SHORT VERSION

Benson's
Microbiological
Applications

Laboratory Manual in General Microbiology

Thirteenth Edition

Alfred Brown
Emeritus Professor, Auburn University

Heidi Smith
Front Range Community College

BENSON'S MICROBIOLOGICAL APPLICATIONS: LABORATORY MANUAL IN GENERAL MICROBIOLOGY, SHORT VERSION, THIRTEENTH EDITION

Published by McGraw-Hill Education, 2 Penn Plaza, New York, NY 10121. Copyright © 2015 by McGraw-Hill Education. All rights reserved. Printed in the United States of America. Previous editions © 2012, 2009, and 2007. No part of this publication may be reproduced or distributed in any form or by any means, or stored in a database or retrieval system, without the prior written consent of McGraw-Hill Education, including, but not limited to, in any network or other electronic storage or transmission, or broadcast for distance learning.

Some ancillaries, including electronic and print components, may not be available to customers outside the United States.

 This book is printed on acid-free paper.

1 2 3 4 5 6 7 8 9 0 QVS/QVS 1 0 9 8 7 6 5 4

ISBN 978–0–07–340241–3
MHID 0–07–340241–9

Senior Vice President, Products & Markets: *Kurt L. Strand*
Vice President, General Manager, Products & Markets: *Marty Lange*
Vice President, Content Production & Technology Services: *Kimberly Meriwether David*
Managing Director: *Michael S. Hackett*
Brand Manager: *Amy Reed*
Director of Development: *Rose M. Koos*
Product Developer: *Darlene M. Schueller*
Digital Product Analyst: *John J. Theobald*
Executive Marketing Manager: *Patrick E. Reidy*
Director, Content Production: *Terri Schiesl*
Content Project Manager: *Laura L. Bies*
Buyer: *Nichole Birkenholz*
Designer: *Margarite Reynolds*
Cover Image: *Science Photo Library RF/Getty Images*
Senior Content Licensing Specialist: *Lori Hancock*
Compositor: *Lachina Publishing Services*
Typeface: *11/12 Times LT Std*
Printer: *Quad/Graphics*

Some of the laboratory experiments included in this text may be hazardous if materials are handled improperly or if procedures are conducted incorrectly. Safety precautions are necessary when you are working with chemicals, glass test tubes, hot water baths, sharp instruments, and the like, or for any procedures that generally require caution. Your school may have set regulations regarding safety procedures that your instructor will explain to you. Should you have any problems with materials or procedures, please ask your instructor for help.

The Internet addresses listed in the text were accurate at the time of publication. The inclusion of a website does not indicate an endorsement by the authors or McGraw-Hill Education, and McGraw-Hill Education does not guarantee the accuracy of the information presented at these sites.

About the Authors

Alfred Brown

Emeritus Professor of
Microbiology
Auburn University

B.S. Microbiology, California
State College, Long Beach

Ph.D. Microbiology, UCLA

Research My research has focused on the physiology of the purple nonsulfur bacteria. This has involved how bacteriochlorophyll and photosynthetic membrane synthesis are coordinated. Herbicides, such as atrazine, have been used to determine the binding site for ubiquinone in photosynthetic electron transport. Binding occurs on the L-subunit, a protein in the photosynthetic reaction center. Resistance to atrazine involves a single amino acid change in the L-subunit that prevents the herbicide from binding to the protein and inhibiting electron transport. This is comparable to how atrazine inhibits electron transport in plants and how resistance to these herbicides develops in weed populations. My laboratory also investigated how the sulfonylurea herbicides inhibit acetolactate synthase, a crucial enzyme in the pathway for branched-chain amino acids. Most recently, I and my graduate students consulted for a company that manufactures roofing shingles. Because of the presence of calcium carbonate in shingles, cyanobacteria can easily grow on their surface, causing problems of contamination. The resulting discoloration caused by these bacteria on shingles has caused significant financial losses to the industry. My laboratory isolated various species of cyanobacteria involved in the problem and taxonomically characterized them. We also tested possible growth inhibitors that might be used in their control.

Teaching Dr. Brown has taught various courses in microbiology over a teaching career that spans more than 30 years. Courses have included general microbiology, medical microbiology, microbial physiology, applied and environmental microbiology, photosynthesis, microbiological methods, and graduate courses, such as biomembranes. In 2008, Dr. Brown retired from the Auburn University faculty as an emeritus professor of microbiology. At present, he continues to work on this manual and travel extensively.

Administration During his tenure at Auburn University, Dr. Brown served as the director of the University Electron Microscope Facility. He also served as the chair of the Department of Botany and Microbiology and the chair of the Department of Biological Sciences.

Heidi Smith

Front Range Community College

B.A. Biology/Pre-Medicine,
Taylor University, IN

M.S. Biology, Bowling Green
State University, OH

Heidi Smith is the lead faculty member for microbiology at Front Range Community College in Fort Collins, CO, and teaches a variety of biology courses each semester including microbiology, anatomy/physiology, and biotechnology. Heidi has also served as the director of the Honors Program at the college for five years, working with a group of faculty to build the program from the ground up.

Student success is a strategic priority at FRCC and a personal passion of Heidi's, and she continually works to develop professionally in ways that help her do a better job of reaching this important goal. Throughout the past few years, Heidi has had the opportunity to collaborate with faculty all over the country in developing digital tools, such as LearnSmart, LearnSmart Labs, and Connect, to facilitate student learning and measure learning outcomes. This collaborative experience and these tools have revolutionized her approach to teaching and have dramatically affected student performance in her courses, especially microbiology hybrid courses where content is delivered partially online.

Heidi is an active member of the American Society for Microbiology and has presented instructional technology and best online and face-to-face teaching practices on numerous occasions at the annual conference for undergraduate educators. She also served as a member of the ASM Task Force on Curriculum Guidelines for Undergraduate Microbiology Education, assisting in the identification of core microbiology concepts as a guide to undergraduate instruction.

Contents

Indicates a LearnSmart Lab™ activity is available for all or part of this exercise. For more information, visit mhhe.com/lslabsmicro.

Preface

Benson's Microbiological Applications has been the "gold standard" of microbiology laboratory manuals for over 30 years. This manual has a number of attractive features that resulted in its adoption in universities, colleges, and community colleges for a wide variety of microbiology courses. These features include user-friendly diagrams that students can easily follow, clear instructions, and an excellent array of reliable exercises suitable for beginning or advanced microbiology courses.

In revising the lab manual for the thirteenth edition, we have tried to maintain the proven strengths of the manual and further enhance it. We have updated the introductory material of the fungi, protozoa, and algae to reflect changes in scientific information. Finally, the names of microorganisms used in the manual are consistent with those used by the American Type Culture Collection. This is important for those users who rely on the ATCC for a source of cultures.

Guided Tour Through a Lab Exercise

Learning Outcomes

Each exercise opens with Learning Outcomes, which list what a student should be able to do after completing the exercise.

> ### Learning Outcomes
>
> After completing this exercise, you should be able to
>
> 1. Prepare a negative stain of bacterial cells using the slide-spreading or loop-spreading techniques.
> 2. Use the negative stain to visualize cells from your teeth and mouth.
> 3. Discern different morphological types of bacterial cells in a negative stain.

Introduction

The introduction describes the subject of the exercise or the ideas that will be investigated. It includes all of the information needed to perform the laboratory exercise.

> **Bacteriophages** are viruses that infect bacterial cells. They were first described by Twort and d'Herelle in 1915 when they both noted that bacterial cultures spontaneously cleared and the bacteria-free liquid that remained could cause new cultures of bacteria to also clear. Because it appeared that the cultures were being "eaten" by some unknown agent, d'Herelle coined the term *bacteriophage,* which means "bacterial eater." Like all viruses, bacteriophages, or phages, for short,

First and Second Periods

In many cases, instructions are presented for two or more class periods so you can proceed through an exercise in an appropriate fashion.

> ### 🕒 First Period
>
> (Inoculations and Incubation)
>
> Since six microorganisms and three kinds of media are involved in this experiment, it will be necessary for economy of time and materials to have each student work with only three organisms. The materials list for this

> ### 🕒 Second Period
>
> (Culture Evaluations and Spore Staining)
>
> Remove the lid from the GasPak jar. If vacuum holds the inner lid firmly in place, break the vacuum by sliding the lid to the edge. When transporting the plates and tubes to your desk *take care not to agitate the FTM tubes*. The position of growth in the medium can be easily changed if handled carelessly.

Materials Needed

This section lists the laboratory materials that are required to complete the exercise.

> ### Materials
>
> - microscope slides
> - broth cultures of *Staphylococcus, Streptococcus,* and *Bacillus*
> - Bunsen burner
> - wire loop
> - marking pen
> - slide holder (clothespin)

Procedures

The procedures and methods provide a set of detailed instructions for accomplishing the planned laboratory activities.

> ### Scrub Procedure
>
> The two members of the class who are chosen to perform the surgical scrub will set up their materials near a sink for convenience. As one student performs the scrub, the other will assist in reading the instructions and providing materials as needed. The basic steps,

Illustrations

Illustrations provide visual instructions for performing steps in procedures or are used to identify parts of instruments or specimens.

Figure 1.2 **The compound microscope.**
© Charles D. Winters/Science Source.

Figure 13.1 **Procedure for demonstration of a capsule.**

Laboratory Reports

A Laboratory Report to be completed by students immediately follows most of the exercises. These Laboratory Reports are designed to guide and reinforce student learning and provide a convenient place for recording data. These reports include various types of review activities, tables for recoding observations and experimental results, and questions dealing with the analysis of such data.

As a result of these activities, students will increase their skills in gathering information by observation and experimentation. By completing all of the assessments in the Laboratory Reports, students will be able to determine if they accomplished all of the learning outcomes.

Digital Tools for Your Lab Course

McGraw-Hill Connect®

McGraw-Hill Connect allows instructors and students to use art and animations for assignments and lectures. A robust set of questions and activities are presented and aligned with the lab manual's exercises. As an instructor, you can edit existing questions and author entirely new problems. Track individual student performance—by question, assignment, or in relation to the class overall—with detailed grade reports. Instructors also have access to a variety of new resources, including assignable and gradable lab questions from the lab manual, additional pre- and post-lab activities, case study activities, interactive questions based on atlas images, lab skill videos, and more. In addition, digital images, PowerPoint slides, and instructor resources are available through Connect. Visit www.mcgrawhillconnect.com.

Digital Lecture Capture

Tegrity Campus™ is a service that allows class time to be any time by automatically capturing every lecture in a searchable video format for students to review at their convenience. Educators know that the more students can see, hear, and experience class resources, the better they learn. Help turn all your students' study time into learning moments by supplying them with your lecture videos.

New! LearnSmart Labs™

Fueled by LearnSmart—the most widely used and intelligent adaptive learning resource—LearnSmart Labs is a highly realistic and adaptive simulated lab experience that allows students to "do and think" like scientists.

LearnSmart Labs brings meaningful scientific exploration to students by giving them an adaptive environment in which to practice the scientific method, enabling them to make mistakes safely and to think critically about their findings. As a result, students come to lab better prepared and more likely to engage in the learning process.

The LearnSmart Labs questions, learning resources, and simulations were created with lab manual exercises in mind. This pairing gives you flexibility to implement technology in your lab course in a way that works for you. Whether you go all digital or try a hybrid approach, LearnSmart Labs will help your students learn the techniques AND the reasons why they're important.

This revolutionary technology is available only from McGraw-Hill Education as part of the LearnSmart Advantage series.

Visit www.learnsmartadvantage.com to learn more.

New! LearnSmart Prep™

LearnSmart Prep is an adaptive tool that prepares students for college level work. LearnSmart Prep identifies the concepts each student doesn't know or fully understand and provides learning resources to teach essential concepts so he or she enters the classroom prepared for college level work. Visit www.learnsmart advantage.com to learn more.

Website

Visit the website at www.mhhe.com/benson13e for additional resources, including access to two additional exercises, Slime Mold Culture and White Blood Cell Study. For instructors, a password-protected instructor's manual and image library are available. The instructor's manual provides: (1) a list of equipment and supplies needed for performing all of the experiments, (2) procedures for setting up the experiments, and (3) answers to all the questions for the Laboratory Reports. The image library contains figures from the laboratory manual. Please contact your sales representative for additional information.

Electronic Book—GO GREEN!

Green . . . it's on everybody's mind these days. It's not only about saving trees; it's also about saving money. If you or your students are ready for an alternative, McGraw-Hill eBooks offer a cheaper and eco-friendly alternative to traditionally printed textbooks and laboratory manuals. This laboratory manual is available as an eBook at www.CourseSmart.com. At CourseSmart your students can take advantage of significant savings off the cost of a print textbook, reduce their impact on the environment, and gain access to powerful web tools for learning. CourseSmart eBooks can be viewed online or downloaded to a computer. The eBooks allow students to do full text searches, add highlighting and notes, and share notes with classmates. Contact your McGraw-Hill sales representative or visit www.Course Smart.com to learn more.

Personalize Your Lab

Craft your teaching resources to match the way you teach! With McGraw-Hill Create™, www.mcgrawhill create.com, you can easily rearrange chapters, combine material from other content sources, and quickly upload content you have written like your course syllabus or teaching notes. Find the content you need in Create by searching through thousands of leading McGraw-Hill textbooks. Arrange your book to fit your teaching style. Create even allows you to personalize your book's appearance by selecting the cover and adding your name, school, and course information. Order a Create book and you'll receive a complimentary print review copy in 3–5 business days or a complimentary electronic review copy (eComp) via email in minutes. Go to www.mcgrawhillcreate.com today and register to experience how McGraw-Hill Create empowers you to teach *your* students *your* way.

Student Resources

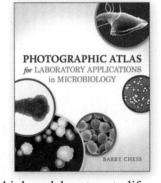

Photographic Atlas for Laboratory Applications in Microbiology (0-07-737159-3), prepared by Barry Chess at Pasadena City College, can be packaged with this laboratory manual. This beautifully prepared photo atlas contains more than 300 color photos that bring the microbiology laboratory to life. The photo atlas is divided into eight major sections: staining techniques; cultural and biochemical tests; bacterial colonial morphology; bacterial microscopic morphology; fungi; protists; helminths; and hematology and serology. A picture is worth a thousand words, and this is definitely the case with this beautifully prepared atlas. Contact your McGraw-Hill sales representative for additional information and packaging options.

Annual Editions: Microbiology 10/11 (0-07-738608-6) is a series of over 65 volumes, each designed to pro-

vide convenient, inexpensive access to a wide range of current articles from some of the most respected magazines, newspapers, and journals published today. *Annual Editions* are updated on a regular basis through a continuous monitoring of over 300 periodical sources. The articles selected are authored by prominent scholars, researchers, and commentators writing for a general audience. The *Annual Editions* volumes have a number of common organizational features designed to make them particularly useful in the classroom: a general introduction; an annotated table of contents; a topic guide; an annotated listing of selected World Wide Web sites; and a brief overview for each section. Visit www.mhhe.com/cls for more details.

Changes to the Thirteenth Edition

- The thirteenth edition has a beautiful new design, with a striking color palette. Different-colored bars are used for Periods, Materials, Procedures, and Results.
- Each part opener now features a photograph depicting the theme of the exercises.
- For exercises that require several periods to complete, a clock icon has been included in the heading to indicate that results are required for the period and that further procedures may be necessary for the exercise.
- All cautions and warnings are denoted with a red bar to call attention to hazards associated with the exercise.
- A LearnSmart Lab icon has been included by the exercise title for those exercises where a Learn-Smart Lab activity is available for all or part of that exercise.

Part 2, Survey of Microorganisms

- Exercise 5, Microbiology of Pond Water—Protists, Algae, and Cyanobacteria, has been revised to reflect changes in the taxonomy of these organisms. More emphasis is given to their role in disease and their importance in the environment and ecosystems.
- Exercise 6, Ubiquity of Bacteria, is now focused on the diversity and ubiquity of bacteria in the environment and how they are cultured. The morphology of bacterial cells has been moved to Exercise 11, Simple Staining.
- The Fungi: Molds and Yeasts, Exercise 7, has been revised to reflect changes in their taxonomy. Greater emphasis is given to their role in human disease, their importance as foods and in food production, and their role in environmental processes such as mineralization and the turnover of organic materials in ecosystems. The relationship between fungi that form a mycelium and yeasts is discussed.

Part 4, Staining and Observation of Microorganisms

- Exercises 10–16 on staining reactions have been reorganized and modified. Exercises have photographs in larger formats and some new photographs to show expected results. Also, each exercise now has its own set of results and questions, thus eliminating the necessity for students to record results and search for photos in later exercises.
- Exercise 14, Gram Staining, now includes new enlarged photomicrographs depicting gram-positive and gram-negative cells and align clearly with the procedure. More background on the history, importance, and theoretical basis of the Gram staining method has also been included. A corrected figure for the Gram stain steps is included as well. Procedures have slightly changed so that students have control bacteria on each slide to assist them in determining the success of their staining technique. The Laboratory Report has been enhanced, asking students to apply the concepts of cell shape and arrangement when evaluating their stained results.

Part 5, Culture Methods

- Enumeration of Bacteria: The Standard Plate Count, Exercise 19, now has photos of a set of serial dilution plates of a bacterial sample.

Part 6, Bacterial Viruses

- The steps in the infection of a bacterial cell by a bacteriophage have been revised in Exercise 21, Determination of a Bacteriophage Titer.

Part 7, Environmental Influences and Control of Microbial Growth

- Exercise 24, Effects of Oxygen on Growth, has been revised to include the discussion of how oxygen influences growth and how it defines various classes of bacteria. Photographs showing the growth patterns of aerobes, anaerobes, microaerophiles, and facultative anaerobes in thioglycollate broth have replaced a diagram.
- A description of the organisms associated with human skin has been added to Exercise 30, Evaluation of Alcohol: Its Effectiveness as an Antiseptic.
- Antimicrobic Sensitivity Testing: The Kirby-Bauer Method, Exercise 31, has been updated with new information concerning health worker–acquired infections and the problem of antibiotic-resistant bacteria. New photographs of Kirby-Bauer plates showing sensitivity and resistance have been added, as well as photos showing how to measure the zone of inhibition.

Part 8, Identification of Unknown Bacteria

- Exercise 36, Physiological Characteristics: Oxidation and Fermentation Tests, now includes new photos of fermentation reactions (Durham tubes), the MRVP test, the citrate test, and the catalase.
- Added to Exercise 37, Physiological Characteristics: Hydrolytic and Degradative Reactions, are new photos of starch, casein, and fat hydrolysis.
- Physiological Characteristics: Multiple Test Media, Exercise 38, has new photos for SIM medium showing motility and hydrogen sulfide production. Enhanced photos of litmus milk reactions, including stormy fermentation, have also been added.
- The introductory material and separation outlines for Exercise 39, Use of *Bergey's Manual*, have been updated to reflect the current edition of Bergey's and the different volumes (both determinative and systematic). An additional challenge was added to this exercise aimed at teaching students how to use a table of test results and construct a flow chart to determine the identity of an unknown bacterium. A new lab report was added to this exercise for both the original procedure and the additional challenge.

Part 11, Medical Microbiology

- Exercise 51, The Staphylococci: Isolation and Identification, includes new photos for Gram stain of staph, coagulase test, methyl green DNase test agar, and novobiocin sensitivity of Staphylococcus epidermidis.
- The Streptococci and Enterococci: Isolation and Identification, Exercise 52, includes a new photo of Gram stain of strep.
- Exercise 53, Gram-Negative Intestinal Pathogens, includes new photos of lactose fermenters on McConkey and Eosin methylene blue agar.
- A Synthetic Epidemic, Exercise 54, has two new figures added to the introductory material, highlighting the two different categories of epidemics and the concept of herd immunity. Procedure B and its corresponding lab report section were revised to illustrate the concept of herd immunity.

Part 12, Immunology and Serology

- Exercise 56, Slide Agglutination Test for *S. aureus*, has a new photo of the C-reactive protein agglutination test showing positive and negative results.

Acknowledgements

Our deepest gratitude to Judy Kaufman for her thorough review of the exercises and to Jill Kolodsick for her assistance in preparing the final draft of the manuscript and for her recommendations concerning the exercises.

We also wish to express our gratitude to Lisa Burgess for her excellent photographic contributions to the manual. The many new photos she provided will greatly improve the manual by providing clarity to many exercises.

The updates and improvements in this edition were guided by the helpful reviews and survey responses from the following instructors. Their input was critical to the decisions that shaped this edition.

Debra Albright *University of Phoenix*
Lorrie Burnham *San Bernardino Valley College*
Jessica Cofield *Craven Community College*
Iris Cook *Westchester Community College*
Adrienne Cottrell-Yongye *Indian River State College*
Amy Warenda Czura *Suffolk County Community College Eastern Campus*
Angela Edwards *Trident Technical College*
Mary Farone *Middle Tennessee State University*
Deborah Faurot *University of Kansas*
Rebecca Giorno-McConnell *Louisiana Tech University*
Dale Harrington *Caldwell Community College and Technical Institute, Watauga Campus*
Betsy Hogan *Trident Technical College*
Jessica Jean *Capital Community College*
Mustapha Lahrach *Hillsborough Community College SouthShore Campus*
Jeffrey D. Leblond *Middle Tennessee State University*

Steven McConnell *Southeast Community College*
Thomas M. McNeilis *Dixie State College of Utah*
Caroline McNutt *Schoolcraft College*
Lathika Moragoda *Henry Ford Community College*
Christian Nwamba *Wayne County Community College District*
Teresa P. Palos *El Camino College*
Anita Denise Patterson *Northeast Alabama Community College*
Charles Pumpuni *Northern Virginia Community College*
Yilei Qian *Indiana University South Bend*
Pushpa Samkutty *Southern University*
Austin Smith *Jones County Junior College*
Davood Soleymani *California State University—Dominguez Hills*
Jamie Welling *South Suburban College*
Narinder Whitehead *Capital Community College*

We would like to thank all the people at McGraw-Hill for their tireless efforts and support with this project. They are professional and competent and always a pleasure to work with on this manual. A special and deep thanks to Darlene Schueller, our product developer. Once again, she kept the project focused, made sure we met deadlines and made suggestions that improved the manual in many ways. Thanks as well to Amy Reed, brand manager; Laura Bies, content project manager; Nichole Birkenholz, buyer; and Margarite Reynolds, designer, and many who worked "behind the scenes."

Basic Microbiology Laboratory Safety

Every student and instructor must focus on the need for safety in the microbiology laboratory. While the lab is a fascinating and exciting learning environment, there are hazards that must be acknowledged and rules that must be followed to prevent accidents and contamination with microbes. The following guidelines will provide every member of the laboratory section the information required to assure a safe learning environment.

Microbiological laboratories are special, often unique environments that may pose identifiable infectious disease risks to persons who work in or near them. Infections have been contacted in the laboratory throughout the history of microbiology. Early reports described laboratory-associated cases of typhoid, cholera, glanders, brucellosis, and tetanus, to name a few. Recent reports have documented laboratory-acquired cases in laboratory workers and health-care personnel involving *Bacillus anthracis, Bordetella pertussis, Brucella, Burkholderia pseudomallei, Campylobacter, Chlamydia,* and toxins from *Clostridium tetani, Clostridium botulinum,* and *Corynebacterium diphtheriae.* While we have a greater knowledge of these agents and antibiotics with which to treat them, safety and handling still remain primary issues.

The "Biohazard" symbol must be affixed to any container or equipment used to store or transport potentially infectious materials.

Courtesy of the Centers for Disease Control.

The term "containment" is used to describe the safe methods and procedures for handling and managing microorganisms in the laboratory. An important laboratory procedure practiced by all microbiologists that will guarantee containment is **aseptic technique,** which prevents workers from contaminating themselves with microorganisms, ensures that others and the work area do not become contaminated, and also ensures that microbial cultures do not become unnecessarily contaminated with unwanted organisms. Containment involves personnel and the immediate laboratory and is provided by good microbiological technique and the use of appropriate safety equipment. Containment also guarantees that infectious agents do not escape from the laboratory and contaminate the environment external to the lab. Containment, therefore, relies on good microbiological technique and laboratory protocol as well as elements of laboratory design.

Biosafety Levels (BSL)

The recommended biosafety level(s) for handling microorganisms represent the potential of the agent to cause disease and the conditions under which the agent should be safely handled. The Centers for Disease Control classifies organisms into levels and sets guidelines for handling and safety measures required. These levels take into account many factors such as virulence, pathogenicity, antibiotic resistance patterns, vaccine and treatment availability, and other factors. The recommended biosafety levels are as follows:

1. **BSL 1**—agents not known to cause disease in healthy adults; standard microbiological practices (SMP) apply; no safety equipment required; sinks required. Examples: *Bacillus subtilis, Micrococcus luteus.*
2. **BSL 2**—agents associated with human disease; standard microbiological practices apply plus limited access, biohazard signs, sharps precautions, and a biosafety manual required. Biosafety cabinet (BSC) used for aerosol/splash generating operations; lab coats, gloves, face protection required; contaminated waste is autoclaved.

All microorganisms used in the exercises in this manual are classified as BSL 1 or BSL 2. Examples: *Staphylococcus aureus, Streptococcus pyogenes.* **Note:** Although some of the organisms that students will culture and work with are classified as BSL 2, these organisms are laboratory strains that do not pose the same threat of infection as primary isolates of the same organism taken from patients in clinical samples. Hence, these laboratory strains can, in most cases, be handled using normal procedures and equipment found in the vast majority of student teaching laboratories. However, it should be emphasized that many bacteria are opportunistic pathogens, and therefore all microorganisms should be handled by observing proper techniques and precautions.

3. **BSL 3**—indigenous/exotic agents that may have serious or lethal consequences and with a potential for aerosol transmission. BSL 2 practices plus controlled access; decontamination of all waste and lab clothing before laundering; determination of baseline antibody titers to agents; biosafety cabinets used for all specimen manipulations; respiratory protection used as needed; physical separation from access corridors; double door access; negative airflow into the lab; exhaust air not recirculated. Examples: *Mycobacterium tuberculosis* and vesicular stomatitis virus (VSV).
4. **BSL 4**—dangerous/exotic agents of a life-threatening nature or unknown risk of transmission; BSL 3 practices plus clothing change before entering the laboratory; shower required before leaving the lab; all materials decontaminated on

exit; positive pressure personnel suit required for entry; separated/isolated building; dedicated air supply/exhaust and decontamination systems. Examples: Ebola and Lassa viruses.

Each of the biosafety levels consist of combinations of laboratory practices and techniques, safety equipment, and laboratory facilities. Each combination is specifically appropriate for the operations performed and the documented or suspected routes of transmission of the infectious agents. Common to all biosafety levels are standard practices, especially aseptic technique. Refer to the Biosafety Levels table on page xv for a list of common organisms.

Standard Laboratory Rules and Practices

1. Students should store all books and materials not used in the laboratory in areas or receptacles designated for that purpose. Only necessary materials such as a lab notebook, the laboratory manual, and pen/pencil should be brought to the student work area.
2. Eating, drinking, chewing gum, and smoking are not allowed in the laboratory. Students must also avoid handling contact lenses or applying makeup while in the laboratory.
3. Safety equipment:
   a. Some labs will require that lab coats be worn in the laboratory at all times. Others may make this optional or not required. Lab coats can protect a student from contamination by microorganisms that he/she is working with and prevent contamination from stains and chemicals. At the end of the laboratory session, lab coats are usually stored in the lab in a manner prescribed by the instructor. Lab coats, gloves, and safety equipment should not be worn outside of the laboratory unless properly decontaminated first.
   b. You may be required to wear gloves while performing the lab exercises. This is especially important if you have open wounds. They protect the hands against contamination by microorganisms and prevent the hands from coming in direct contact with stains and other reagents.
   c. Face protection/safety glasses may be required by some instructors while you are performing experiments. Safety glasses can prevent materials from coming in contact with the eyes. They must be worn especially when working with ultraviolet light to prevent eye damage because they block out UV rays. If procedures involve the potential for splash/aerosols, face protection should be worn.
   d. Know the location of eye wash and shower stations in the event of an accident that requires the use of this equipment. Also know the location of first aid kits.
4. Sandals or open-toe shoes are not to be worn in the laboratory. Accidental dropping of objects or cultures could result in serious injury or infection.
5. Students with long hair should tie the hair back to avoid accidents when working with Bunsen burners/open flames. Long hair can also be a source of contamination when working with cultures.
6. Before beginning the activities for the day, work areas should be wiped down with the disinfectant that is provided for that purpose. Likewise, when work is finished for the day, the work area should be treated with disinfectant to ensure that any contamination from the exercise performed is destroyed. Avoid contamination of the work surface by not placing contaminated pipettes, loops/needles, or swabs on the work surface. Dispose of contaminated paper towels used for swabbing in the biohazard container.
7. Use extreme caution when working with open flames. The flame on a Bunsen burner is often difficult to see when not in use. Caution is imperative when working with alcohol and open flames. Alcohol is highly flammable, and fires can easily result when using glass rods that have been dipped in alcohol. **Always make sure the gas is turned off before leaving the laboratory.**
8. Any cuts or injuries on the hands must be covered with band-aids to prevent contamination. If you injure or cut yourself during the laboratory, notify the laboratory instructor immediately.
9. Pipetting by mouth is prohibited in the lab. All pipetting must be performed with pipette aids. Be especially careful when inserting glass pipettes into pipette aids as the pipette can break and cause a serious injury.
10. Know the location of exits and fire extinguishers in the laboratory.
11. Most importantly, read the exercise and understand the laboratory protocol before coming to laboratory. In this way you will be familiar with potential hazards in the exercise.
12. When working with microfuges, be familiar with their safe operation and make sure that all microfuge tubes are securely capped before centrifuging.
13. When working with electrophoresis equipment, follow the directions carefully to avoid electric shock.
14. If you have any allergies or medical conditions that might be complicated by participating in the laboratory, inform the instructor. Women who are pregnant should discuss the matter of enrolling in

the lab with their family physician and the laboratory instructor.

15. Unless directed to do so, do not subculture any unknown organisms isolated from the environment as they could be potential pathogens.

16. Avoid handling personal items such as cell phones, calculators, and cosmetics while performing the day's exercise.

17. You may be required to sign a safety agreement stating that you have been informed about safety issues and precautions and the hazardous nature of microorganisms that you may handle during the laboratory course.

18. Avoid wearing dangling jewelry to lab.

Disposal of Biological Wastes

Dispose of all contaminated materials properly and in the appropriate containers:

1. Biohazard containers—biohazard containers are to be lined with clear autoclave bags; disposable petri plates, used gloves, and any materials such as contaminated paper towels should be discarded in these containers; no glassware, test tubes, or sharp items are to be disposed of in biohazard containers.

2. Sharps containers—sharps, slides, coverslips, broken glass, disposable pipettes, and Pasteur pipettes should be discarded in these containers. If instructed to do so, you can discard contaminated swabs, wooden sticks, and microfuge tubes in the sharps containers.

3. Discard shelves, carts, bins, etc.—contaminated culture tubes and glassware used to store media and other glassware should be placed in these areas for decontamination and washing.

4. Trash cans—any noncontaminated materials, paper, or trash should be discarded in these containers. Under no circumstances should laboratory waste be disposed of in trash cans.

Discard other materials as directed by your instructor. This may involve placing materials such as slides contaminated with blood in disinfectant baths before these materials can be discarded.

Emergencies

Surface Contamination

1. Report all spills immediately to the laboratory instructor.

2. Cover the spill with paper towels and saturate the paper towels with disinfectant.

3. Allow the disinfectant to act for at least 20 minutes.

4. Remove any glass or solid material with forceps or scoop and discard the waste in an appropriate manner.

Personnel Contamination

1. Notify lab instructor.

2. Clean exposed area with soap/water, eye wash (eyes) or saline (mouth).

3. Apply first aid.

Biosafety Levels for Selected Infectious Agents

BIOSAFETY LEVEL (BSL)	TYPICAL RISK	ORGANISM
BSL 1	Not likely to pose a disease risk to healthy adults.	*Achromobacter denitrificans* *Alcaligenes faecalis* *Bacillus cereus* *Bacillus subtilis* *Corynebacterium pseudodiphtheriticum* *Enterococcus faecalis* *Micrococcus luteus* *Neisseria sicca* *Proteus vulgaris* *Pseudomonas aeruginosa* *Staphylococcus epidermidis* *Staphylococcus saprophyticus*
BSL 2	Poses a moderate risk to healthy adults; unlikely to spread throughout community; effective treatment readily available.	*Enterobacter aerogenes* *Escherichia coli* *Klebsiella pneumoniae* *Mycobacterium phlei* *Salmonella enterica var. Typhimurium* *Shigella flexneri* *Staphylococcus aureus* *Streptococcus pneumoniae* *Streptococcus pyogenes*
BSL 3	Can cause disease in healthy adults; may spread to community; effective treatment readily available.	*Blastomyces dermatitidis* *Chlamydia trachomatis* *Coccidioides immitis* *Coxiella burnetii* *Francisella tularensis* *Histoplasma capsulatum* *Mycobacterium bovis* *Mycobacterium tuberculosis* *Pseudomonas mallei* *Rickettsia canadensis* *Rickettsia prowazekii* *Yersinia pestis*
BSL 4	Can cause disease in healthy adults; poses a lethal risk and does not respond to vaccines or antimicrobial therapy.	Filovirus *Herpesvirus simiae* Lassa virus Marburg virus

Microorganisms Used or Isolated in the Lab Exercises in This Manual

ORGANISM	GRAM STAIN AND MORPHOLOGY	HABITAT	BSL	LAB EXERCISE
Alcaligenes faecalis ATCC 8750	Negative rod	Decomposing organic material, feces	1	26, 39
Bacillus coagulans ATCC 7050	Positive rod	Spoiled food, silage	1	49
Bacillus megaterium ATCC 14581	Positive rod	Soil, water	1	11, 12, 14, 15, 28, 48
Bacillus subtilis ATCC 23857	Positive rod	Soil, decomposing organic matter	1	24, 37
Candida glabrata ATCC 200918	Yeast	Human oral cavity	1	26
Chromobacterium violaceum ATCC 12472	Negative rod	Soil and water; opportunistic pathogen in humans	1	9
Citrobacter freundii ATCC 8090	Negative rod	Humans, animals, soil water; sewage opportunistic pathogen	1	53
Clostridium beijerinckii ATCC 25752	Positive rod	Soil	1	24
Clostridium sporogenes ATCC 3584	Positive rod	Soil, animal feces	1	24, 49
Corynebacterium xerosis ATCC 373	Positive rods, club-shaped	Conjunctiva, skin	1	11
Enterobacter aerogenes ATCC 13048	Negative rods	Feces of humans and animals	2	24, 36, 39, 45
Enterococcus faecalis ATCC 19433	Positive cocci in pairs, short chains	Water, sewage, soil, dairy products	2	24, 39, 52, 57
Enterococcus faecium ATCC 19434	Positive cocci in pairs, short chains	Feces of humans and animals	2	52, 57
Escherichia coli ATCC 11775	Negative rods	Sewage, intestinal tract of warm-blooded animals	2	8, 9, 14, 19, 21, 22, 24, 25, 26, 27, 29, 31, 36, 37, 38, 39, 45, 48, 49
Geobacillus stearothermophilus ATCC 12980	Gram-positive rods	Soil, spoiled food	1	25, 49
Halobacterium salinarium ATCC 33170	Gives gram-negative reaction; rods	Salted fish, hides, meats	1	27
Klebsiella pneumoniae ATCC 13883	Negative rods	Intestinal tract of humans; respiratory and intestinal pathogen in humans	2	13, 39

Microorganisms Used or Isolated in the Lab Exercises in This Manual (continued)

ORGANISM	GRAM STAIN AND MORPHOLOGY	HABITAT	BSL	LAB EXERCISE
Lactococcus lactis ATCC 19435	Positive cocci in chains	Milk and milk products	1	11
Micrococcus luteus ATCC 12698	Positive cocci that occur in pairs	Mammalian skin	1	9, 17, 29, 39
Moraxella catarrhalis ATCC 25238	Negative cocci that often occur in pairs with flattened sides	Pharynx of humans	1	14
Mycobacterium smegmatis ATCC 19420	Positive rods; may be Y-shaped or branched	Smegma of humans	1	16
Proteus vulgaris ATCC 29905	Negative rods	Intestines of humans, and animals; soil and polluted waters	1	17, 31, 37, 38, 39, 53
Pseudomonas aeruginosa ATCC 10145	Negative rods	Soil and water; opportunistic pathogen in humans	1	14, 31, 32, 36, 39
Saccharomyces cerevisiae ATCC 18824	Yeast	Fruit, used in beer, wine, and bread	1	26
Salmonella enterica subsp. *enterica* serovar *Typhimurium* ATCC 700720D-5	Negative rods	Most frequent agent of *Salmonella* gastroenteritis in humans	2	39, 53, 55
Serratia marcescens ATCC 13880	Negative rods	Opportunistic pathogen in humans	1	9, 25, 39, 54
Shigella flexneri ATCC 29903	Negative rods	Pathogen of humans	2	53
Staphylococcus aureus ATCC 12600	Positive cocci, irregular clusters	Skin, nose, GI tract of humans, pathogen	2	9, 11, 12, 14, 16, 23, 24, 26, 27, 28, 29,31, 32, 36, 37, 38, 39, 48, 51, 52, 56
Staphylococcus epidermidis ATCC 14990	Positive cocci that occur in pairs and tetrads	Human skin, animals; opportunistic pathogen	1	39, 51
Staphylococcus saprophyticus ATCC 15305	Positive cocci that occur singly and in pairs	Human skin; opportunistic pathogen in the urinary tract	1	51
Streptococcus agalactiae ATCC 13813	Positive cocci; occurs in long chains	Upper respiratory and vaginal tract of humans, cattle; pathogen	2	52, 57
Streptococcus bovis ATCC 33317	Positive cocci; pairs and chains	Cattle, sheep, pigs; occasional pathogen in humans	2	52, 57
Streptococcus dysagalactiae subsp. *equisimilis* ATCC 43078	Positive cocci in chains	Mastitis in cattle	2	52

Microorganisms Used or Isolated in the Lab Exercises in This Manual (continued)

ORGANISM	GRAM STAIN AND MORPHOLOGY	HABITAT	BSL	LAB EXERCISE
Streptococcus mitis ATCC 49456	Positive cocci in pairs and chains	Oral cavity of humans	2	52
Streptococcus mutans ATCC 25175D-5	Positive cocci in pairs and chains	Tooth surface of humans, causes dental caries	2	52
Streptococcus pneumoniae ATCC 33400D-5	Positive cocci in pairs	Human pathogen	2	52
Streptococcus pyogenes ATCC 12344	Positive cocci in chains	Human respiratory tract; pathogen	2	52, 57
Streptococcus salivarius ATCC 19258	Positive cocci in short and long chains	Tongue and saliva	2	52
Thermoanaerobacterium thermosaccharolyticum ATCC 7956	Negative rods; single cells or pairs	Soil, spoiled canned foods	1	49

Microscopy

© JGI/Blend Images LCC RF

Although there are many kinds of microscopes available to the microbiologist today, only four types will be described here for our use: the brightfield, darkfield, phase-contrast, and fluorescence microscopes. If you have had extensive exposure to microscopy in previous courses, this unit may not be of great value to you; however, if the study of microorganisms is a new field of study for you, there is a great deal of information that you need to acquire about the proper use of these instruments.

Microscopes in a college laboratory represent a considerable investment and require special care to prevent damage to the lenses and mechanical parts. A microscope may be used by several people during the day and moved from the work area to storage, which results in a much greater chance for damage to the instrument than if the microscope were used by only a single person.

The complexity of some of the more expensive microscopes also requires that certain adjustments be made periodically. Knowing how to make these adjustments to get the equipment to perform properly is very important. An attempt is made in the five exercises of this unit to provide the necessary assistance for getting the most out of the equipment.

Microscopy should be as fascinating to the beginner as it is to the professional of long standing; however, only with intelligent understanding can the beginner approach the achievement that occurs with years of experience.

Brightfield Microscopy

Learning Outcomes

After completing this exercise, you should be able to

1. Identify the basic components of a brightfield microscope and understand the function of each component in proper specimen observation.
2. Examine a specimen using the low-power, high-dry, and oil immersion lenses.
3. Understand the proper use, care, and storage of a microscope.

A microscope that allows light rays to pass directly to the eye without being deflected by an intervening opaque plate in the condenser is called a **brightfield microscope.** This is the conventional type of instrument encountered by students in beginning courses in biology; it is also the first type to be used in this laboratory.

All brightfield microscopes have certain things in common, yet they differ somewhat in mechanical operation. Similarities and differences of various makes are discussed in this exercise so that you will know how to use the instrument that is available to you. Before attending the first laboratory session in which the microscope is used, read over this exercise and answer all the questions on the Laboratory Report. Your instructor may require that the Laboratory Report be handed in prior to doing any laboratory work.

Figure 1.1 **The microscope should be held firmly with both hands while being carried.**

Care of the Instrument

Microscopes represent considerable investment and can be damaged easily if certain precautions are not observed. The following suggestions cover most hazards.

Transport When carrying your microscope from one part of the room to another, use both hands to hold the instrument, as illustrated in figure 1.1. If it is carried with only one hand and allowed to dangle at your side, there is always the danger of collision with furniture or some other object. And, *under no circumstances should one attempt to carry two microscopes at one time.*

Clutter Keep your workstation uncluttered while doing microscopy. Keep unnecessary books and other materials away from your work area. A clear work area promotes efficiency and results in fewer accidents.

Electric Cord Microscopes have been known to tumble off of tabletops when students have entangled a foot in a dangling electric cord. Don't let the electric cord on your microscope dangle in such a way as to risk foot entanglement.

Lens Care At the beginning of each laboratory period, check the lenses to make sure they are clean. At the end of each lab session, be sure to wipe any immersion oil off the immersion lens if it has been used. More specifics about lens care are provided on page 6.

Dust Protection In most laboratories dustcovers are used to protect the instruments during storage. If one is available, place it over the microscope at the end of the period.

3

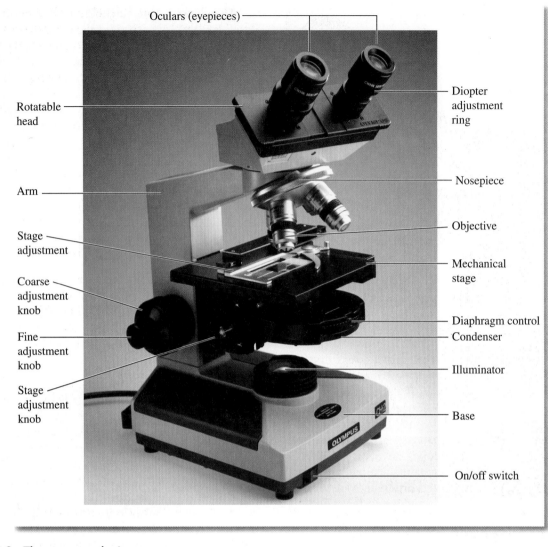

Oculars (eyepieces)

Rotatable head

Arm

Stage adjustment

Coarse adjustment knob

Fine adjustment knob

Stage adjustment knob

Diopter adjustment ring

Nosepiece

Objective

Mechanical stage

Diaphragm control

Condenser

Illuminator

Base

On/off switch

Figure 1.2 The compound microscope.
© Charles D. Winters/Science Source.

Components

Before we discuss the procedures for using a microscope, let's identify the principal parts of the instrument as illustrated in figure 1.2.

Framework All microscopes have a basic frame structure, which includes the **arm** and **base.** To this framework all other parts are attached. On many of the older microscopes the base is not rigidly attached to the arm as is the case in figure 1.2; instead, a pivot point is present that enables one to tilt the arm backward to adjust the eye point height.

Stage The horizontal platform that supports the microscope slide is called the **stage.** Note that it has a clamping device, the **stage adjustment,** which is used for holding and moving the slide around on the stage. Note, also, the location of the **stage adjustment knobs** in figure 1.2.

Light Source In the base of most microscopes is positioned some kind of light source. Ideally, the lamp should have a **light intensity control** to vary the intensity of light. The microscope in figure 1.2 has a knurled wheel on the right side of its base to regulate the voltage supplied to the lightbulb.

Most microscopes have some provision for reducing light intensity with a **neutral density filter.** Such a filter is often needed to reduce the intensity of light below the lower limit allowed by the voltage control. On microscopes such as the Olympus CX41, one can simply place a neutral density filter over the light source in the base. On some microscopes a filter is built into the base.

Lens Systems All compound microscopes have three lens systems: the oculars, the objectives, and the condenser. Figure 1.3 illustrates the light path through these three systems.

The object is now magnified 1000 times its actual size. The third lens system is the **condenser,** which is located under the stage. It collects and directs the light from the lamp to the slide being studied. Unlike the ocular and objective lenses, the condenser lens does not affect the magnifying power of the compound microscope. The condenser can be moved up and down by a knob under the stage. A **diaphragm** within the condenser regulates the amount of light that reaches the slide. Microscopes that lack a voltage control on the light source rely entirely on the diaphragm for controlling light intensity. On the Olympus CX41 microscope in figure 1.2, the diaphragm is controlled by turning a knurled ring. On some microscopes, a diaphragm lever is present. Figure 1.3 illustrates the location of the condenser and diaphragm.

Focusing Knobs The concentrically arranged **coarse adjustment** and **fine adjustment knobs** on the side of the microscope are used for bringing objects into focus when studying an object on a slide. On some microscopes, these knobs are not positioned concentrically as shown here.

Ocular Adjustments On binocular microscopes, one must be able to change the distance between the oculars and to make diopter changes for eye differences. On most microscopes, the interocular distance is changed by simply pulling apart or pushing together the oculars.

To make diopter adjustments, one focuses first with the right eye only. Without touching the focusing knobs, diopter adjustments are then made on the left eye by turning the knurled **diopter adjustment ring** (figure 1.2) on the left ocular until a sharp image is seen. One should now be able to see sharp images with both eyes.

Resolution

It would appear that the magnification of a microscope is only limited by the magnifying power of a lens system. However, in reality the limit for most light microscopes is 1000×, which is set by an intrinsic property of lenses called **resolving power.** The resolving power of a lens is its ability to completely separate two objects in a microscopic field. The resolving power is given by the formula $d = 0.5 \, \lambda/NA$. The limit of resolution, d, or the distance between the two objects, is a function of two properties: the wavelength of the light used to observe a specimen, λ, and a property of lenses called the **numerical aperture** or NA. Numerical aperture is a mathematical expression that describes how the condenser lens concentrates and focuses the light rays from the light source. Its value is maximized when the light rays are focused into a cone of light that then passes through the specimen into the objective lens. However, because some light is refracted or bent as it passes from glass into air, the refracted light rays are

Figure 1.3 **The light pathway of a microscope.**

The **ocular,** or eyepiece, is a complex piece, located at the top of the instrument, that consists of two or more internal lenses and usually has a magnification of 10×. Most modern microscopes (figure 1.2) have two ocular (binocular) lenses.

Three or more **objectives** are usually present. Note that they are attached to a rotatable **nosepiece,** which makes it possible to move them into position over a slide. Objectives on most laboratory microscopes have magnifications of 10×, 40×, and 100×, designated as **low-power, high-dry,** and **oil immersion,** respectively. Some microscopes will have a fourth objective for rapid scanning of microscopic fields that is only 4×.

The total magnification of a compound microscope is determined by multiplying the power of the ocular lens times the power of the objective lens used. Thus, the magnification of a microscope in which the oil immersion lens is being used is:

$$10 \times 100 = 1000$$

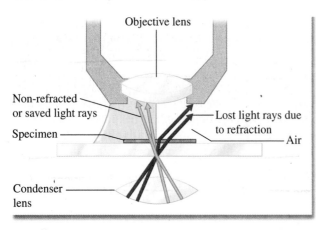

Figure 1.4 Immersion oil, having the same refractive index as glass, prevents light loss due to refraction.

Figure 1.5 When oculars are removed for cleaning, cover the ocular opening with lens tissue. A blast from an air syringe or gas canister removes dust and lint.

lost, and as a result the numerical aperture is diminished (figure 1.4). The greater the loss of refracted light, the lower the numerical aperture. The final result is that the resolving power is greatly reduced.

For any light microscope, the limit of resolution is about 0.2 μm. This means that two objects closer than 0.2 μm would not be seen as two distinct objects. Because bacterial cells are about 1 μm, the cells can be resolved by the light microscope, but that is not the case for internal structures in bacterial cells that are smaller than 0.2 μm.

In order to maximize the resolving power from a lens system, the following should be considered:

- A **blue filter** should be placed over the light source because the shorter wavelength of the resulting light will provide maximum resolution.
- The condenser should be kept at the highest position that allows the maximum amount of light to enter the objective lens and therefore limit the amount of light lost due to refraction.
- The diaphragm should not be stopped down too much. While closing the diaphragm improves the contrast, it also reduces the numerical aperture.
- **Immersion oil** should be used between the slide and the 100× objective lens. This is a special oil that has the same refractive index as glass. When placed between the specimen and objective lens, the oil forms a continuous lens system that limits the loss of light due to refraction.

The bottom line is that for magnification to increase, resolution must also increase. Thus, a greater magnification cannot be achieved simply by adding a stronger ocular lens.

Lens Care

Keeping the lenses of your microscope clean is a constant concern. Unless all lenses are kept free of

dust, oil, and other contaminants, they cannot achieve the degree of resolution that is intended. Consider the following suggestions for cleaning the various lens components:

Cleaning Tissues Only lint-free, optically safe tissues should be used to clean lenses. Tissues free of abrasive grit fall in this category. Booklets of lens tissue are most widely used for this purpose. Although several types of boxed tissues are also safe, *use only the type of tissue that is recommended by your instructor* (figure 1.5).

Solvents Various liquids can be used for cleaning microscope lenses. Green soap with warm water works very well. Xylene is universally acceptable. Alcohol and acetone are also recommended, but often with some reservations. Acetone is a powerful solvent that could possibly dissolve the lens mounting cement in some objective lenses if it were used too liberally. When it is used it should be used sparingly. Your instructor will inform you as to what solvents can be used on the lenses of your microscope.

Oculars The best way to determine if your eyepiece is clean is to rotate it between the thumb and forefinger as you look through the microscope. A rotating pattern will be evidence of dirt.

If cleaning the top lens of the ocular with lens tissue fails to remove the debris, one should try cleaning the lower lens with lens tissue and blowing off any excess lint with an air syringe or gas canister. *Whenever the ocular is removed from the microscope, it is imperative that a piece of lens tissue be placed over the open end of the microscope as illustrated in figure 1.5.*

Objectives Objective lenses often become soiled by materials from slides or fingers. A piece of lens tissue moistened with green soap and water, or one of the acceptable solvents mentioned on page 6, will usually remove whatever is on the lens. Sometimes a cotton swab with a solvent will work better than lens tissue. At any time that the image on the slide is unclear or cloudy, assume at once that the objective you are using is soiled.

Condenser Dust often accumulates on the top surface of the condenser; thus, wiping it off occasionally with lens tissue is desirable.

Figure 1.6 **The slide must be properly positioned as the retainer lever is moved to the right.**

Procedures

If your microscope has three objectives, you have three magnification options: (1) low-power, or $100\times$ total magnification, (2) high-dry magnification, which is $400\times$ total with a $40\times$ objective, and (3) $1000\times$ total magnification with a $100\times$ oil immersion objective.

Whether you use the low-power objective or the oil immersion objective will depend on how much magnification is necessary. Generally speaking, however, it is best to start with the low-power objective and progress to the higher magnifications as your study progresses. Consider the following suggestions for setting up your microscope and making microscopic observations.

Low-Power Examination The main reason for starting with the low-power objective is to enable you to explore the slide to look for the object you are planning to study. Once you have found what you are looking for, you can proceed to higher magnifications. Use the following steps when exploring a slide with the low-power objective:

1. Position the slide on the stage with the material to be studied on the *upper* surface of the slide. Figure 1.6 illustrates how the slide must be held in place by the mechanical stage retainer lever.
2. Turn on the light source, using a *minimum* amount of voltage. If necessary, reposition the slide so that the stained material on the slide is in the *exact center* of the light source.
3. Check the condenser to see that it has been raised to its highest point.
4. If the low-power objective is not directly over the center of the stage, rotate it into position. Be sure that as you rotate the objective into position it clicks into its locked position.
5. Turn the coarse adjustment knob to lower the objective until it stops. A built-in stop will prevent the objective from touching the slide.
6. While looking down through the ocular (or oculars), bring the object into focus by turning the fine adjustment focusing knob. Don't readjust the

coarse adjustment knob. If you are using a binocular microscope, it will also be necessary to adjust the interocular distance and diopter adjustment to match your eyes.
7. For optimal viewing, it is necessary to focus the condenser and adjust it for maximum illumination. This procedure should be performed each time the objective lens is changed. Raise the iris diaphragm to its highest position. Close the iris diaphragm until the edges of the diaphragm image appear fuzzy. Lower the condenser using its adjustment knob until the edges of the diaphragm are brought into sharp focus. You should now clearly see the sides of the diaphragm expand beyond the field of view. Refocus the specimen using the fine adjustment. Note that as you close the iris diaphragm to reduce the light intensity, the contrast improves and the depth of field increases. **Depth of field** is defined as the range of distance in front of and behind a focused image within which other objects will appear clear and sharply defined.
8. Once an image is visible, move the slide about to search out what you are looking for. The slide is moved by turning the knobs that move the mechanical stage.
9. Check the cleanliness of the ocular, using the procedure outlined earlier.
10. Once you have identified the structures to be studied and wish to increase the magnification, you may proceed to either high-dry or oil immersion magnification. However, before changing objectives, *be sure to center the object you wish to observe.*

High-Dry Examination To proceed from low-power to high-dry magnification, all that is necessary is to rotate the high-dry objective into position and open up the diaphragm somewhat. It may be necessary to make a minor adjustment with the fine adjustment knob to sharpen up the image, but *the coarse adjustment knob should not be touched.*

Good quality modern microscopes are usually both **parfocal** and **parcentral.** This means that the image will remain both centered and in focus when changing from a lower-power to a higher-power objective lens.

When increasing the lighting, be sure to open up the diaphragm first instead of increasing the voltage on your lamp; the reason is that *lamp life is greatly extended when used at low voltage.* If the field is not bright enough after opening the diaphragm, feel free to increase the voltage. A final point: Keep the condenser at its highest point.

Oil Immersion Techniques The oil immersion lens derives its name from the fact that a special mineral oil is interposed between the specimen and the $100\times$ objective lens. As stated previously, this reduces light refraction and maximizes the numerical aperture to improve the resolution. The use of oil in this way enhances the resolving power of the microscope. Figure 1.4 reveals this phenomenon.

With parfocal objectives one can go directly to oil immersion from either low-power or high-dry. On some microscopes, however, going from low-power to high-power and then to oil immersion is better. Once the microscope has been brought into focus at one magnification, the oil immersion lens can be rotated into position without fear of striking the slide.

Before rotating the oil immersion lens into position, however, a drop of immersion oil must be placed on the slide. An oil immersion lens should never be used without oil. Incidentally, if the oil appears cloudy, it should be discarded.

When using the oil immersion lens, more light is necessary to adequately visualize an image. Opening the diaphragm increases the resolving power of the microscope at higher magnifications. Thus, the iris diaphragm must be opened wider when using the oil immersion lens. Also, do not forget to refocus the condenser when moving from lower-power to higher-power objectives. Some microscopes also employ blue or green filters on the lamp housing to enhance resolving power.

Since the oil immersion lens will be used extensively in all bacteriological studies, it is of paramount importance that you learn how to use this lens properly. Using this lens takes a little practice due to the difficulties usually encountered in manipulating the lighting. It is important for all beginning students to appreciate that the working distance of a lens, the distance between the lens and microscope slide, decreases significantly as the magnification of the lens increases (table 1.1). Hence, the potential for damage to the oil immersion lens because of a collision with the microscope slide is very great. A final comment of importance: At the end of the laboratory period remove all immersion oil from the lens tip with lens tissue.

Table 1.1 Relationship of Working Distance to Magnification

LENS	MAGNIFICATION	FOCAL LENGTH (mm)	WORKING DISTANCE (mm)
Low-power	10×	16.0	7.7
High-dry	40×	4.0	0.3
Oil immersion	100×	1.8	0.12

Putting It Away

When you take a microscope from the cabinet at the beginning of the period, you expect it to be clean and in proper working condition. The next person to use the instrument after you have used it will expect the same consideration. A few moments of care at the end of the period will ensure these conditions. Check over the following list of items at the end of each period before you return the microscope to the cabinet.

1. Remove the slide from the stage.
2. If immersion oil has been used, wipe it off the lens and stage with lens tissue. Also, make sure that no immersion oil is on the 40× objective. This lens often becomes contaminated with oil as a result of mistakes made by beginning students. (Do not wipe oil off slides you wish to keep. Simply put them into a slide box and let the oil drain off.)
3. Rotate the low-power objective into position.
4. If the microscope has been inclined, return it to an erect position.
5. If the microscope has a built-in movable lamp, raise the lamp to its highest position.
6. If the microscope has a long attached electric cord, wrap it around the base.
7. Adjust the mechanical stage so that it does not project too far on either side.
8. Replace the dustcover.
9. If the microscope has a separate transformer, return it to its designated place.
10. Return the microscope to its correct place in the cabinet.

Laboratory Report

Before the microscope is to be used in the laboratory, answer all the questions on Laboratory Report 1. Preparation on your part prior to going to the laboratory will greatly facilitate your understanding. Your instructor may wish to collect this report at the *beginning of the period* on the first day that the microscope is to be used in class.

Laboratory Report

Student: _____

Date: _____ Section: _____

1 Brightfield Microscopy

A. Short-Answer Questions

1. Describe the position of your hands when carrying the microscope to and from your laboratory bench.

   _____

   _____

2. Differentiate between the limit of resolution of the typical light microscope and that of the unaided human eye.

   _____

   _____

3. (a) What two adjustments can be made to the condenser? (b) What effect do these adjustments have on the image?

   _____

   _____

4. Why are condenser adjustments generally preferred over the use of the light intensity control?

   _____

   _____

5. When using the oil immersion lens, what four procedures can be implemented to achieve the maximum resolution?

   _____

   _____

   _____

6. Why is it advisable to start first with the low-power lens when viewing a slide?

   _____

   _____

7. Why is it necessary to use oil in conjunction with the oil immersion lens and not with the other objectives?

   _____

   _____

8. What is the relationship between the working distance of an objective lens and its magnification power?

   _____

   _____

B. Matching Questions

Match the lens (condenser, high-dry, low-power, ocular, or oil immersion) to its description. Choices may be used more than once.

1. This objective lens provides the highest magnification.
2. This objective lens provides the second-highest magnification.
3. This objective lens provides the lowest magnification.
4. This objective lens has the shortest working distance.
5. The coarse focus knob should be adjusted only when using this objective lens.
6. This lens collects and focuses light from the lamp onto the specimen on the slide.
7. This lens, also known as the eyepiece, often comes in pairs.
8. Diopter adjustments can be made to this lens.
9. A diaphragm is used to regulate light passing through this lens.

C. True-False

1. Acetone is the safest solvent for cleaning an objective lens.
2. Only lint-free, optically safe tissue should be used to wipe off microscope lenses.
3. The total magnification capability of a light microscope is only limited by the magnifying power of the lens system.
4. The coarse focus knob can be used to adjust the focus when using any of the objective lenses.
5. Once focus is achieved at one magnification, a higher-power objective lens can be rotated into position without fear of striking the slide.

D. Multiple Choice

Select the answer that best completes the following statements.

1. The resolving power of a microscope is a function of
   a. the magnifying power of the lenses.
   b. the numerical aperture of the lenses.
   c. the wavelength of light.
   d. Both (a) and (b) are correct.
   e. Both (b) and (c) are correct.

2. The coarse and fine focus knobs adjust the distance between
   a. the objective and ocular lenses.
   b. the ocular lenses.
   c. the ocular lenses and your eyes.
   d. the stage and the condenser lens.
   e. the stage and the objective lens.

3. A microscope that maintains focus when the objective magnification is increased is called
   a. binocular.
   b. myopic.
   c. parfocal.
   d. refractive.
   e. resolute.

ANSWERS

Matching

1. _____

2. _____

3. _____

4. _____

5. _____

6. _____

7. _____

8. _____

9. _____

True-False

1. _____

2. _____

3. _____

4. _____

5. _____

Multiple Choice

1. _____

2. _____

3. _____

4. The total magnification achieved when using a 100× oil immersion lens with 10× binocular eyepieces is
   a. 10×.
   b. 100×.
   c. 200×.
   d. 1000×.
   e. 2000×.

5. The most useful adjustment for increasing image contrast in low-power magnification is
   a. closing down the diaphragm.
   b. closing one eye.
   c. opening up the diaphragm.
   d. placing a drop of oil on the slide.
   e. using a blue filter.

6. Before the oil immersion lens is rotated into place, you should
   a. center the object of interest in the preceding lens.
   b. lower the stage with use of the coarse focus adjustment knob.
   c. place a drop of oil on the slide.
   d. Both (a) and (c) are correct.
   e. All are correct.

4. _____

5. _____

6. _____

Darkfield Microscopy

Learning Outcomes

After completing this exercise, you should be able to

1. Visualize a specimen using a darkfield microscope.

2. Understand the difference between the illumination systems for the darkfield microscope and the brightfield microscope.

Delicate transparent living organisms can be more easily observed with darkfield microscopy than with conventional brightfield microscopy. This method is particularly useful when one is attempting to identify spirochaetes in an exudate from a syphilitic lesion. Figure 2.1 illustrates the appearance of these organisms under such illumination. This effect may be produced by placing a darkfield stop below the regular condenser or by replacing the condenser with a specially constructed one.

Another application of darkfield microscopy is in the fluorescence microscope. Although fluorescence may be seen without a dark field, it is greatly enhanced with this application.

To achieve the darkfield effect it is necessary to alter the light rays that approach the objective in such a way that only oblique rays strike the objects being viewed. The obliquity of the rays must be so extreme

that if no objects are in the field, the background is completely light-free. Objects in the field become brightly illuminated by the rays that are reflected up through the lens system of the microscope.

Although there are several different methods for producing a dark field, only two devices will be described here: the star diaphragm and the cardioid condenser. The availability of equipment will determine the method to be used in this laboratory.

The Star Diaphragm

One of the simplest ways to produce the darkfield effect is to insert a star diaphragm into the filter slot of the condenser housing, as shown in figure 2.2. This device has an opaque disk in the center that blocks the central rays of light. Figure 2.3 reveals the effect of this stop on the light rays passing through the condenser. If such a device is not available, one can be made by cutting round disks of opaque paper of different sizes that are cemented to transparent celluloid disks that will fit into the slot. If the microscope normally has a diffusion disk in this slot, it is best to replace it with rigid clear celluloid or glass.

An interesting modification of this technique is to use colored celluloid stops instead of opaque paper. Backgrounds of blue, red, or any color can be produced in this way.

Figure 2.1 **Darkfield image of *Treponema pallidum*, the bacterium that causes syphilis.**
Centers for Disease Control and Prevention/Susan Lindsley

Figure 2.2 **The insertion of a star diaphragm into the filter slot of the condenser will produce a dark field suitable for low magnifications.**
©McGraw-Hill Education. Auburn University Photographic Services

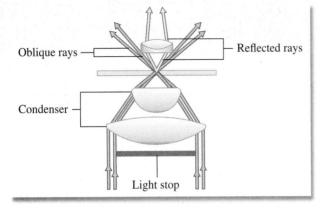

Figure 2.3 **The star diaphragm allows only peripheral light rays to pass up through the condenser. This method requires maximum illumination.**

Figure 2.4 **A cardioid condenser provides greater light concentration for oblique illumination than the star diaphragm.**

In setting up this type of darkfield illumination, it is necessary to keep these points in mind:

1. Limit this technique to the study of large organisms that can be seen easily with low-power magnification. *Good resolution with higher-powered objectives is difficult with this method.*
2. Keep the diaphragm wide open and use as much light as possible. If the microscope has a voltage regulator, you will find that the higher voltages will produce better results.
3. Be sure to center the stop as precisely as possible.
4. Move the condenser up and down to produce the best effects.

The Cardioid Condenser

The difficulty that results from using the star diaphragm or opaque paper disks with high-dry and oil immersion objectives is that the oblique rays are not as carefully metered as is necessary for the higher magnifications. Special condensers such as the cardioid or paraboloid types must be used. Since the cardioid type is the most frequently used type, its use will be described here.

Figure 2.4 illustrates the light path through such a condenser. Note that the light rays entering the lower element of the condenser are reflected first off a convex mirrored surface and then off a second concave surface to produce the desired oblique rays of light. Once the condenser has been installed in the microscope, the following steps should be followed to produce ideal illumination.

Materials

- slides and cover glasses of excellent quality (slides of 1.15–1.25 mm thickness and No. 1 cover glasses)

1. Adjust the upper surface of the condenser to a height just below stage level.
2. Place a clear glass slide in position over the condenser.
3. Focus the 10× objective on the top of the condenser until a bright ring comes into focus.

4. Center the bright ring so that it is concentric with the field edge by adjusting the centering screws on the darkfield condenser. If the condenser has a light source built into it, it will also be necessary to center it as well to achieve even illumination.
5. Remove the clear glass slide.
6. If a funnel stop is available for the oil immersion objective lens, remove the oil immersion objective and insert the funnel stop. (This stop serves to reduce the numerical aperture of the oil immersion objective to a value that is less than that of the condenser.)
7. Place a drop of immersion oil on the upper surface of the condenser and place the slide on top of the oil. The following preconditions in slide usage must be adhered to:

   - Slides and cover glasses should be optically perfect. Scratches and imperfections will cause annoying diffractions of light rays.
   - Slides and cover glasses must be free of dirt or grease of any kind.
   - A cover glass should always be used.

8. If the oil immersion lens is to be used, place a drop of oil on the cover glass.
9. If the field does not appear dark and lacks contrast, return to the 10× objective and check the ring concentricity and light source centration. If contrast is still lacking after these adjustments, the specimen is probably too thick.
10. If sharp focus is difficult to achieve under oil immersion, try using a thinner cover glass and adding more oil to the top of the cover glass and bottom of the slide.

Laboratory Report

This exercise may be used in conjunction with Part 2 when studying the various types of organisms. After reading over this exercise and doing any special assignments made by your instructor, answer the questions in Laboratory Report 2 about darkfield microscopy.

2 Darkfield Microscopy

A. Short-Answer Questions

1. For which types of specimens is darkfield microscopy preferred over brightfield microscopy?

    _____

    _____

    _____

2. If a darkfield condenser causes all light rays to bypass the objective, where does the light come from that makes an object visible in a dark field?

    _____

    _____

    _____

3. What advantage does a cardioid condenser have over a star diaphragm?

    _____

    _____

    _____

4. If accessories for darkfield microscopy were not available, how would you construct a simple star diaphragm?

    _____

    _____

    _____

Phase-Contrast Microscopy

Learning Outcomes

After completing this exercise, you should be able to

1. Understand how the phase-contrast microscope takes advantage of density differences in cell constituents to produce an image.
2. Properly use a phase-contrast microscope to obtain an image of living cells.
3. Align a phase-contrast microscope.

Figure 3.1 **Phase-contrast image of an endospore-forming bacterium.**
©Stephen Durr

If one tries to observe cells without the benefit of staining, very little contrast or detail can be seen because the cells appear transparent against the aqueous medium in which they are usually suspended. Staining increases the contrast between the cell and its surrounding medium, allowing the observer to see more cellular detail, including some inclusions and various organelles. However, staining usually results in cell death, which means we are unable to observe living cells or their activities, and staining can also lead to undesirable artifacts.

A microscope that is able to differentiate the transparent protoplasmic structures and enhance the contrast between a cell and its surroundings, without the necessity of staining, is the **phase-contrast microscope.** This microscope was developed by the Dutch physicist Frits Zernike in the 1930s. For his discovery of phase-contrast microscopy, he was awarded the Nobel Prize in 1953. Today it is the microscope of choice for viewing living cells and their activities such as motility. Figure 3.1 illustrates a phase-contrast image of an endospore-forming bacterium. In this exercise, you will learn to use the phase-contrast microscope and observe the activities of living cells.

To understand how a phase-contrast microscope works, it is necessary to review some of the physical properties of light and how it interacts with matter such as biological material. Light energy can be represented as a waveform that has both an amplitude and a characteristic wavelength (illustration 1, figure 3.2). Some objects can reduce the amplitude of a light wave, and they would appear as dark objects in a microscope. In contrast, light can pass through matter without affecting the amplitude, and these objects would

appear transparent in a microscope. However, as light passes through some of the transparent objects, it can be slowed down by $\frac{1}{4}$ wavelength, resulting in a **phase shift** of the light's wavelength (illustration 2, figure 3.2). For a cell, the phase shift without a reduction in amplitude results in the cell having a different refractive index than its surroundings. However, the phase shifts caused by biological material are usually too small to be seen as contrast differences in a brightfield microscope. Therefore, in a brightfield microscope, cells appear transparent rather than opaque against their surroundings. Since biological material lacks any appreciable contrast, it becomes necessary to stain cells with various dyes in order to study them. However, Zernike took advantage of the $\frac{1}{4}$ wavelength phase shift to enhance the small contrast differences in the various components that comprise a cell, making them visible in his microscope. This involved manipulating the light rays that were shifted and those that were unchanged as they emerged from biological material.

Two Types of Light Rays

Light rays passing through a transparent object emerge as either direct or diffracted rays. Those rays that pass

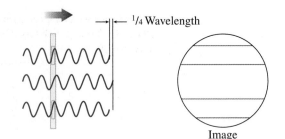

AMPLITUDE OBJECTS

(1) The extent to which the amplitude of light rays is diminished determines the darkness of an object in a microscopic field.

PHASE OBJECTS

(2) Note that the retardation of light rays without amplitude diminution results in transparent phase objects.

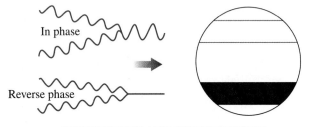

DIRECT AND DIFFRACTED RAYS

(3) A light ray passing through a slit or transparent object emerges as a direct ray with several orders of diffracted rays. The diffracted rays are $\frac{1}{4}$ wavelength out of phase with the direct ray.

COINCIDENCE AND INTERFERENCE

(4) Note that when two light rays are in phase they will unite to produce amplitude summation. Light rays in reverse phase, however, cancel each other (interference) to produce dark objects.

Figure 3.2   The utilization of light rays in phase-contrast microscopy.

straight through unaffected by the medium are called **direct rays.** They are unaltered in amplitude and phase. The rays that are bent because they are retarded by the medium (due to density differences) emerge from the object as **diffracted rays.** It is these specific light rays that are retarded $\frac{1}{4}$ wavelength. Illustration 3, figure 3.2, shows these two types of light rays.

If the direct and diffracted light waves are brought into exact phase with each other, the result is **coincidence** with the resultant amplitude of the converged waves being the sum of the two waves. This increase in amplitude will produce increased brightness of the object in the field. In contrast, if two light waves of equal amplitude are in reverse phase ($\frac{1}{2}$ wavelength off), their amplitudes will cancel each other to produce a dark object. This is called **interference.** Illustration 4, figure 3.2, shows these two conditions.

Phase-Contrast Microscope

In constructing his first phase-contrast microscope, Zernike experimented with various configurations of diaphragms and various materials that could be used to retard or advance the direct light rays. Figure 3.3 illustrates the optical system of a typical modern phase-contrast microscope. It differs from a conventional brightfield microscope by having (1) a different type of diaphragm and (2) a phase plate.

The diaphragm consists of an **annular stop** that allows only a hollow cone of light rays to pass through the condenser to the specimen on the slide. The phase plate is a special optical disk located on the objective lens near its rear focal plane. It has a phase ring that advances or retards the direct light rays $\frac{1}{4}$ wavelength.

Note in figure 3.3 that the direct rays converge on the phase ring to be advanced or retarded $\frac{1}{4}$ wavelength. These rays emerge as solid lines from the object on the slide. This ring on the phase plate is coated with a material that will produce the desired phase shift. The diffracted rays, on the other hand, which have already been retarded $\frac{1}{4}$ wavelength by the phase object on the slide, completely miss the phase ring and are not affected by the phase plate. It should be clear, then, that depending on the type of phase-contrast microscope, the convergence of diffracted and direct rays on the image plane will result in either a brighter image (*amplitude summation*) or a darker image (*amplitude interference* or *reverse phase*). The former is referred to as **bright-phase**

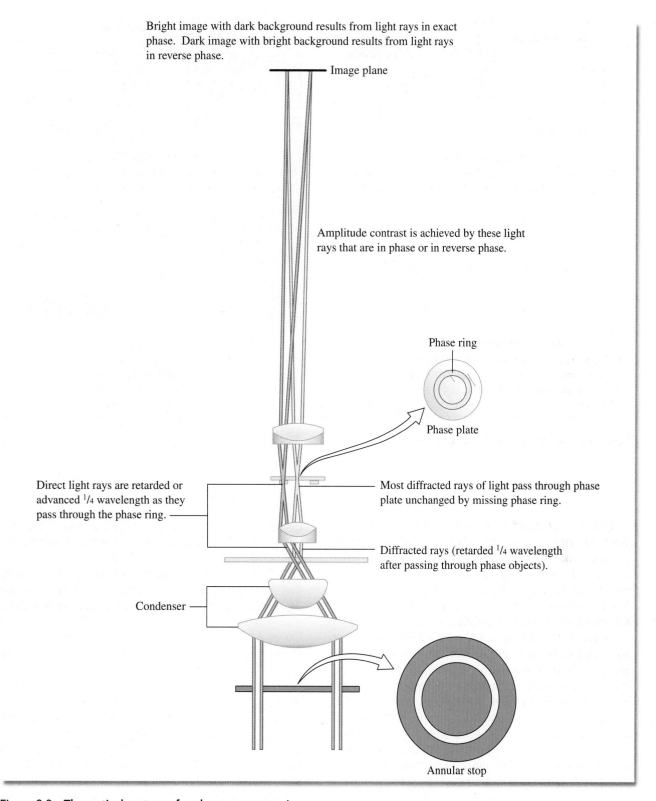

Bright image with dark background results from light rays in exact phase. Dark image with bright background results from light rays in reverse phase.

Image plane

Amplitude contrast is achieved by these light rays that are in phase or in reverse phase.

Phase ring

Phase plate

Direct light rays are retarded or advanced $1/4$ wavelength as they pass through the phase ring.

Most diffracted rays of light pass through phase plate unchanged by missing phase ring.

Diffracted rays (retarded $1/4$ wavelength after passing through phase objects).

Condenser

Annular stop

Figure 3.3   **The optical system of a phase-contrast microscope.**

microscopy; the latter as ***dark-phase*** microscopy. The apparent brightness or darkness, incidentally, is proportional to the square of the amplitude; thus, the image will be four times as bright or dark as one seen through a brightfield microscope.

It should be added here that the phase plates of some microscopes have coatings to change the phase of the diffracted rays. In any event, the end result will be the same: to achieve coincidence or interference of direct and diffracted rays.

Microscope Adjustments

If the annular stop under the condenser of a phase-contrast microscope can be moved out of position, this instrument can also be used for brightfield studies. Although a phase-contrast objective has a phase ring attached to the top surface of one of its lenses, the presence of that ring does not impair the resolution of the objective when it is used in the brightfield mode. It is for this reason that manufacturers have designed phase-contrast microscopes in such a way that they can be quickly converted to brightfield operation.

To make a microscope function efficiently in both phase-contrast and brightfield situations, one must master the following procedures:

- lining up the annular ring and phase rings so that they are perfectly concentric,
- adjusting the light source so that maximum illumination is achieved for both phase-contrast and brightfield usage, and
- being able to shift back and forth easily from phase-contrast to brightfield modes. The following suggestions should be helpful in coping with these problems.

Alignment of Annulus and Phase Ring

Unless the annular ring below the condenser is aligned perfectly with the phase ring in each objective lens, good phase-contrast imagery cannot be achieved. Figure 3.4 illustrates the difference between nonalignment and alignment. If a microscope has only one phase-contrast objective, there will be only one annular stop that has to be aligned. If a microscope has two or more phase objectives, there must be a substage unit with separate annular stops for each phase objective, and the alignment procedure must be performed separately for each objective and its annular stop.

Since the objective cannot be moved once it is locked in position, all adjustments are made to the annular stop. On some microscopes the adjustment may be made with tools, as illustrated in figure 3.5. On other microscopes, in figure 3.6, the annular rings are moved into position with special knobs on the substage unit. Since the method of adjustment varies from one brand of microscope to another, one has to follow the instructions provided by the manufacturer. Once the adjustments have been made, they are rigidly set and needn't be changed unless someone inadvertently disturbs them.

To observe ring alignment, one can replace the eyepiece with a **centering telescope** as shown in figure 3.7. With this unit in place, the two rings can be brought into sharp focus by rotating the focusing ring on the telescope. Refocusing is necessary for each objective and its matching annular stop. Some manufacturers, such as American Optical, provide

Figure 3.4 **The image on the right illustrates the appearance of the rings when perfect alignment of phase ring and annulus diaphragm has been achieved.**

Figure 3.5 **Alignment of the annulus diaphragm and phase ring is accomplished with a pair of Allen-type screwdrivers on this American Optical microscope.**

an aperture viewing unit (figure 3.8), which enables one to observe the rings without using a centering telescope. Zeiss microscopes have a unit called the *Optovar,* which is located in a position similar to the American Optical unit that serves the same purpose.

Light Source Adjustment

For both brightfield and phase-contrast modes, it is essential that optimum lighting be achieved. This is no great problem for a simple setup such as the American Optical instrument shown in figure 3.9. For multiple-phase objective microscopes, however, there are many more adjustments that need to be made.

Figure 3.6   **Alignment of the annulus and phase ring is achieved by adjusting the two knobs as shown.**
©McGraw-Hill Education. Auburn University Photographic Services

Figure 3.8   **Some microscopes have an aperture viewing unit that can be used instead of a centering telescope for observing the orientation of the phase ring and annular ring.**

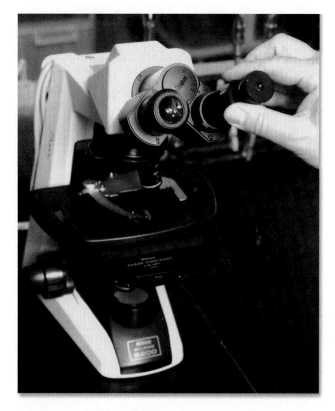

Figure 3.7   **If the ocular of a phase-contrast microscope is replaced with a centering telescope, the orientation of the phase ring and annular ring can be viewed.**
©McGraw-Hill Education. Auburn University Photographic Services

Figure 3.9   **The annular stop on this American Optical microscope is located on a slideway. When pushed in, the annular stop is in position.**

A few suggestions that highlight some of the problems and solutions follow:

1. Since blue light provides better images for both phase-contrast and brightfield modes, make certain that a blue filter is placed in the filter holder that is positioned in the light path. If the microscope has no filter holder, placing the filter over the light source on the base will help.

2. Brightness of field under phase-contrast is controlled by adjusting the voltage or the iris diaphragm on the base. Considerably more light is required for phase-contrast than for brightfield since so much light is blocked out by the annular stop.

3. The evenness of illumination on some microscopes, seen on these pages, can be adjusted by removing the lamp housing from the microscope and focusing the light spot on a piece of translucent

white paper. For the detailed steps in this procedure, one should consult the instruction manual that comes with the microscope. Light source adjustments of this nature are not necessary for the simpler types of microscopes.

Working Procedures

Once the light source is correct and the phase elements are centered you are finally ready to examine slide preparations. Keep in mind that from now on most of the adjustments described earlier should not be altered; however, if misalignment has occurred due to mishandling, it will be necessary to refer back to the alignment procedures. The following guidelines should be adhered to in all phase-contrast studies:

- Use only optically perfect slides and cover glasses (no bubbles or striae in the glass).
- Be sure that slides and cover glasses are completely free of grease or chemicals.
- Use wet mount slides instead of hanging drop preparations. The latter leave much to be desired. Culture broths containing bacteria or protozoan suspensions are ideal for wet mounts.
- In general, limit observations to living cells. In most instances, stained slides are not satisfactory.

The first time you use phase-contrast optics to examine a wet mount, follow these suggestions:

1. Place the wet mount slide on the stage and bring the material into focus, *using brightfield optics* at low-power magnification.

2. Once the image is in focus, switch to phase optics at the same magnification. Remember, it is necessary to place in position the matching annular stop.
3. Adjust the light intensity, first with the base diaphragm and then with the voltage regulator. In most instances, you will need to increase the amount of light for phase-contrast.
4. Switch to higher magnifications, much in the same way you do for brightfield optics, except that you have to rotate a matching annular stop into position.
5. If an oil immersion phase objective is used, add immersion oil to the top of the condenser as well as to the top of the cover glass.
6. Don't be disturbed by the "halo effect" that you observe with phase optics. Halos are normal.

Laboratory Report

This exercise may be used in conjunction with Part 2 in studying various types of organisms. Organelles in protozoans and algae will show up more distinctly than with brightfield optics. After reading this exercise and doing any special assignments made by your instructor, answer the questions in Laboratory Report 3.

Laboratory Report

Student: _____

Date: _____ Section: _____

3 Phase-Contrast Microscopy

A. Short-Answer Questions

1. Staining of cells is often performed to enhance images acquired by brightfield microscopy. Phase-contrast microscopy does not require cell staining. Why is this advantageous?

   _____

   _____

2. As light passes through a transparent object, how are direct and diffracted light rays produced? How much phase shift occurs?

   _____

   _____

3. How do coincidence and interference of light rays differ? What type of image does each produce? How does that contribute to a sharper image?

   _____

   _____

   _____

4. Differentiate between bright-phase and dark-phase microscopy in terms of phase shift.

   _____

   _____

   _____

5. Which two items can be used to check the alignment of the annulus and phase ring?

   _____

   _____

B. Multiple Choice

Select the answer that best completes the following statements.

1. A phase-contrast microscope differs from a brightfield microscope by having a
   a. blue filter in the ocular lens.
   b. diaphragm with an annular stop.
   c. phase plate in the objective lens.
   d. Both (b) and (c) are correct.
   e. All are correct.

ANSWERS

Multiple Choice

1. _____

2. Which of the following would be best observed for a bacterial cell using phase-contrast microscopy?
   a. Motility of cells
   b. Bacterial nucleoid
   c. Cell wall
   d. Glycocalyx

3. Amplitude summation occurs in phase-contrast optics when both direct and diffracted rays are
   a. in phase.
   b. in reverse phase.
   c. off $\frac{1}{4}$ wavelength.
   d. None of these are correct.

4. The phase-contrast microscope is best suited for observing
   a. living organisms in an uncovered drop on a slide.
   b. stained slides with cover glasses.
   c. living organisms in hanging drop slide preparations.
   d. living organisms on a slide with a cover glass.

2. _____

3. _____

4. _____

Microscopic Measurements

Learning Outcomes

After completing this exercise, you should be able to

1. Calibrate an ocular micrometer using a stage micrometer.

2. Use an ocular micrometer to accurately measure the dimensions of stained cells.

Figure 4.2 **Stage micrometer is positioned by centering the small glass disk over the light source.**

Figure 4.1 **The ocular micrometer with retaining ring is inserted into the base of the eyepiece.**

With an ocular micrometer properly installed in the eyepiece of your microscope, it is a simple matter to measure the size of microorganisms that are seen in the microscopic field. An **ocular micrometer** consists of a circular disk of glass that has graduations engraved on its upper surface. These graduations appear as shown in illustration B, figure 4.4. On some microscopes one has to disassemble the ocular so that the disk can be placed on a shelf in the ocular tube between the two lenses. On most microscopes, however, the ocular micrometer is simply inserted into the bottom of the ocular, as shown in figure 4.1. Before one can use the micrometer, it is necessary to calibrate it for each of the objectives by using a stage micrometer.

The principal purpose of this exercise is to show you how to calibrate an ocular micrometer for the various objectives on your microscope.

Calibration Procedure

The distance between the lines of an ocular micrometer is an arbitrary value that has meaning only if the ocular micrometer is calibrated for the objective that is being used. A **stage micrometer** (figure 4.2), also known as an *objective micrometer,* has lines inscribed on it that are exactly 0.01 mm (10 μm) apart. Illustration C, figure 4.4, shows these graduations.

To calibrate the ocular micrometer for a given objective, it is necessary to superimpose the two scales and determine how many of the ocular graduations coincide with one graduation on the scale of the stage micrometer. Illustration A in figure 4.4 shows how the two scales appear when they are properly aligned in the microscopic field. In this case, seven ocular divisions match up with one stage micrometer division of 0.01 mm to give an ocular value of 0.01/7, or 0.00143 mm. Since there are 1000 micrometers in 1 millimeter, these divisions are 1.43 μm apart.

With this information known, the stage micrometer is replaced with a slide of organisms (figure 4.3) to be measured. Illustration D, figure 4.4, shows how a field of microorganisms might appear with the ocular micrometer in the eyepiece. To determine the size of an organism, then, it is a simple matter to count the graduations and multiply this number by the known distance between the graduations. When calibrating the objectives of a microscope, proceed as follows.

Figure 4.3   **After calibration is completed, the stage micrometer is replaced with a slide for measurements.**

Stage micrometer

Ocular micrometer

(a) View showing the alignment of stage micrometer graduations (top bold lines) with ocular micrometer graduations (bottom lines). One space of the stage micrometer (0.01 mm) aligns with 7 graduations of the ocular micrometer. Therefore, each graduation of the ocular micrometer $\frac{0.01}{7} = 0.0014$ mm, or 1.4 microns.

(b) View showing appearance of ocular micrometer graduations. Spacing is arbitrary.

(c) Appearance of stage micrometer graduations. Lines are exactly 0.01 mm (10 micrometers) apart.

(d) On the basis of the calibration calculations in view **(a)** above, what is the total length of the yeast cell and bud in this view?

Figure 4.4   **Calibration of ocular micrometer.**

Materials

- ocular micrometer or eyepiece that contains a micrometer disk
- stage micrometer

1. If eyepieces are available that contain ocular micrometers, replace the eyepiece in your microscope with one of them. If it is necessary to insert an ocular micrometer in your eyepiece, find out from your instructor whether it is to be inserted below the bottom lens or placed between the two lenses within the eyepiece. In either case, great care must be taken to avoid dropping the eyepiece or reassembling the lenses incorrectly. *Only with your instructor's prior approval shall eyepieces be disassembled.* Be sure that the graduations are on the upper surface of the glass disk.

2. Place the stage micrometer on the microscope stage and center it exactly over the light source.

3. With the low-power (10×) objective in position, bring the graduations of the stage micrometer into focus, *using the coarse adjustment knob. Reduce the lighting.* Note: If the microscope has an automatic stop, do not use it as you normally would for regular microscope slides. The stage micrometer slide is too thick to allow it to function properly.

4. Rotate the eyepiece until the graduations of the ocular micrometer lie parallel to the lines of the stage micrometer.

5. If the **low-power objective** is the objective to be calibrated, proceed to step 8.

6. If the **high-dry objective** is to be calibrated, swing it into position and proceed to step 8.

7. If the **oil immersion lens** is to be calibrated, place a drop of immersion oil on the stage micrometer, swing the oil immersion lens into position, and bring the lines into focus; then, proceed to the next step.

8. Move the stage micrometer laterally until the lines at one end coincide. Then look for another line on the ocular micrometer that coincides *exactly* with one on the stage micrometer. Occasionally, one stage micrometer division will include an even number of ocular divisions, as shown in illustration A of figure 4.4. In most instances, however, several stage graduations will be involved. In this case, divide the number of stage micrometer divisions by the number of ocular divisions that coincide. The figure you get will be that part of a stage micrometer division that is seen in an ocular division. This value must then be multiplied by 0.01 mm to get the amount of each ocular division.

Example: 3 divisions of the stage micrometer line up with 20 divisions of the ocular micrometer.

$$\text{Each ocular division} = \tfrac{3}{20} \times 0.01$$
$$= 0.0015 \text{ mm}$$
$$= 1.5 \ \mu\text{m}$$

Replace the stage micrometer with slides of organisms to be measured.

Measuring Assignments

Organisms such as protozoans, algae, fungi, and bacteria in the next few exercises may need to be measured. If your instructor requires that measurements be made, you will be referred to this exercise.

Later on you will be working with unknowns. In some cases measurements of the unknown organisms will be pertinent to identification.

If trial measurements are to be made at this time, your instructor will make appropriate assignments. **Important:** Remove the ocular micrometer from your microscope at the end of the laboratory period.

Laboratory Report

Answer the questions in Laboratory Report 4.

Laboratory Report

Student: _____

Date: _____ Section: _____

4 Microscopic Measurements

A. Short-Answer Questions

1. How do the graduations differ between ocular and stage micrometers?

    _____

    _____

2. If 13 ocular divisions line up with two divisions of the stage micrometer, what is the diameter (μm) of a cell that spans 16 ocular divisions?

    _____

    _____

3. Why must the entire calibration procedure be performed for each objective?

    _____

    _____

Survey of Microorganisms

Microorganisms abound in the environment. Eukaryotic microbes such as protozoa, algae, diatoms, and amoebas are plentiful in ponds and lakes. Bacteria are found associated with animals, occur abundantly in the soil and in water systems, and have even been isolated from core samples taken from deep within the earth's crust. Bacteria are also present in the air where they are distributed by convection currents that transport them from other environments. The Archaea, modern-day relatives of early microorganisms, occupy some of the most extreme environments such as acidic-volcanic hot springs, anaerobic environments devoid of any oxy-

CDC/Dr. Lucille K. Georg

gen, and lakes and salt marshes excessively high in sodium chloride. Cyanobacteria are photosynthetic prokaryotes that can be found growing in ponds and lakes, on limestone rocks, and even on the shingles that protect the roofs of our homes. Fungi are a very diverse group of microorganisms that are found in most common environments. For example, they degrade complex molecules in the soil, thus contributing to its fertility. Sometimes, however, they can be nuisance organisms; they form mildew in our bathroom showers and their spores cause allergies. The best way of describing the distribution of microorganisms is to say that they are ubiquitous, or found everywhere.

Intriguing questions to biologists are how are the various organisms related to one another and where do the individual organisms fit in an evolutionary scheme? Molecular biology techniques have provided a means to analyze the genetic relatedness of the organisms that comprise the biological world and determine where the various organisms fit into an evolutionary scheme. By comparing the sequence of ribosomal RNA molecules, coupled with biochemical data, investigators have developed a phylogenetic tree that illustrates the current thinking on the placement of the various organisms into such a scheme. This evolutionary scheme divides the biological world into three domains.

Domain Bacteria These organisms have a prokaryotic cell structure. They lack organelles such as mitochondria and chloroplasts, are devoid of an organized nucleus with a nuclear membrane, and possess 70S ribosomes that are inhibited by many broad-spectrum antibiotics. The vast majority of

© Roberto Osti Illustrations. Used by permission.

organisms are enclosed in a cell wall composed of peptidoglycan. The bacteria and cyanobacteria are members of this domain.

Domain Eukarya   Organisms in this domain have a eukaryotic cell structure. They contain membrane-bound organelles such as mitochondria and chloroplasts, an organized nucleus enclosed in a nuclear membrane, and 80S ribosomes that are not inhibited by broad-spectrum antibiotics. Plants, animals, and microorganisms such as protozoa, algae, and fungi belong in this domain. Plants have cell walls composed of cellulose and fungi have cell walls composed of chitin. In contrast, animal cells lack a cell wall structure.

Domain Archaea   The Archaea exhibit the characteristics of both the Bacteria and Eukarya. These organisms are considered to be the relatives of ancient microbes that existed during Archaean times. Like their bacterial counterparts, they possess a simple cell structure that lacks organelles and an organized nucleus. They have 70S ribosomes like bacteria, but the protein makeup and morphology of their ribosomes are more similar to eukaryotic ribosomes. Like eukaryotes, the ribosomes in Archaea are not sensitive to antibiotics. They have a cell wall but its structure is not composed of peptidoglycan. The principal habitats of these organisms are extreme environments such as volcanic hot springs, environments with excessively high salt, and environments devoid of oxygen. Thus, they are referred to as "extremophiles." The acido-thermophiles, the halobacteria, and the methanogens (methane bacteria) are examples of the Archaea.

In the exercises of Part 2, you will have the opportunity to study some of these organisms. In pond water, you may see amoebas, protozoa, various algae, diatoms, and cyanobacteria. You will sample for the presence of bacteria by exposing growth media to various environments. The fungi will be studied by looking at cultures and preparing slides of these organisms. Because the Archaea occur in extreme conditions and also require specialized culture techniques, it is unlikely that you will encounter any of these organisms.

Microbiology of Pond Water—Protists, Algae, and Cyanobacteria

Learning Outcomes

After completing this exercise, you should be able to

1. Prepare wet mounts of samples from aquatic environments.
2. Identify various eukaryotic microorganisms belonging to the algae and protozoa groups, and differentiate them from the prokaryotic cyanobacteria.
3. Differentiate between the prokaryotic cyanobacteria and the eukaryotic algae.

In this exercise you will study the diverse organisms that can occur in pond water. This will include eukaryotic organisms classified as algae and the protozoa, as well as the prokaryotic cyanobacteria. Illustrations and descriptions are provided to assist you in identifying some of these organisms. In addition, you may also observe small nematodes, insect larvae, microcrustaceans, rotifers, and other invertebrates. Your instructor may supply you with additional materials to identify these organisms, as they are not covered in this manual.

The purpose of this exercise is to familiarize you with the diversity of organisms that occur in a pond habitat. Most of your identifications will involve comparisons of morphological characteristics and differences. To study these organisms, you will make wet mounts of various samples of pond water. This is achieved by pipetting a drop of pond water onto a microscope slide and covering the drop with a coverslip. To achieve the best results, use the following guidelines:

1. Clean the slide and cover glass with soap and water.
2. Insert the pipette into the bottom of the sample bottle to obtain the maximum number of organisms. Fewer will occur at mid-depth.
3. Remove filamentous algae using forceps. Avoid using too much material.
4. Explore the slide first with the low-power objective. Reduce the illumination using the iris diaphragm to provide better contrast.

5. When you find an organism of interest, switch to the high-dry objective and adjust the illumination appropriately.
6. Refer to figures 5.1 to 5.7 and the accompanying text to help in your identification.
7. Record your observations on the Laboratory Report.

Materials

- bottles of pond water
- microscope slides and cover glasses
- Pasteur pipettes, pipette aids, and forceps
- marking pen
- additional reference books
- prepared slides of protozoa, amoeba, and algae

Survey of Organisms

An impressive variety of protists and cyanobacteria will likely be encountered during this laboratory exercise. You will be asked to identify and categorize these organisms based on their morphological characteristics. Traditionally, such morphological characteristics were used to construct formal classification schemes; however, genetic analyses have demonstrated that such classification schemes do not necessarily represent evolutionary relationships between organisms. Presently, there is a lively debate among protistologists as to which taxonomic scheme (and there are many) of protists should be accepted worldwide. Because of this lack of consensus among scientists about how to classify protists, and because true evolutionary relationships cannot be determined simply by observing organisms, you will use an informal system based on morphology to categorize the organisms that you encounter in this exercise. Table 5.1 will help you understand some of the major morphological groups of the organisms that you may see. **Please keep in mind that these are not formally recognized taxonomic groups,** but they are useful for identifying and categorizing organisms in the laboratory based on specific physical traits.

Eukaryotes

Protozoa

The Protozoa are protists, which means "first animals." They were some of the original microorganisms

Table 5.1 Classification of Organisms

PROKARYOTES (DOMAIN BACTERIA)	EUKARYOTES (DOMAIN EUKARYA)

Cyanobacteria (figure 5.7)	Protozoa
	    Diplomonads and Parabasalids
	    Euglenozoans
	        Euglenids (figure 5.2, illustrations 1, 2)
	        Kinetoplastids
	    Alveolates
	        Ciliates
	        Dinoflagellates (figure 5.3, illustrations 10, 11)
	        Apicomplexans
	    Stramenophiles
	        Diatoms (figure 5.6)
	        Golden algae (figure 5.3, illustrations 1–4)
	    Amoebozoans
	        Gymnamoebas (figure 5.1, illustration 6)
	        Entamoebas
	Algae
	    Red and green algae
	        Unicellular red algae
	        Unicellular green algae (figures 5.2, 5.3, 5.4)

observed by Leeuwenhoek over 300 years ago when he studied samples of rain water with his simple microscopes. Protozoa were initially classified as unicellular, colorless organisms, lacking a cell wall and were further subdivided based on their modes of motility. However, the application of modern genetic analysis has shown that these organisms are a diverse group that have little relationship to one another. They are mostly heterotrophic, unicellular organisms that lack a rigid cellulose cell wall characteristic of algae or a chitinous cell wall characteristic of fungi. Their specialized structures for motility such as cilia, flagella, or pseudopodia were important in previous classification schemes, but genetic analysis subdivides them into the following groups with the accompanying characteristics:

Diplomonads and Parabasalids   These organisms contain two nuclei and organelles called **mitosomes,** which are reduced mitochondria that lack electron transport components and Krebs cycle constituents. *Giardia lamblia* is a diplomonad that causes giardiasis, a diarrheal disease acquired by campers drinking contaminated water from lakes and streams. The parabasalids are characterized by the presence of a **parabasal body** that is associated with the Golgi apparatus. They lack mitochondria but instead contain hydrogenosomes that carry out anaerobic respiration, releasing hydrogen gas. Some inhabit the guts of termites where they digest cellulose for the insect. They are also serious pathogens of vertebrates, causing infections in the intestinal or urogenital tracts. *Trichomonas vaginalis* is a parabasalid that causes a sexually transmitted disease in humans.

Euglenozoans   These organisms are a diverse group that are unicellular and contain a unique crystalline rod associated with their flagellum. Some are pathogens while others are free living. They are comprised of the kinetoplastids and the euglenids.

- **Kinetoplastids**
  The kinetoplastids have a single large mitochondrion in the cell that contains a kinetoplast, which is a large mass of DNA. Some are found in aquatic habitats while others are very serious pathogens in humans. The trypanosomes are responsible for several diseases in humans, such as African sleeping sickness caused by *Trypanosoma brucei*, Chagas' disease caused by *Trypanosoma cruzi*, and leishmaniasis caused by *Leishmania major*. Trypanosomes are crescent-shaped and possess a single flagellum that originates at a basal body and is enclosed by part of the cytoplasmic membrane, thus forming an undulating structure over the cell body.

- **Euglenids**
  The euglenids can grow either heterotrophically or phototrophically because they contain chloroplasts. They are nonpathogens that occur primarily in aquatic habitats, where many subsist on bacteria. Food is taken in by the process of **phagocytosis** when the cytoplasmic membrane surrounds the food and brings it into the cell for digestion. *Euglena* (figure 5.2, illustrations 1 and 2) is a member of this group.

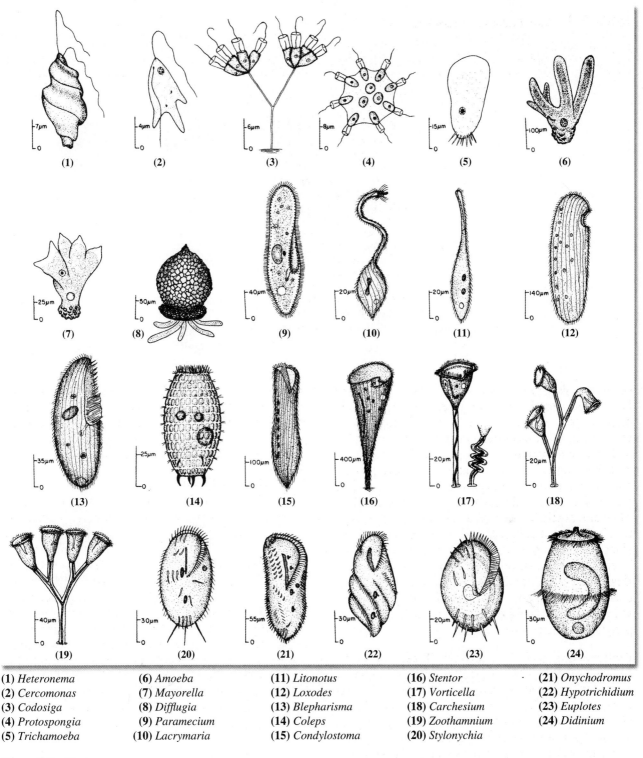

Figure 5.1   Protozoa.

(1) *Heteronema*	**(6)** *Amoeba*	**(11)** *Litonotus*	**(16)** *Stentor*	**(21)** *Onychodromus*
(2) *Cercomonas*	**(7)** *Mayorella*	**(12)** *Loxodes*	**(17)** *Vorticella*	**(22)** *Hypotrichidium*
(3) *Codosiga*	**(8)** *Difflugia*	**(13)** *Blepharisma*	**(18)** *Carchesium*	**(23)** *Euplotes*
(4) *Protospongia*	**(9)** *Paramecium*	**(14)** *Coleps*	**(19)** *Zoothamnium*	**(24)** *Didinium*
(5) *Trichamoeba*	**(10)** *Lacrymaria*	**(15)** *Condylostoma*	**(20)** *Stylonychia*	

Alveolates   The alveolates are comprised of the ciliates, the dinoflagellates, and the apicomplexans. Members of this group contain sacs called **alveoli** associated with the cytoplasmic membrane that may function in maintaining the osmotic balance of the cell. The apicomplexans are obligate parasites.

- **Ciliates**
  Ciliates are characterized by having two kinds of nuclei, *micronuclei* and *macronuclei*. Micronuclei genes function in sexual reproduction, whereas genes encoding for cellular functions are associated with the macronuclei.

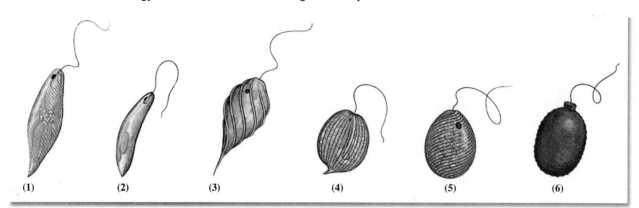

(1) *Euglena* (700×)   **(2)** *Euglena* (700×)   **(3)** *Phacus* (1000×)   **(4)** *Phacus* (350×)   **(5)** *Lepocinclis* (350×)   **(6)** *Trachelomonas* (1000×)

Figure 5.2   Flagellated euglenids.
(Courtesy of the U.S. Environmental Protection Agency, Office of Research & Development, Cincinnati, OH 45268)

An example of a ciliate is *Paramecium* (figure 5.1, illustration 9). Reproduction occurs by the process of conjugation, in which two *Paramecium* cells fuse and exchange their micronuclei. The ciliates are usually covered with cilia that are responsible for motility of the cell. Cilia are short, hairlike structures that have the same structure as flagella, that is, a 9 + 2 arrangement of microtubules. Motility occurs when cilia beat in a coordinated fashion to propel the cell forward or backward. Cilia also function in digestion. They line the cytostome or oral groove of the organism to direct food such as bacteria to the cell mouth, where it is enclosed in a vacuole and taken into the cell by phagocytosis. Many *Paramecium* harbor endosymbiotic prokaryotes in their cytoplasm or in the macronucleus; for example, methanogens produce hydrogen gas in the hydrogenosome of the *Paramecium*. Some species can attach to solid surfaces—for example, *Vorticella* and *Zoothamnium* (figure 5.1, illustrations 17 and 19). Ciliates can be pigmented: *Stentor* (figure 5.1, illustration 16), blue; *Blepharisma* (figure 5.1, illustration 13), pink; and *Paramecium busaria*, green (figure 5.1, illustration 9). The green color is due to the presence of endosymbiotic algal cells. The ciliates are important inhabitants of the forestomach of ruminant animals such as cattle. In the rumen (stomach), they degrade cellulose and starch, which can be used by the animal for nutrition. They also feed on rumen bacteria to maintain their numbers in the rumen. The ciliate *Balantidium coli* is an intestinal pathogen of domestic animals that occasionally infects humans. It causes an intestinal disease similar to *Entamoeba histolytica*.

- **Dinoflagellates**
The dinoflagellates (figure 5.3, illustrations 10, 11) are characterized by the presence of two flagella of different lengths. The flagella encircle the cell and when they beat, they cause the cell to whirl or spin. These organisms occur in both marine and freshwater habitats. *Gonyaulax* is a dinoflagellate that occurs in marine coastal waters. Some species are bioluminescent and can cause luminescence when waves break on the shore or boats move through the water. The organisms produce "red tides" caused by xanthophyll in the cells. Blooms of these organisms can occur naturally or by increased pollution in coastal waters. These blooms can result in fish kills and poisoning in humans when contaminated shellfish are consumed. *Gonyaulax* produces a neurotoxin, called **saxitoxin,** that is responsible for the symptoms of dizziness, numbness of the lips, and difficulty breathing.

- **Apicomplexans**
The apicomplexans are obligate parasites in animals and humans. In humans they are the cause of malaria (*Plasmodium* spp.), toxoplasmosis (*Toxoplasma gondii*), and coccidiosis (*Eimeria*). They produce resting stages called *sporozoites* that facilitate transmission of the pathogen. The cells contain structures called *apicoplasts*, which are vestiges of chloroplasts that have degenerated. They contain no photosynthetic pigments, but they do have some genes that encode for fatty acid and heme synthesis.

Stramenophiles   The stramenophiles are composed of the oomycetes (water molds), the diatoms, the golden algae, and the multicellular paeophytes, which includes the seaweeds. The unicelluar stramenophiles have flagella with short, hairlike extensions. The name is derived from this feature. *Stramen* is Latin for straw and *pilos* means hair.

- **Diatoms**
Diatoms are phototrophic freshwater and marine organisms that are unique in the biological

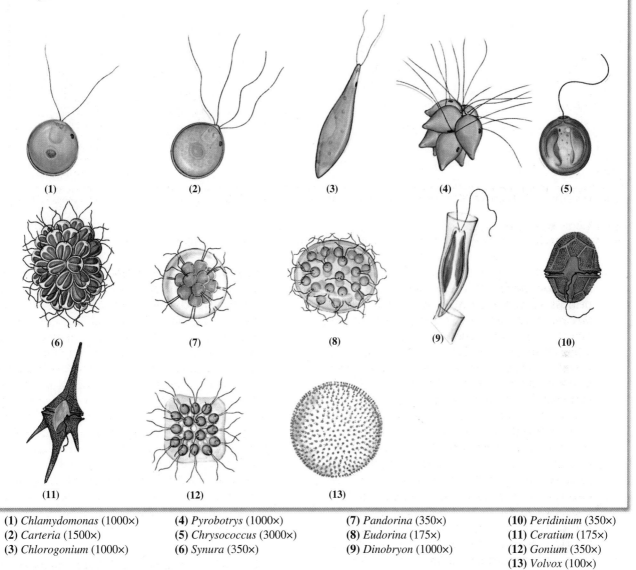

(1)   (2)   (3)   (4)   (5)

(6)   (7)   (8)   (9)   (10)

(11)   (12)   (13)

(1) *Chlamydomonas* (1000×)	**(4)** *Pyrobotrys* (1000×)	**(7)** *Pandorina* (350×)	**(10)** *Peridinium* (350×)
(2) *Carteria* (1500×)	**(5)** *Chrysococcus* (3000×)	**(8)** *Eudorina* (175×)	**(11)** *Ceratium* (175×)
(3) *Chlorogonium* (1000×)	**(6)** *Synura* (350×)	**(9)** *Dinobryon* (1000×)	**(12)** *Gonium* (350×)
			(13) *Volvox* (100×)

Figure 5.3   **Flagellated green algae.**
(Courtesy of the U.S. Environmental Protection Agency, Office of Research & Development, Cincinnati, OH 45268)

world because they synthesize a cell wall composed of silica. The external portion of the wall is called a *frustule* and is very diverse in its shape. The frustule consists of two halves, the epitheca and the hypotheca. It is analogous to a box with a lid, with the epitheca fitting over the hypotheca. Pores in the frustule called *areolae* function as passageways for gases and nutrients. The frustules remain after the death of the organism and are referred to as diatomaceous earth, which is used in many applications such as a polishing abrasive in toothpaste. They do not decay when the organism dies, and hence they are a preserved fossil record of diatoms. Based on fossil records of the frustules, it is estimated that diatoms were present on the earth about 200 million years ago. *Nitzschia* is a common diatom (figure 5.6, illustration 7).

- **Golden Algae**
The *chrysophytes* or golden (brown) algae are classified with the stramenophiles because genetic analysis has determined that they are more closely related to the diatoms and oomycetes (water molds) than to the other unicellular algae. They are inhabitants of fresh and marine waters. Some species are also chemoorganotrophs, which means they can derive food from the transport of organic compounds across the cytoplasmic membrane instead of carrying out photosynthesis. They are motile by means of two flagella. Their gold or brown color is the result of the carotenoid, *fucoxanthin*. They also contain chlorophyll *c* rather than the chlorophyll *a* found in other algae. *Dinobryon* is a colonial golden alga that occurs in fresh water (figure 5.3, illustration 9).

Amoebozoa The amoebozoa occur in both terrestrial and aquatic habitats. They are often found in pond water. This group is composed of the gymnamoebas, the entamoebas, the slime molds, and the cellular slime molds. In this exercise we will only focus on the gymnamoebas and mention the entamoebas because of their importance as pathogens.

- **Gymnamoebas**

  These amoebozoa are primarily free living, occurring in aquatic habitats and in the soil. They achieve motility by amoeboid movement in which pseudopodia are extended and the cell cytoplasm streams into the tip of the pseudopodium. Microfilaments associated with the cytoplasmic membrane aid in the overall process. Pseudopodia are also utilized by the amoebas for entrapping and surrounding food by phagocytosis. They can range in size from 15μ to above 700μ. An example is shown in figure 5.1, illustration 6.

- **Entamoebas**

  These amoebas are strictly parasites of invertebrates and vertebrates. An important pathogen in humans is *Entamoeba histolytica,* which causes amoebic dysentery. The organism is transmitted in a cyst form by fecal contamination of water and food. In the intestine it causes ulceration, resulting in a bloody diarrhea.

Algae The algae are a diverse group of organisms that obtain their carbon requirements from oxygenic photosynthesis in which carbon dioxide is fixed into cellular materials and water is split to evolve oxygen. They are typically smaller and less complex in their structure than land plants, but they are similar to plants because they possess photosynthetic pigments such as chlorophyll and carotenoids that harvest light energy from the sun. They range in size from microscopic unicellular forms to the seaweeds that form giant kelp beds in the oceans. The microscopic algae form filaments or colonies that are comprised of several individuals loosely held together in an organized fashion. Many of the unicellular algae are motile by means of a flagellum. Reproduction can occur by both asexual and sexual mechanisms.

The algae are diverse in their ecology. They occur primarily in aquatic habitats such as freshwater lakes and streams and in the oceans, where they are important members of the phytoplankton. Algae also occur in terrestrial habitats. Some grow on snow, imparting a pink color. They enter into unique symbiotic relationships with fungi called lichens, which are found on trees and rocks. Although unique morphologically, the algae are related genetically to other protists.

Red and Green Algae The red and green algae belong to the *rhodophytes* and *chlorophytes*, respectively. They may be unicellular (figure 5.3, illustrations 1–5), colonial (figure 5.3, illustrations 6, 7, 8, 12, 13), or filamentous (figure 5.4). Many seaweeds are included in the red algae. Agar used in microbiological media is extracted from a seaweed, *Gelidium*, a member of the red algae.

- **Unicellular Red Algae**

  The unicellular red algae occur primarily in marine habitats, with a few freshwater varieties. Cells may have one or more flagella. They contain chloroplasts, which carry out photosynthesis. Their primary photosynthetic pigment is chorophyll *a*; however, they are unique because they lack chorophyll *b*. In addition, they also have phycobiliproteins, which assist in harvesting light. Their red coloration is in fact due to the presence of *phycoerythrin*, one of the phycobiliproteins. These pigments are also found in the cyanobacteria.

- **Unicellular Green Algae**

  The majority of algae observed in fresh water such as ponds and lakes are green algae and hence, these will be the ones that you will most likely encounter in this exercise. Like green plants, they contain chlorophyll *a* and *b*. They store starch granules as an energy reserve. *Chlamydomonas* is an excellent example of a unicellular green alga (figure 5.3, illustration 1). This organism has been the object of extensive studies, especially involving photosynthesis. Colonial forms of green algae include *Pandorina, Eudorina, Gonium,* and Volvox (figure 5.3, illustrations 7, 8, 12, 13). *Vaucheria* and *Tribonema* (figure 5.4, illustrations 5 and 6) are not green algae but have been recently reclassified as Tribophyceae based on genetic analysis.

  A unique group of green algae are the "desmids," whose cells are composed of two halves, called semi-cells (figure 5.4, illustration 12, and figure 5.5, illustrations 16–20). The two halves of the cell are separated by a constriction called the isthmus. Some of the organisms such as *Spirogyra* form a filament, whereas others such as *Desmidium* form a pseudofilament.

Prokaryotes

Cyanobacteria

The cyanobacteria first appeared on the earth some 2.7 billion years ago. They were critical to the evolution of life on the earth because they were the first to carry out oxygenic photosynthesis in which oxygen was released into the atmosphere. The accumulation of oxygen over time paved the way for the development of aerobic bacteria and eukaryotic organisms that required oxygen for respiratory metabolism.

(1) *Rhizoclonium* (175×)
(2) *Cladophora* (100×)
(3) *Bulbochaete* (100×)
(4) *Oedogonium* (350×)
(5) *Vaucheria* (100×)
(6) *Tribonema* (300×)
(7) *Chara* (3×)
(8) *Batrachospermum* (2×)
(9) *Microspora* (175×)
(10) *Ulothrix* (175×)
(11) *Ulothrix* (175×)
(12) *Desmidium* (175×)
(13) *Mougeotia* (175×)
(14) *Spirogyra* (175×)
(15) *Zygnema* (175×)
(16) *Stigeoclonium* (300×)
(17) *Draparnaldia* (100×)

Figure 5.4 **Filamentous algae.**
(Courtesy of the U.S. Environmental Protection Agency, Office of Research & Development, Cincinnati, OH 45268)

Prior to the cyanobacteria's appearance, the earth was anaerobic, and only organisms capable of anaerobic metabolism could survive. Evidence also suggests that these bacteria may have invaded a primitive cell and established an endosymbiotic relationship, thus giving rise to chloroplasts in algae and plants.

Cyanobacteria are extremely diverse and occur in a variety of environments, from the tropics to the poles. They primarily occur in aquatic habitats such as the oceans, lakes, and streams. The cyanobacteria play a critical role in the global carbon cycle, accounting, for 20–30% of the carbon fixed on the earth by

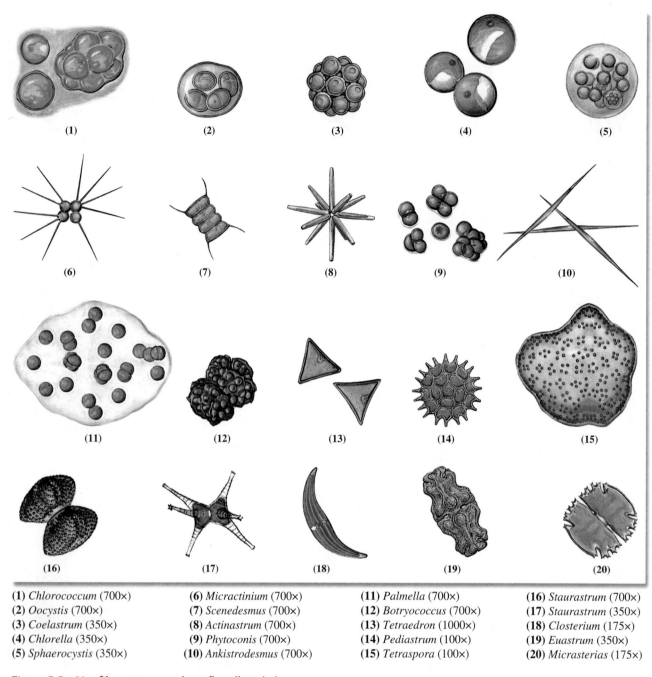

(1)   (2)   (3)   (4)   (5)

(6)   (7)   (8)   (9)   (10)

(11)   (12)   (13)   (14)   (15)

(16)   (17)   (18)   (19)   (20)

(1) *Chlorococcum* (700×)
(2) *Oocystis* (700×)
(3) *Coelastrum* (350×)
(4) *Chlorella* (350×)
(5) *Sphaerocystis* (350×)
(6) *Micractinium* (700×)
(7) *Scenedesmus* (700×)
(8) *Actinastrum* (700×)
(9) *Phytoconis* (700×)
(10) *Ankistrodesmus* (700×)
(11) *Palmella* (700×)
(12) *Botryococcus* (700×)
(13) *Tetraedron* (1000×)
(14) *Pediastrum* (100×)
(15) *Tetraspora* (100×)
(16) *Staurastrum* (700×)
(17) *Staurastrum* (350×)
(18) *Closterium* (175×)
(19) *Euastrum* (350×)
(20) *Micrasterias* (175×)

Figure 5.5   Nonfilamentous and nonflagellated algae.
(Courtesy of the U.S. Environmental Protection Agency, Office of Research & Development, Cincinnati, OH 45268)

photosynthesis. This, coupled with their ability to fix nitrogen, makes them primary producers in the environment. They comprise part of the phytoplankton in the oceans and are therefore extremely important in food chains in these habitats.

Their taxonomy is not completely settled at this time. They can be divided into five morphological groups: (1) unicellular, dividing by binary fission; (2) unicellular, dividing by multiple fission; (3) filamentous, containing nitrogen-fixing heterocysts; (4) filamentous, containing no heterocysts; and (5) branching filamentous species. Over 1000 species have been described. They were once classified as "blue-green algae" primarily because they contain chlorophyll *a* and release oxygen from photosynthesis. However, they are prokaryotes because they lack an organized nucleus or chloroplast and do not contain other organelles. They have 70S ribosomes and contain peptidoglycan in their cell walls. They differ from the green sulfur and purple sulfur bacteria because the latter contain bacteriochlorophyll and carry out anoxygenic photosynthesis in which oxygen is not produced. Figure 5.7 illustrates only a random few that are frequently seen.

(1)   (2)   (3)   (4)   (5)   (6)
(7)   (8)   (9)   (10)   (11)
(12)   (13)   (14)   (15)   (16)   (17)
(18)   (19)   (20)   (21)   (22)   (23)

(1) *Diatoma* (1000×) **(7)** *Nitzschia* (1500×) **(13)** *Synedra* (175×) **(19)** *Asterionella* (175×)
(2) *Gomphonema* (175×) **(8)** *Pinnularia* (175×) **(14)** *Melosira* (750×) **(20)** *Asterionella* (750×)
(3) *Cymbella* (175×) **(9)** *Cyclotella* (1000×) **(15)** *Surirella* (350×) **(21)** *Navicula* (750×)
(4) *Cymbella* (1000×) **(10)** *Tabellaria* (175×) **(16)** *Stauroneis* (350×) **(22)** *Stephanodiscus* (750×)
(5) *Gomphonema* (2000×) **(11)** *Tabellaria* (1000×) **(17)** *Fragillaria* (750×) **(23)** *Meridion* (750×)
(6) *Cocconeis* (750×) **(12)** *Synedra* (350×) **(18)** *Fragillaria* (750×)

Figure 5.6   Diatoms.
(Courtesy of the U.S. Environmental Protection Agency, Office of Research & Development, Cincinnati, OH 45268)

The previous designation of these organisms as "blue-green" was somewhat misleading as many are black, purple, red, and various shades of green, including blue-green. The varying colors are due to the differing proportions of photosynthetic pigments present in the cells, which include chlorophyll *a*, carotenoids, and phycobiliproteins. The latter pigments consist of allophycocyanin, phycocyanin, and phycoerythrin that combine with protein molecules to form **phycobili-proteins,** which serve as light-harvesting molecules

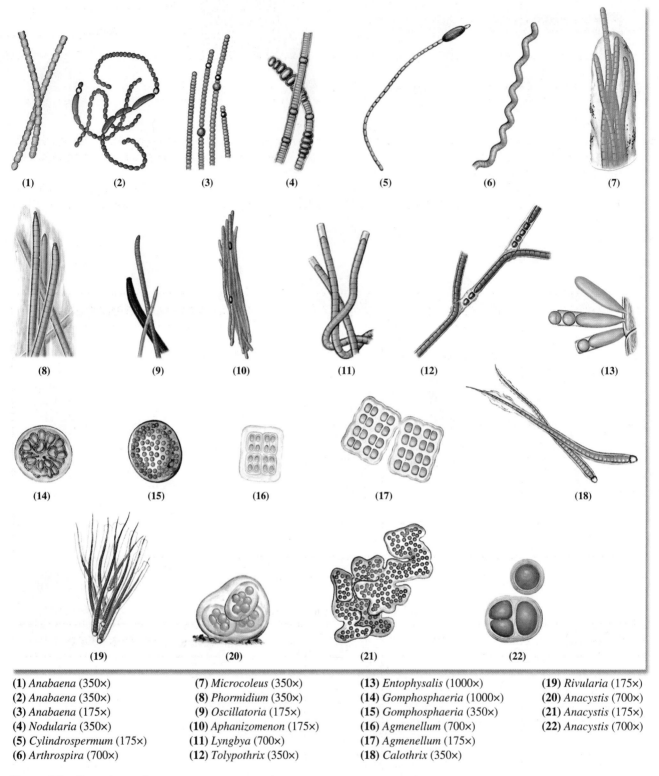

(1)   (2)   (3)   (4)   (5)   (6)   (7)

(8)   (9)   (10)   (11)   (12)   (13)

(14)   (15)   (16)   (17)   (18)

(19)   (20)   (21)   (22)

(1) *Anabaena* (350×)

(2) *Anabaena* (350×)

(3) *Anabaena* (175×)

(4) *Nodularia* (350×)

(5) *Cylindrospermum* (175×)

(6) *Arthrospira* (700×)

(7) *Microcoleus* (350×)

(8) *Phormidium* (350×)

(9) *Oscillatoria* (175×)

(10) *Aphanizomenon* (175×)

(11) *Lyngbya* (700×)

(12) *Tolypothrix* (350×)

(13) *Entophysalis* (1000×)

(14) *Gomphosphaeria* (1000×)

(15) *Gomphosphaeria* (350×)

(16) *Agmenellum* (700×)

(17) *Agmenellum* (175×)

(18) *Calothrix* (350×)

(19) *Rivularia* (175×)

(20) *Anacystis* (700×)

(21) *Anacystis* (175×)

(22) *Anacystis* (700×)

Figure 5.7 **Cyanobacteria.**

(Courtesy of the U.S. Environmental Protection Agency, Office of Research & Development, Cincinnati, OH 45268)

for the photosystems. Interestingly, these structures are also found in the red algae. Because an organized chloroplast is not present in these bacteria, their pigments and photosystems are organized in **thylakoids,** which are parallel arrays of stacked membranes.

Laboratory Report

Answer the questions in Laboratory Report 5.

5 Microbiology of Pond Water—Protists, Algae, and Cyanobacteria

A. Results

In this study of freshwater microorganisms, record your observations in the following tables. The number of organisms to be identified will depend on the availability of time and materials. Your instructor will indicate the number of each type that should be recorded.

Record the genus of each identifiable organism. Also, indicate the group to which the organism belongs. Microorganisms that you cannot identify should be sketched in the space provided. It is not necessary to draw those that are identified.

1. **Protozoa**

GENUS	GROUP	BOTTLE NO.	SKETCHES OF UNIDENTIFIED

2. **Algae**

GENUS	GROUP	BOTTLE NO.	SKETCHES OF UNIDENTIFIED

3.   **Cyanobacteria**

GENUS	BOTTLE NO.	SKETCHES OF UNIDENTIFIED

B. Short-Answer Questions

1. In which domains are algae, protozoa, and cyanobacteria classified?

   _____

   _____

2. (a) Name one similarity between algae and plants. (b) Name one difference.

   _____

   _____

3. Compare and contrast the three mechanisms of motility displayed by protozoa.

   _____

   _____

4. (a) What organisms were formerly known as "blue-green algae"? (b) Why are these organisms not algae?

   _____

   _____

5. What makes "red tides" red?

   _____

   _____

6. What is the genus of the causative agent of malaria? In what group does it belong?

   _____

   _____

7. What are mitosomes? In what organisms do they occur?

   _____

   _____

8. What is a kinetoplast? What pathogenic organisms contain this structure?

_____

_____

9. In the alveolates, what structure may control osmotic balance?

_____

_____

10. What function do the micronuclei play in the ciliates?

_____

_____

11. What is a frustule, and what unique compound comprises this structure?

_____

_____

12. Besides chlorophyll, what additional light-harvesting pigments are found in the cyanobacteria?

_____

_____

C. Fill-in-the-Blanks Question

1. For each type of organism, place a check mark in the box to indicate whether the cellular characteristic or function is present.

CHARACTERISTIC OR FUNCTION	PROTOZOA	ALGAE	CYANOBACTERIA
Nucleus			
Flagella			
Pseudopodia			
Cilia			
Photosynthetic pigment(s)			
Chloroplasts			
Cell wall			

Ubiquity of Bacteria

Learning Outcomes

After completing this exercise, you should be able to

1. Culture bacteria from natural sources.
2. Appreciate that the kinds of bacteria and their numbers will vary for the different environments sampled.

Figure 6.1   Nutrient agar plate exposed to the environment.
© Andreas Reh RF

Bacteria are the most widely distributed organisms in the biosphere. The major portion of biomass on the earth is made up from bacteria and other microbial life. Bacteria occur as part of the normal flora of humans and animals, where they occupy sites on surfaces such as the skin and intestinal tract. In humans it is estimated that the normal flora outnumber our own cells by approximately tenfold. In plants they are found on leaf surfaces and in nodules on roots, where they form a symbiotic partnership with the plant to fix nitrogen from the atmosphere. They are also responsible for a variety of diseases in both plants and animals. We depend on bacteria to break down our sewage and waste. They are important in the soil, where they are active in the nitrogen and carbon cycles, thus contributing to soil fertility. Bacteria are abundant in the oceans, where they contribute to cycles and mineralization. In the oceans, some have unique metabolic capabilities that allow them to grow next to superheated hydrothermal vents and utilize hydrogen sulfide and other gases emitted from volcanic activity associated with these vents. Bacteria have been isolated from core samples taken from deep within the earth's crust. They form colorful blooms in sulfur hot springs found in places such as Yellowstone National Park. In essence, they have been found in almost every place humans have searched for them. Thus, they are almost ubiquitous in their distribution on the earth.

When Robert Koch first studied bacteria in his laboratory, one of the first challenges was to devise a method to grow bacteria in culture so that populations could be separated into individual species. Initial approaches used pieces of vegetables such as potatoes and carrots, or the addition of gelatin to meat broths. It was found that bacterial cells would grow as visible, discrete **colonies** on a solid medium. A **colony** is a visible mass of cells usually resulting from the division of a single cell (figure 6.1), and the number of cells in a single colony can exceed one

billion (10^9). However, a colony can arise from more than one cell, for example when a chain of cells such as streptococci grow on a nutrient medium. The cultivation of bacterial cells was vastly improved in Koch's laboratory when Frau Hesse, the wife of an early coworker of Koch, suggested the use of agar-agar as a solidifying agent. A major advantage was that agar media were easily manipulated, and agar, unlike gelatin, was not degraded by the pathogenic bacteria Koch was studying. As a solidifying agent, agar could be added to rich broths to form a solid medium on which isolated colonies would develop after inoculation. The purity of cultures was further improved when R. J. Petri, a worker in Koch's lab, introduced a covered dish that protected the nutrient surface of media from contamination by bacteria in the environment, especially the air. These methods were extremely important in the accomplishments of Koch's lab because they provided a means for separating and culturing the pathogens that they were studying, including anthrax caused by *Bacillus anthracis*. Microbiologists still use agar and the petri dish in the cultivation of bacteria today.

No single medium exists that will support the growth of all bacteria, owing to the diverse metabolic capabilities and requirements of bacteria. However, many bacteria will grow on extracts of meat or vegetables that have been solidified with agar, an example being nutrient agar. During this laboratory period, you will be provided with sterile bacteriological media that you will expose to the environment in various ways. Any bacterial cells that occur in the environment and are deposited on the agar medium or in the broth will subsequently undergo cell division to produce colonies or cause the broth to become turbid. The idea is to appreciate that bacteria are ubiquitous in the environment and can be spread by convection currents in the air. It is important to also understand that the morphology of individual bacterial colonies that develop on an agar medium usually differ from one species to another. This may involve different aspects of the colony such as the regularity of its edge, pigmentation, the configuration and texture of its surface, and its elevation. Examples of these characteristics are given in Exercise 35, figure 35.4 on page 250.

You will be provided with three kinds of sterile bacteriological media that you will expose to the environment in various ways. To ensure that your exposures cover as wide a spectrum as possible, specific assignments will be made for each student. This may involve the use of a swab to remove bacteria from an object or surface; in other instances a petri plate containing a nutrient medium will be exposed to the air in different environments. A number will designate your assignment, as detailed in the chart below.

Materials

per student:
- 1 tube of nutrient broth
- 1 petri plate of trypticase soy agar (TSA)
- 1 sterile cotton swab
- Sharpie marking pen

per two or more students:
- 1 petri plate of blood agar

1. Scrub down your desktop with a disinfectant (see Exercise 8, Aseptic Technique).
2. Expose your TSA plate according to your assignment in the chart below. *Label the bottom* of your plate with your initials, your assignment number, and the date.

3. Moisten a sterile swab by immersing it into a tube of nutrient broth and expressing most of the broth out of it by pressing the swab against the inside wall of the tube.
4. Rub the moistened swab over a part of your body such as a finger or ear, or some object such as a doorknob or telephone mouthpiece, and return the swab to the tube of broth. It may be necessary to break off the stick end of the swab so that you can replace the cap on the tube.
5. Label the tube with your initials and the source of the bacteria.
6. Expose the blood agar plate by coughing onto it. Label the bottom of the plate with the initials of the individuals who cough onto it. Be sure to date the plate also.
7. Incubate the plates and tube at 37°C for 48 hours.

Evaluation

After 48 hours' incubation, examine the tube of nutrient broth and two plates. Shake the tube vigorously without wetting the cap. Is it cloudy or clear? Compare it with an uncontaminated tube of broth. What is the significance of cloudiness? Do you see any colonies growing on the blood agar plate? Are the colonies all the same size and color? If not, what does this indicate? Group together a set of TSA plates representing all nine types of exposure. Record your results on the Laboratory Report.

Your instructor will indicate whether these tubes and plates are to be used for making slides in Exercise 11 (Simple Staining). If the plates and tubes are to be saved, containers will be provided for their storage in the refrigerator. Place the plates and tubes in the designated containers.

Laboratory Report

Record your results in Laboratory Report 6.

EXPOSURE METHOD FOR TSA PLATE	STUDENT NUMBER
1. To the air in laboratory for 30 minutes	1, 10, 19, 28
2. To the air in room other than laboratory for 30 minutes	2, 11, 20, 29
3. To the air outside of building for 30 minutes	3, 12, 21, 30
4. Blow dust onto exposed medium	4, 13, 22, 31
5. Moist lips pressed against medium	5, 14, 23, 32
6. Fingertips pressed lightly on medium	6, 15, 24, 33
7. Several coins pressed temporarily on medium	7, 16, 25, 34
8. Hair combed over exposed medium (10 strokes)	8, 17, 26, 35
9. Optional: Any method not listed above	9, 18, 27, 36

6 Ubiquity of Bacteria

A. Results

1. After examining your TSA and blood agar plates, record your results in the following table and on a similar table that your instructor has drawn on the chalkboard. With respect to the plates, we are concerned with a quantitative evaluation of the degree of contamination and differentiation as to whether the organisms are bacteria or molds. Quantify your recording as follows:

0	no growth		
+	1 to 10 colonies	+ + +	51 to 100 colonies
+ +	11 to 50 colonies	+ + + +	over 100 colonies

   After shaking the tube of broth to disperse the organisms, look for cloudiness (turbidity). If the broth is clear, no bacterial growth occurred. Record no growth as 0. If the tube is turbid, record + in the last column.

STUDENT INITIALS	PLATE EXPOSURE METHOD		COLONY COUNTS		BROTH	
	TSA	Blood Agar	Bacteria	Mold	Source	Result

Use the class results to answer the following questions.

2. Using the number of colonies as an indicator, which habitat sampled by the class appears to contain the most bacteria? _____

3. Why do you suppose this habitat contains such a high microbial count? _____

_____

4. a. Were any plates completely lacking in colonies?

_____

   b. Do you think that the habitat sampled was really sterile?

_____

   c. If your answer to *b* is *no,* then how can you account for the lack of growth on the plate?

_____

   d. If your answer to *b* is *yes,* defend it:

_____

B. **Short-Answer Questions**

1. In what ways do the macroscopic features of bacterial colonies differ from those of molds?

_____

2. Why is the level of contamination measured as number of colonies rather than size of colonies?

_____

3. Should one be concerned to find bacteria on the skin? How about molds? Explain.

_____

4. How can microbial levels be controlled on the skin? On surfaces in the environment? In the air?

_____

5. Compare the following features of bacteria to those of eukaryotic microorganisms:

   a. size.

_____

   b. organization of genetic material.

_____

   c. ribosomes.

_____

   d. cell wall.

_____

   e. respiration and photosynthesis.

_____

   f. motility mechanisms.

_____

The Fungi:
Molds and Yeasts

Learning Outcomes

After completing this exercise, you should be able to

1. Differentiate between yeast and fungi.
2. Identify some common fungi based on their microscopic morphology and colony characteristics.
3. Identify the fruiting structures of yeasts and molds based on their microscopic morphology.

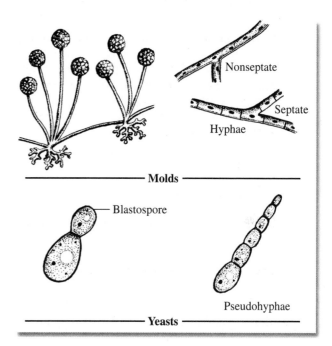

Figure 7.1 **Structural differences between molds and yeasts.**

The fungi comprise a large and diverse group of saprophytic, eukaryotic organisms that obtain their nutritional needs from degrading organic materials in the environment. This group includes the molds, mushrooms, and yeast. Unlike plants, which have cell walls composed of cellulose, fungal cell walls are composed of chitin, a polymer of the carbohydrate N-acetyl-glucosamine. They mostly occur in terrestrial environments, although some are found exclusively in aquatic habitats. They secrete **exoenzymes** that break down polysaccharides and proteins into their monomeric components of sugars, peptides, and amino acids, which are utilized for nutrition. They are responsible for the decay of dead plant and animal material, and thus they are essential for the mineralization and recycling of organic and biological material in the environment. For example, the wood-rotting fungi play an almost exclusive role in the degradation and turnover of dead trees. Wood has a complex structure composed of cellulose and lignin, a polymer of phenolic compounds. Fungi are unique in their capacity to degrade the compounds that comprise wood. They can grow under widely varying environmental conditions, with the ability to withstand low pH values and temperatures up to 62°C. Because of their unique metabolic capabilities and diverse growth capacities, they are often nuisance organisms. They are responsible for significant economic losses of foods such as fruits and vegetables. They especially thrive in moist habitats such as bathrooms, where they are responsible for the formation of mildew on surfaces.

Molds are fungi that form colonies composed of microscopic, rounded, intertwining filaments called **hyphae** (hypha, singular, figure 7.1). There are two different structural types of hyphae in fungi. One type is characterized by individual cells that are separated by the formation of *septa* or cross-walls, with each cell having a typical eukaryotic structure. In this type of hyphae, pores in the septa allow streaming of cytoplasm between compartments. A second type of hyphae is *coenocytic,* in which cross-walls are not present to form individual cells. Repeated nuclear divisions in coenocytic hyphae result in nuclei that are transported by cytoplasmic streaming throughout the fungal colony. Growth of a mold colony occurs at the ends of hyphae, called apical tips. Hyphae cover the surface of a substrate to form a branching, filamentous network called a **mycelium.** A visible mold colony is a mat formed by the hyphae that make up this substrate mycelium. A mold colony often has a characteristic pigmentation that is useful in identification. Structures that produce spores such as sporangiophores and conidiophores arise when branches of the substrate mycelium form and undergo cellular differentiation to form these specialized reproductive structures.

Yeasts are fungi that do not ordinarily form hyphae. The primary mode of reproduction for yeasts such as *Saccharomyces cerevisiae* is by the process of budding, whereby a bud forms and eventually separates from the main cell to form a new yeast cell. However, sometimes buds may not separate but instead form a chain of cells called **pseudohyphae** (figure 7.1). Some pathogenic fungi are **dimorphic.** In tissue, they occur as yeast cells and reproduce by budding. However, when cultured onto nutrient media, they form mycelia and typical sporulation structures. An example is *Histoplasma capsulatum.*

Fungi are important pathogens in humans, where they can cause minor infections of hair, skin, and nails or serious deep mycoses in internal organs that result in death. The **dermatophytes** such as *Trichophyton* infect hair, skin, and nails in humans to cause diseases such as athlete's foot and ringworm. A number of different fungi cause subcutaneous mycoses, which are characterized by lesions that develop at the site of inoculation. Two examples are *Sporothrix*, which causes sporotrichosis, and *Madurella*, which usually infects the feet to cause edema and swelling in a disease called mycetoma. The true pathogenic fungi cause mycoses that may involve all internal organs of the body. Many are acquired by inhaling infective spores into the respiratory tract. These include the following: *Blastomyces*, North American blastomycosis; *Coccidioides*, coccidioidomycosis; *Histoplasma*, histoplasmosis; and *Paracoccidioidomycosis*, South American blastomycosis. Progressive infections by *Histoplasma* can involve the liver, spleen, and lymph nodes. Fungi that cause deep mycoses are usually dimorphic because they produce mycelia when grown at 25°C on culture media but produce budding yeast forms when growing in tissue at 37°C.

Yeasts are also responsible for infections in humans. *Candida albicans* causes infections in various body sites. This organism can infect the mouth and tongue, especially in newborns, to cause **thrush.** The latter results when *Candida* is introduced from the mother's vaginal tract during birth, and because newborns initially lack a normal flora in the mouth, *Candida* can establish a presence and thrive. This organism also causes yeast infections in the vaginal tract as well as the intestinal tract. It can occasionally invade the heart and central nervous system to cause endocarditis and meningitis, respectively. Another species of *Candida, C. glabrata,* is a primary cause of infections in patients with compromised immune systems, such as AIDS patients. *Cryptococcus neoformans* is responsible for various infections involving the skin, skeleton, and the central nervous system. It occurs in approximately 8% of AIDS patients.

Fungi are also important pathogens of plants, where they cause significant economic losses each year. They infect crops such as corn, grains, grasses, and peanuts. Carcinogenic (cancer-causing) aflatoxins are produced by fungi growing in infected peanuts and grains. These toxins are monitored by the U.S. government, which has set allowable limits for human consumption of peanut products. The fungus *Claviceps* infects grain to produce ergot alkaloids that can cause hallucinations and even death. Some have theorized that the Salem witch hunts and trials in colonial Massachusetts involved ergot poisoning from infected grain used to make bread.

Fungi are an important part of our diet as well as producers of certain foods. Fermentation of grapes and grains by yeasts produces beverages such as wine and beer. Roquefort and blue cheese are produced by the growth of *Penicillium* in cheeses made from the milk of cattle, sheep, and goats. Mushrooms such as *Agaricus* and morels, members of the Basidiomycetes, are used in a variety of foods consumed by humans. However, some mushrooms such as *Amanita* are extremely poisonous because they produce very potent toxins that cause death when ingested.

Fungi can form important and beneficial symbiotic relationships with organisms such as plants and cyanobacteria. **Mycorrhizae** are close symbiotic associations between fungi and plant roots. The fungus facilitates the uptake of minerals such as phosphate and water by the plant, and in return, the fungus is supplied with carbohydrates from the plants that are used for nutrition.

Fungi also form symbiotic associations with algae and cyanobacteria called **lichens.** Approximately 85% of these associations involve filamentous or unicellular eukaryotic algae, 10% involve cyanobacteria, and 4% consist of both an alga and a cyanobacterial partner. It is estimated that 20% of all fungi form lichens, presumably because these associations are a reliable way for fungi to obtain fixed carbon from the photosynthetic partner and nitrogen from a cyanobacterial partner. Ascomycetes account for about 98% of fungal partners. Lichens are prevalent in the environment and can be found on the surfaces of trees, rocks, and even the roofs of houses. The fungus anchors the lichen colony to the surface and protects the phototroph from erosion by rain and winds and from drying. The fungal partner also synthesizes lichen acids that dissolve and chelate inorganic nutrients in the environment that are needed by the phototrophic partner. The fungus also can facilitate the uptake of water needed by the phototroph. Lichens are sensitive to many toxic chemicals in the environment. Owing to their sensitivity to compounds such as sulfur dioxide and metals, they are important indicators of pollution such as acid rain.

Many diverse organisms have been traditionally categorized as fungi, including water molds, puff balls, bracket fungi, yeasts, and molds. These organisms do not have a uniform genetic background, and they are believed to have evolved from at least

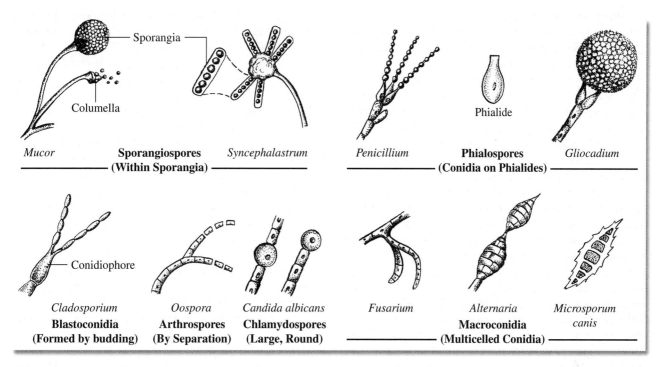

Figure 7.2 **Types of asexual spores seen in fungi.**

two ancestral lines. Classification of fungi is a complex and dynamic process. Traditional classification schemes relied primarily on morphology and reproductive mechanisms to classify fungal groups. However, modern schemes employ genetic analysis to ascertain relatedness between groups and species. Recent genetic analyses of fungi indicate that fungal classification based on morphological characteristics does not reflect evolutionary or phylogenetic relationships between organisms. However, for the beginning laboratory, identification and characterization of fungi is still most easily performed using morphological characteristics. In this exercise, you will examine prepared slides and those made from living cultures of molds and yeasts. The spores of fungi are abundant in the air and can be easily cultured on appropriate media for both macroscopic and microscopic study. Additionally, you will examine various fungi microscopically and attempt to identify them based on their morphological characteristics. Before attempting to identify various molds or yeasts, familiarize yourself with the figures that depict the characteristics of the groups. Note that yeasts differ from molds in that yeasts are unicellular and do not form a true mycelium (figure 7.1).

Fungal Spores: Asexual and Sexual

Fungal spores are essential in the reproduction of fungi and are produced by either asexual or sexual means. Traditional taxonomy relied greatly on the morphological characteristics of fungal spores and spore-related structures. Classification was based on morphological characteristics such as variation in spore size, color, the appearance of spores, and the type of fruiting body. These two types of spores are described next.

Asexual Spores

Asexual spores are produced by mitotic division and differentiation of specialized hyphae that extend above the colony. Types of asexual spores are shown in figure 7.2.

Sporangiospores: Sporangiospores form within a thick-walled sac called a sporangium. They can be either motile or nonmotile.

Conidia: Conidia are nonmotile asexual spores that form on specialized hyphae called conidiophores. They include the following:

Phialospores: These spores are produced on a vase- or flask-shaped cell called a *phialide*. They are found in *Penicillium* and *Gliocladium*.

Blastoconidia: These are present in some filamentous fungi and occur by budding in yeast cells. *Cladosporium* and *Candida* produce blastoconidia.

Arthrospores: Arthrospores form by the fragmentation and formation of crosswalls in preexisting hyphae. They occur in *Geotrichum* and *Galactomyces*.

Chlamydospores: Asexual, thick-walled spores that are round or irregular. They occur in most fungi and may function in survival.

Sexual Spores

Fungi produce sexual spores by the union of two unicellular and genetically distinct gametes to form a diploid cell which undergoes meiosis and mitosis. Sexual reproduction can also occur by the fusion of specialized hyphae called *gametangia*. Sexual spores are resistant to heat, drying, and some chemical compounds but are not as resistant as bacterial endospores. Germination of both kinds of spores, asexual and sexual, produces hyphae and mycelia. The various sexual spores are shown in figure 7.3.

Zygospores are formed by the fusion and genetic exchange between hyphae which have formed gametangia. The hyphae are genetically distinct (+, −). The common bread mold *Rhizopus* forms zygospores.

Ascospores are haploid sexual spores formed in the interior of an oval or elongated structure called an ascus. An example is the fungus *Chaetomium*.

Basidiospores are sexual haploid spores produced externally on a club-shaped *basidium*. Basidospores are produced by mushrooms such as *Agaricus campestris*.

Subdivision of Fungi

The fungi are divided into five groups based on genetic analysis of ribosomal RNA:

Chytridiomycetes

These fungi are primarily aquatic. They may be unicellular or form hyphae. They produce flagellated zoospores, an adaptation of their aquatic habitat. Some are pathogens on reptiles such as *Batrachochytrium dendrobatidis*, which causes an infection in frogs that inhibits respiration across the epidermis of the animal.

Zygomycetes

The zygomycetes produce coenocytic (multinucleate) hyphae. They form sexual zygospores or nonmotile asexual sporangiospores. Some are important in food spoilage, such as *Rhizopus stolonifer*, a common bread mold. They prefer to grow in an atmosphere with high humidity.

Glomeromycetes

These fungi are a small group of about 160 species that almost exclusively form ectomycorrhizae on plants. They may have played a role in the colonization of land habitats by vascular plants. The fungal hyphae penetrate the plant cell wall to form structures called arbuscules, which are swollen vesicles. These structures are thought to aid the plant in the uptake of minerals from the soil.

Ascomycetes

The ascomycetes are a large and diverse group that are represented by single-celled organisms such as *Saccharomyces cerevisiae* but also by filamentous organisms such as *Neurospora crassa*, another type of bread mold. Sexual spores called *ascospores* are produced in saclike structures called asci (figure 7.3). Some asci occur in structures called ascocarps. Asexual reproduction involves the formation of conidiospores. Some ascomycetes are the fungal partners in the ectomycorrhizae on plant roots and in lichens with cyanobacteria.

Basidiomycetes

The basidiomycetes contain more than 30,000 species described to date. Mushrooms and toadstools

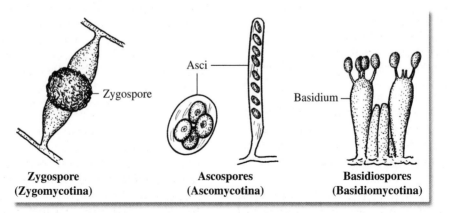

Figure 7.3   Types of sexual spores seen in the fungi.

are found in this of group of fungi. The differentiation between mushrooms and toadstools is that mushrooms, such as *Agaricus campestris*, are edible, whereas toadstools, such as *Amanita*, are considered poisonous. Normally a basidiomycete grows as haploid mycelium in the soil. However, distinct mating strains of mycelium occur which can fuse to form dikaryotic mycelia containing two nuclei per cell. When growth conditions are favorable, for instance, when moisture is abundant, the mycelium can differentiate into a *basidiocarp* which occurs above ground and is the mushroom structure that is normally seen. Gills, which are flat plates beneath the basidiocarp, contain the nuclei that fuse to form a basidium ("small pedestal") with diploid nuclei (figure 7.3). Two meiotic divisions of the diploid nuclei finally yield four haploid nuclei in the basidium. Each nucleus develops into a basidiospore, which can be distributed by winds and other natural means to habitats where new basidiomycete colonies are established to repeat the cycle of growth. One mushroom, *Coprinus*, distributes its basidiospores by producing chitinase that degrades its basidiocarp.

Not all of the fungi described here will be available for study in this exercise. Most will probably belong to zygomycetes and ascomycetes. Follow the directions of your laboratory instructor.

Laboratory Procedures

Several options are provided here for the study of fungi. The procedures to be followed will be outlined by your instructor.

Yeast Study

The organism *Saccharomyces cerevisiae,* which is used in bread making and alcohol fermentation, will be used for this study (figure 7.4). Either prepared slides or living organisms may be used.

Materials

- prepared slides of *Saccharomyces cerevisiae*
- broth cultures of *Saccharomyces cerevisiae*
- methylene blue stain
- microscope slides and coverslips

Prepared Slides   If prepared slides are used, they may be examined under high-dry or oil immersion. Look for typical **blastospores** and **ascospores.** Space is provided on the Laboratory Report for drawing the organisms.

Living Materials   If broth cultures of *Saccharomyces cerevisiae* are available, they should be examined on a wet mount slide with phase-contrast or brightfield

Figure 7.4   *Saccharomyces cerevisiae*, **methylene blue stain.**
©McGraw-Hill Education. Lisa Burgess, photographer

optics. Place two or three loopfuls of the organisms on the slide with a drop of methylene blue stain. Oil immersion will reveal the greatest amount of detail. Look for the **nucleus** and **vacuole.** The nucleus is the smaller body. Draw a few cells on the Laboratory Report.

Mold Study

Examine a petri plate of Sabouraud's agar that has been exposed to the air for about an hour and incubated at room temperature for 3–5 days. This medium has a low pH, which makes it selective for fungi. A good plate will have many different-colored colonies. Note the characteristic "cottony" nature of the colonies. Also, look at the bottom of the plate and observe how the colonies differ in color here. Colony surface color, underside pigmentation, hyphal structure, and the type of spores produced are important phenotypic characteristics used in the identification of fungi.

Figure 7.5 reveals how some common molds appear when grown on Sabouraud's agar. Keep in mind that the appearance of a fungal colony can change appreciably as it gets older. The photographs in figure 7.5 are of colonies that are 10–21 days old.

Conclusive identification cannot be made unless a microscope slide is made to determine the type of hyphae and spores that are present. Figure 7.6 reveals, diagrammatically, the microscopic differences to look for when identifying fungal genera.

Two Options   In making slides from fungal colonies, one can make either wet mounts directly from the colonies by the procedure outlined here or make cultured slides as outlined in Exercise 20. The following steps should be used for making stained slides directly from the colonies. Your instructor will indicate the number of identifications that are to be made.

Mold	Top	Reverse

(1) *Alternaria*

(2) *Aspergillus*

(3) *Cunninghamella*

(4) *Fusarium*

(5) *Helminthosporium*

(6) *Penicillium*

(7) *Paecilomyces*

(8) *Syncephalastrum*

Figure 7.5 Colony characteristics of some of the more common molds.

1. Place an uncovered plate on a dissecting microscope and examine the colony. Look for hyphal structures and spore arrangement. Increase the magnification if necessary to more clearly see spores. Ignore white colonies as they are usually young and have not begun the sporulation process.
2. Consult figures 7.5 and 7.6 to make a preliminary identification based on colony characteristics and low-power magnification of hyphae and spores.
3. Make a wet mount slide by transferring a small amount of mycelium with a scalpel, dissecting needle, or toothpick to a drop of lactophenol cotton blue. Gently tease apart the mycelium with the dissecting needles. Cover the specimen with a coverslip and examine with the low-power objective. Look for hyphae that have spore structures. Go to the high-dry objective to discern more detail about the spores. Compare your specimen to figure 7.6 and see if you can identify the culture based on microscopic morphology.
4. Repeat the procedure for each colony.

Note: An alternative procedure that preserves the fruiting structures is the **cellophane tape method.** Place 1–2 drops of lactophenol cotton blue on a microscope slide. Using a piece of *clear* cellophane tape slightly smaller than the length of the slide, gently touch the surface of a fungal colony with the sticky side of the tape. Transfer the tape containing the material from the fungal colony to the lactophenol cotton blue stain and press the tape onto the slide, making sure that the culture material is in the stain. Observe with the low-power and high-dry lens.

Prepared Slides If prepared slides are available, first examine them with high-dry objective and then switch to the oil immersion for greater detail. Look for typical structures such as sporangiospores, phialospores, and conidiospores. Space is provided on the Laboratory Report for drawing the organisms.

Laboratory Report

After recording your results on the Laboratory Report, answer all the questions.

Materials

- fungal cultures on Sabouraud's agar
- microscope slides and coverslips
- lactophenol cotton blue stain
- sharp-pointed scalpels or dissecting needles
- prepared slide of a fungal culture

(1)   (2)   (3)   (4)   (5)

(6)   (7)   (8)   (9)   (10)

(11)   (12)   (13)   (14)   (15)   (16)

(17)   (18)   (19)   (20)   (21)

(1) *Penicillium*– bluish-green; brush arrangement of phialospores.

(2) *Aspergillus*– bluish-green with sulfur-yellow areas on the surface. *Aspergillus niger* is black.

(3) *Verticillium*– pinkish-brown, elliptical microconidia.

(4) *Trichoderma*– green, resemble *Penicillium* macroscopically.

(5) *Gliocadium*– dark-green; conidia (phialospores) borne on phialides, similar to *Penicillium*; grows faster than *Penicillium*.

(6) *Cladosporium (Hormodendrum)*– light green to grayish surface; gray to black back surface; blastoconidia.

(7) *Pleospora*– tan to green surface with brown to black back; ascospores shown are produced in sacs borne within brown, flask-shaped fruiting bodies called pseudothecia.

(8) *Scopulariopsis*– light-brown; rough-walled microconidia.

(9) *Paecilomyces*– yellowish-brown, elliptical microconidia.

(10) *Alternaria*– dark greenish-black surface with gray periphery; black on reverse side; chains of macroconidia.

(11) *Bipolaris*– black surface with grayish periphery; macroconidia shown.

(12) *Pullularia*– black, shiny, leathery surface; thick-walled; budding spores.

(13) *Diplosporium*– buff-colored wooly surface; reverse side has red center surrounded by brown.

(14) *Oospora (Geotrichum)*– buff-colored surface; hyphae break up into thin-walled rectangular arthrospores.

(15) *Fusarium*– variants, of yellow, orange, red, and purple colonies; sickle-shaped macroconidia.

(16) *Trichothecium*– white to pink surface; two-celled conidia.

(17) *Mucor*– a zygomycete; sporangia with a slimy texture; spores with dark pigment.

(18) *Rhizopus*– a zygomycete; spores with dark pigment.

(19) *Syncephalastrum*– a zygomycete; sporangiophores bear rod-shaped sporangioles, each containing a row of spherical spores.

(20) *Nigrospora*– conidia black, globose, one-celled, borne on a flattened, colorless vesicle at the end of a conidiophore.

(21) *Montospora*– dark gray center with light gray periphery; yellow-brown conidia.

Figure 7.6   **Microscopic appearance of some of the more common molds.**

7 The Fungi: Molds and Yeasts

A. Results

1. Yeast Study

Draw a few representative cells of *Saccharomyces cerevisiae* in the appropriate circles below. Blastospores (buds) and ascospores, if seen, should be shown and labeled.

Prepared Slide   **Living Cells**

2. Mold Study

In the following table, list the genera of molds identified in this exercise. Under colony description, give the approximate diameter of the colony, its topside color and backside (bottom) color. For microscopic appearance, make a sketch of the organism as it appears on the slide preparation.

GENUS	COLONY DESCRIPTION	MICROSCOPIC APPEARANCE (DRAWING)

B. Short-Answer Questions

1. What does the term coenocytic mean?

   _____

   _____

2. What criteria are the basis for traditional classification schemes? What modern approach to classification has shown that traditional schemes do not apply?

   _____

   _____

3. What distinguishes the cell walls of fungi from those of plants?

   _____

   _____

4. What is one of the fungi that is responsible for infections of skin and nails?

   _____

   _____

5. What does the term "dimorphic" refer to? Give an example of an organism that is dimorphic and what disease it causes.

   _____

   _____

6. How do zygospores differ from conidiospores?

   _____

   _____

7. What foods are produced by fungi?

   _____

   _____

8. What is considered to be the difference between mushrooms and toadstools?

   _____

   _____

9. Why might have fungi been theorized to be involved in the Salem witch trials?

   _____

10. How are fungi important in the production of foods?

   _____

11. What are the ectomycorrhizae?

   _____

12. What components of wood must be degraded for its turnover by the wood-rotting fungi?

   _____

Manipulation of Microorganisms

One of the most critical techniques that any beginning student in microbiology must learn is aseptic technique. This technique insures that an aseptic environment is maintained when handling microorganisms. This means two things:

1. no contaminating microorganisms are introduced into cultures or culture materials and
2. the microbiologist is not contaminated by cultures that are being manipulated.

Aseptic technique is crucial in characterizing an unknown organism. Multiple transfers must often be made from a stock culture to various test media. It is imperative that only the desired organism is transferred each time and that no foreign bacteria are introduced during the transfer. Aseptic technique is also obligatory in isolating and purifying bacteria from a mixed source of organisms. The

© Getty RF

streak-plate and pour-plate techniques provide a means to isolate an individual species. And once an organism is in pure culture and stored as a stock culture, aseptic technique insures that the culture remains pure when it is necessary to retrieve the organism.

Individuals who work with pathogenic bacteria must be sure that any pathogen that is being handled is not accidently released, causing harm to themselves or to coworkers. Failure to observe aseptic technique can obviously pose a serious threat to many.

In the following exercises, you will learn the techniques that allow you to handle and manipulate cultures of microorganisms. Once you have mastered these procedures, you will be able to make transfers of microorganisms from one kind of medium to another with confidence. You will also be able to isolate an organism from a mixed culture to obtain a pure isolate. It is imperative that you have a good grasp of these procedures, as they will be required over and over in the exercises in this manual.

Aseptic Technique

Learning Outcomes

After completing this exercise, you should be able to

1. Aseptically transfer a bacterial culture from one broth tube to a new broth tube.

2. Aseptically transfer a bacterial culture from an agar slant to a new agar slant.

3. Aseptically transfer a bacterial colony from an agar plate to an agar slant.

The use of aseptic technique insures that no contaminating organisms are introduced into culture materials when the latter are inoculated or handled in some manner. It also insures that organisms that are being handled do not contaminate the handler or others who may be present. And its use means that no contamination remains after you have worked with cultures.

As you work with these procedures, with time, they will become routine and second nature to you. You will automatically know that a set of procedures outlined below will be used when dealing with cultures of microorganisms. This may involve the transfer of a broth culture to a plate for streaking, or inoculating an isolated colony from a plate onto a slant culture to prepare a stock culture. It may also involve inoculating many tubes of media and agar plates from a stock culture in order to characterize and identify an unknown bacterium. Ensuring that only the desired organism is transferred in each inoculation is of paramount importance in the identification process. The general procedure for aseptic technique follows.

Work Area Disinfection The work area is first treated with a disinfectant to kill any microorganisms that may be present. This process destroys vegetative cells and viruses but may not destroy endospores.

Loops and Needles The transfer of cultures will be achieved using inoculating loops and needles. These implements must be sterilized before transferring any culture. A loop or needle is sterilized by inserting it into a Bunsen burner or incinerator flame until it is red-hot. This will incinerate any contaminating organisms that may be present. Allow the loop to cool completely before picking up inoculum. This will ensure that viable cells are transferred.

Culture Tube Flaming and Inoculation Prior to inserting a cooled loop or needle into a culture tube, the cap is removed and the mouth of the tube may be flamed. If the tube is a broth tube, the loop is inserted into the tube and twisted several times to ensure that the organisms on the loop are delivered to the liquid. If the tube is an agar slant, the surface of the slant is inoculated by drawing the loop up the surface of the slant from the bottom of the slant to its top. For stab cultures, a needle is inserted into the agar medium by stabbing it into the agar. After the culture is inoculated, the mouth of the tube may be reflamed and the tube is recapped.

Petri Plate Inoculations Loops are used to inoculate or streak petri plates. The plate cover is raised and held diagonally over the plate to protect the surface from any contamination in the air. The loop containing the inoculum is then streaked gently over the surface of the agar. It is important not to gouge or disturb the surface of the agar with the loop. The cover is replaced and the loop is flamed.

Final Flaming of the Loop or Needle After the inoculation is complete, the loop or needle is flamed to destroy any organisms that remain on these implements. The loop or needle is then returned to its receptacle for storage. It should never be placed on the desk surface.

Final Disinfection of the Work Area When all work for the day is complete, the work area is treated with disinfectant to insure that any organism that might have been deposited during any of the procedures is killed.

To gain some practice in aseptic transfer of bacterial cultures, three simple transfers will be performed in this exercise:

1. broth culture to broth tube
2. agar slant culture to an agar slant and
3. agar plate to an agar slant.

Transfer from Broth Culture to Another Broth

Do a broth tube to broth tube inoculation using the following technique. Figure 8.1 illustrates the procedure for removing organisms from a culture, and figure 8.2 shows how to inoculate a tube of sterile broth.

Materials

- broth culture of *Escherichia coli*
- tubes of sterile nutrient broth
- inoculating loop
- Bunsen burner or incinerator
- disinfectant for desktop and paper towels
- marking pen

1. Prepare your desktop by swabbing down its surface with a disinfectant. Use a sponge or paper towels.

2. With a marking pen, label a tube of sterile nutrient broth with your initials and *E. coli.*

3. Sterilize your inoculating loop by flaming it *until it becomes bright red.* The entire wire must be heated. See illustration 1, figure 8.1.

4. Using your free hand, gently shake the tube to disperse the culture (illustration 2, figure 8.1).

5. Grasp the tube cap with the little finger of your hand holding the inoculating loop and remove it from the tube. Flame the mouth of the tube,

(1) Inoculating loop is heated until it is red-hot.

(2) Organisms in culture are dispersed by shaking tube.

(3) Tube enclosure is removed and mouth of tube is flamed.

(4) A loopful of organisms is removed from tube.

(5) Loop is removed from culture and tube mouth is flamed.

(6) Tube enclosure is returned to tube.

Figure 8.1   Procedure for removing organisms from a broth culture with inoculating loop.

as shown in illustration 3, figure 8.1. Note: if an incinerator is used, the tube is not flamed.

6. Insert the inoculating loop into the culture (illustration 4, figure 8.1).

7. Remove the loop containing the culture, flame the mouth of the tube again (illustration 5, figure 8.1), and recap the tube (illustration 6). Place the culture tube back on the test-tube rack.

8. Grasp a tube of sterile nutrient broth with your free hand, carefully remove the cap with your little finger, and flame the mouth of this tube (illustration 1, figure 8.2).

9. Without flaming the loop, insert it into the sterile broth, inoculating it (illustration 2, figure 8.2). To disperse the organisms into the medium, move the loop back and forth in the tube.

10. Remove the loop from the tube and flame the mouth (illustration 3, figure 8.2). Replace the cap on the tube (illustration 4, figure 8.2).

11. Sterilize the loop by flaming it (illustration 5, figure 8.2). Return the loop to its container.

12. Incubate the culture you just inoculated at 37°C for 24–48 hours.

(1) Cap is removed from sterile broth and tube mouth is flamed.

(2) Sterile, cooled loop is inserted into tube of sterile broth.

(3) Loop is removed from broth and tube mouth is flamed.

(4) Tube enclosure is returned to tube.

(5) Loop is flamed and returned to receptacle.

Figure 8.2 **Procedure for inoculating a nutrient broth.**

Transfer of Bacteria from a Slant

To inoculate a sterile nutrient agar slant from an agar slant culture, use the following procedure. Figure 8.3 illustrates the entire process.

Materials

- agar slant culture of *E. coli*
- sterile nutrient agar slant
- inoculating loop
- Bunsen burner or incinerator
- marking pen

1. If you have not already done so, prepare your desktop by swabbing down its surface with a disinfectant.
2. With a marking pen, label a tube of nutrient agar slant with your initials and *E. coli.*
3. Sterilize your inoculating loop by holding it over the flame of a Bunsen burner *until it becomes bright red* (illustration 1, figure 8.3). The entire wire must be heated. Allow the loop to cool completely.
4. Using your free hand, pick up the slant culture of *E. coli* and remove the cap using the little finger of the hand that is holding the loop (illustration 2, figure 8.3).
5. Flame the mouth of the tube and insert the cooled loop into the tube. Pick up some of the culture on

the loop (illustration 3, figure 8.3) and remove the loop from the tube.
6. Flame the mouth of the tube (illustrations 4 and 5, figure 8.3) and replace the cap, being careful not to burn your hand. Return tube to rack.
7. Pick up a sterile nutrient agar slant with your free hand, remove the cap with your little finger as before, and flame the mouth of the tube (illustration 6, figure 8.3).
8. Without flaming the loop containing the culture, insert the loop into the tube and gently inoculate the surface of the slant by moving the loop back and forth over the agar surface, while moving up the surface of the slant (illustration 7, figure 8.3). This should involve a type of serpentine or zigzag motion.
9. Remove the loop, flame the mouth of the tube, and recap the tube (illustration 8, figure 8.3). Replace the tube in the rack.
10. Flame the loop, heating the entire wire to red-hot (illustration 9, figure 8.3), allow to cool, and place the loop in its container.
11. Incubate the inoculated agar slant at 37°C for 24–48 hours.

Working with Agar Plates

(Inoculating a slant from a petri plate)

The transfer of organisms from colonies on agar plates to slants or broth tubes is very similar to the

(1) Inoculating loop is heated until it is red-hot.

(2) Cap is removed from slant culture and tube mouth is heated.

(3) Organism is picked up from slant with inoculating loop.

continued

Figure 8.3   Procedure for inoculating a nutrient agar slant from a slant culture.

(4) Mouth of tube is flamed. Inoculating loop is not flamed.

(5) Slant culture is recapped and returned to test-tube rack.

(6) Tube of sterile agar slant is uncapped and mouth is flamed.

(7) Slant surface is streaked with unflamed loop in a zigzag motion.

(8) Tube mouth is flamed, recapped, and incubated.

(9) Loop is flamed red-hot and returned to receptacle.

Figure 8.3 *(continued)*

procedures used in the last two transfers (broth to broth and slant to slant). The following rules should be observed:

Loops and Needles Loops are routinely used when streaking agar plates and slants. When used properly, a loop will not gouge or tear the agar surface. Needles are used in transfers involving stab cultures.

Plate Handling Media in plates must always be protected against contamination. To prevent exposure to air contamination, covers should always be left closed. When organisms are removed from a plate culture, the cover should be only partially opened as shown in illustration 2, figure 8.4.

Flaming Procedures Inoculating loops or needles must be flamed in the same manner that you used when working with previous tubes. One difference when working with plates is that plates are never flamed!

Plate Labeling and Incubation Petri plates containing inoculated media are labeled on the bottom of the plate. Inoculated plates are almost always incubated upside down. This prevents moisture from condensing on the agar surface and spreading the inoculated organisms.

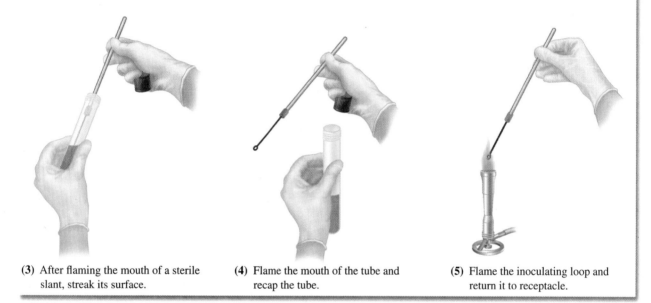

(1) Inoculating loop is heated until it is red-hot.

(2) With free hand, raise the lid of the petri plate just enough to access a colony to pick up a loopful of organisms.

(3) After flaming the mouth of a sterile slant, streak its surface.

(4) Flame the mouth of the tube and recap the tube.

(5) Flame the inoculating loop and return it to receptacle.

Figure 8.4 Procedure for inoculating a nutrient agar slant from an agar plate.

To transfer organisms from a petri plate to an agar slant, use the following procedure:

Materials

- nutrient agar plate with bacterial colonies
- sterile nutrient agar slant
- inoculating loop
- marking pen

1. If you have not done so, swab your work area with disinfectant. Allow area to dry.
2. Label a sterile nutrient agar slant with your name and organism to be transferred.
3. Flame an inoculating loop until it is red-hot (illustration 1, figure 8.4). Allow the loop to cool.
4. As shown in illustration 2, figure 8.4, raise the lid of a petri plate sufficiently to access a colony with your sterile loop.

    Do not gouge the agar with your loop as you pick up organisms. Simply allow the loop to gently glide over the gelatin-like surface of the agar. Do not completely remove the lid while inoculating or removing organisms from the agar plate. This will expose the agar surface to air and potential contamination. Always close the lid once you have removed organisms from the plate.
5. With your free hand, pick up the sterile nutrient agar slant tube. Remove the cap by grasping the cap with the little finger of the hand that is holding the loop.

6. Flame the mouth of the tube and insert the loop into the tube to inoculate the surface of the slant, using a serpentine motion (illustration 3, figure 8.4). Avoid disrupting the agar surface with the loop.
7. Remove the loop from the tube and flame the mouth of the tube. Replace the cap on the tube (illustration 4, figure 8.4).

8. Flame the loop (illustration 5, figure 8.4) and place it in its container.
9. Incubate the nutrient agar slant at 37°C for 24–48 hours.

Results

Examine all three tubes and record your results in Laboratory Report 8.

8 Aseptic Technique

A. Results

1. Were all your transfers successful? _____

2. What evidence do you have that they were successful? _____

    _____

3. What evidence do you have that a transfer is unsuccessful? _____

    _____

B. Short-Answer Questions

1. Provide three reasons why the use of aseptic technique is essential when handling microbial cultures in the laboratory.

    _____

    _____

2. Provide two examples of how heat is used during inoculation of a tube culture.

    _____

    _____

3. How is air contamination prevented when an inoculating loop is used to introduce or take a bacterial sample to/from an agar plate?

    _____

    _____

4. Where should a label be written on an agar plate?

    _____

    _____

5. How should agar plates be incubated? Why?

    _____

    _____

6. Against which organisms are disinfectants effective? Against which type of organism may they not be effective? What disinfectant(s) is used in your laboratory?

    _____

    _____

C. Multiple Choice

Select the answer that best completes the following statements.

1. A disinfectant is used on your work surface
   a. before the beginning of laboratory procedures.
   b. after all work is complete.
   c. after any spill of live microorganisms.
   d. Both (b) and (c) are correct.
   e. All of the above are correct.

2. To retrieve a sample from a culture tube with an inoculating loop, the cap of the tube is
   a. removed and held in one's teeth.
   b. removed and held with the fingers of the loop hand.
   c. removed with the fingers of the loop hand and placed in the fingers of the tube hand.
   d. removed with the fingers of the loop hand and placed on the laboratory bench.
   e. Any of these methods can be used.

3. An inoculating loop or needle is sterilized using heat
   a. by one brief passage.
   b. for exactly 5 minutes.
   c. until the entire wire is bright red.
   d. until the handle is bright red.
   e. until the tip is bright red.

ANSWERS

Multiple Choice

1. _____

2. _____

3. _____

Pure Culture Techniques

Learning Outcomes

After completing this exercise, you should be able to

1. Obtain isolated colonies of a mixed culture using the streak-plate method.

2. Obtain isolated colonies from a bacterial culture using the loop dilution and pour-plate method.

3. Evaluate the purity of your isolated colonies by transferring a single colony to an agar slant and obtaining the growth of a single type of organism.

Figure 9.1 A streak plate demonstrating well-isolated colonies of three different bacteria.

When we try to study the bacterial flora of the body, soil, water, or just about any environment, we realize quickly that bacteria exist in natural environments as mixed populations. It is only in very rare instances that they occur as a single species. Robert Koch, the father of medical microbiology, was one of the first to recognize that if he was going to prove that a particular bacterium causes a specific disease, it would be necessary to isolate the agent from all other bacteria and characterize the pathogen. From his studies on pathogenic bacteria, his laboratory contributed many techniques to the science of microbiology, including the method for obtaining **pure cultures** of bacteria. A pure culture contains only a single kind of an organism, whereas a mixed culture contains more than one kind of organism. A contaminated culture contains a desired organism but also unwanted organisms. With a pure culture, we can study the cultural, morphological, and physiological characteristics of an individual organism.

Several methods for obtaining pure cultures are available to the microbiologist. Two commonly used procedures are the **streak plate** and the **pour plate.** Both procedures involve diluting the bacterial cells in a sample to an end point where a single cell divides, giving rise to a single **pure colony.** The colony is therefore assumed to be the identical progeny of the original cell and can be picked and used for further study of the bacterium.

In this exercise, you will use both the streak-plate and pour-plate methods to separate a mixed culture of bacteria. The bacteria may be differentiated by the characteristics of the colony, such as color, shape, and other colony characteristics. Isolated colonies can then be subcultured and stains prepared to check for purity.

Streak-Plate Method

The streak-plate method is the procedure most often used by microbiologists to obtain pure cultures. It is simple and allows for economy of materials. However, it requires a certain level of skill which is only obtained through practice. Your instructor may want you to try more than one method or only concentrate on one of the quadrant streak-plate procedures. Figure 9.1 illustrates how colonies of a mixed bacterial culture should be spread out and separated on a properly made quadrant streak plate shown in method B of figure 9.2. Good spacing between colonies on the plate is critical so that a single pure colony can be aseptically isolated from quadrant 4 and used for further testing and study. This will insure that you are not working with a mixed or contaminated culture.

Figure 9.2 shows three procedures for producing a streak plate that will yield isolated colonies. By far the most popular and most utilized procedure is the

Quadrant Streak
(Method A)

Quadrant Streak
(Method B)

(1) Streak one loopful of organisms over Area 1 near edge of the plate. Apply the loop lightly. Don't gouge into the medium.

(2) Flame the loop, cool 5 seconds, and make 5 or 6 streaks from Area 1 through Area 2. Momentarily touching the loop to a sterile area of the medium before streaking insures a cool loop.

(3) Flame the loop again, cool it, and make 6 or 7 streaks from Area 2 through Area 3.

(4) Flame the loop again, and make as many streaks as possible from Area 3 through Area 4, using up the remainder of the plate surface.

(5) Flame the loop before putting it aside.

(1) Streak one loopful of organisms back and forth over Area 1, starting at point designated by "s". Apply loop lightly. Don't gouge into the medium.

(2) Flame the loop, cool 5 seconds, and touch the medium in a sterile area momentarily to insure coolness.

(3) Rotate dish 90 degrees while keeping the dish closed. Streak Area 2 with several back and forth strokes, hitting the original streak a few times.

(4) Flame the loop again. Rotate the dish and streak Area 3 several times, hitting the last area several times.

(5) Flame the loop, cool it, and rotate the dish 90 degrees again. Streak Area 4, contacting Area 3 several times and drag out the culture as illustrated.

(6) Flame the loop before putting it aside.

Radiant Streak

(1) Spread a loopful of organisms in a small area near the edge of the plate in Area 1. Apply the loop lightly. Don't gouge into the medium.

(2) Flame the loop and allow it to cool for 5 seconds. Touching a sterile area will insure coolness.

(3) **From the edge** of Area 1 make 7 or 8 straight streaks to the opposite side of the plate.

(4) Flame the loop again, cool it sufficiently, and cross streak over the last streaks, **starting near Area 1.**

(5) Flame the loop before putting it in its receptacle.

Figure 9.2 **Three different streak techniques.**

quadrant streak plate, shown in methods A and B. All of the methods depend upon the physical dilution of cells over the plate surface until a single cell is deposited in an area and grows to produce an isolated bacterial colony. It is important for beginning students to master the streak plate as success in future exercises will depend on using this technique to obtain isolated cultures. This is especially true for the exercises in Part 8 involving the identification of an unknown bacterium.

Materials

- electric hot plate and beaker of water
- Bunsen burner or incinerator
- inoculating loop, thermometer, and marking pen
- 20 ml nutrient agar pour and 1 sterile petri plate
- mixed culture of *Serratia marcescens* or *Escherichia coli*, and *Micrococcus luteus* or *Chromobacterium violaceum*

1. Prepare your tabletop by disinfecting its surface with the disinfectant that is available in the laboratory (Roccal, Zephiran, Betadine, etc.). Use paper towels to scrub it clean.
2. Label the bottom surface of a sterile petri plate with your name and date. Use a marking pen such as a Sharpie.
3. Liquefy a tube of nutrient agar, cool to 50°C, and pour the medium into the bottom of the plate, following the procedure illustrated in figure 9.3. Be sure to flame the neck of the tube prior to pouring to destroy any bacteria around the end of the tube.

   After pouring the medium into the plate, gently rotate the plate so that it becomes evenly distributed, but do not splash any medium up over the sides.

Agar-agar, the solidifying agent in this medium, becomes liquid when boiled and resolidifies at around 42°C. Failure to cool it prior to pouring into the plate will result in condensation of moisture on the cover. Any moisture on the cover is undesirable because it can become deposited on the agar surface, causing the organisms to spread over the surface and thereby defeating the entire isolation procedure.

4. Streak your plate using one of the methods shown in figure 9.2. Method B is the most commonly used procedure.

Caution

Be sure to follow the routine in figure 9.4 for obtaining the organism from culture.

---

5. Incubate the plate in an inverted position for 24 to 48 hours. If the plate is not incubated in an inverted position, condensate from the dish lid will be deposited on the agar surface, allowing organisms to swim and preventing the formation of individual colonies.

(1) Liquefy a nutrient agar pour by boiling for 5 minutes.

(2) Cool down the nutrient agar pour to 50°C. Hold at 50°C for 5 minutes.

(3) Remove the cap from the tube and flame the open end of the tube.

(4) Pour the contents of the tube into the bottom of the petri plate and allow it to solidify.

Figure 9.3 **Procedure for pouring an agar plate for streaking.**

(1) Shake the culture tube from side to side to suspend organisms. Do not moisten cap on tube.

(2) Heat loop and wire to red-hot. Flame the handle slightly also.

(3) Remove the cap and flame the neck of the tube. Do not place the cap down on the table.

(4) After allowing the loop to cool for at least 5 seconds, remove a loopful of organisms. Avoid touching the side of the tube.

(5) Flame the mouth of the culture tube again.

(6) Return the cap to the tube and place the tube in a test-tube rack.

continued

Figure 9.4   **Routine for inoculating a petri plate.**

(7) Streak the plate, holding it as shown. Do not gouge into the medium with the loop.

(8) Flame the loop before placing it down.

Figure 9.4   *(continued)*

Results for isolating colonies using the four procedures are shown in figure 9.5. The quadrant streak method is the most commonly used and gives consistent results.

(a)

(b)

(c)

Figure 9.5   **Spread plates for the three procedures in figure 9.4 for producing isolated colonies: (a) quadrant streak method A; (b) quadrant streak method B; (c) radiant streak.**

Pour-Plate Method

(Loop Dilution)

This method of separating one species of bacteria from another consists of diluting out one loopful of organisms with three tubes of liquefied nutrient agar in such a manner that one of the plates poured will have an optimum number of organisms to provide good isolation. Figure 9.6 illustrates the general procedure. One advantage of this method is that it requires somewhat less skill than that required for a good streak plate; a disadvantage, however, is that it requires more media, tubes, and plates. Proceed as follows to make three dilution pour plates, using the same mixed culture you used for your streak plate.

Materials

- mixed culture of bacteria
- 3 nutrient agar pours
- 3 sterile petri plates
- electric hot plate
- beaker of water
- thermometer
- inoculating loop and marking pen

1. Label the three nutrient agar pours **I, II,** and **III** with a marking pen and place them in a beaker of water on an electric hot plate to be liquefied. To save time, start with hot tap water if it is available.
2. While the tubes of media are being heated, label the bottoms of the three petri plates **I, II,** and **III.**

3. Cool down the tubes of media to 50°C, using the same method that was used for the streak plate.
4. Following the routine in figure 9.6, inoculate tube I with one loopful of organisms from the mixed culture. Note the sequence and manner of handling the tubes in figure 9.7.
5. Inoculate tube II with one loopful from tube I after thoroughly mixing the organisms in tube I by shaking the tube from side to side or by rolling the tube vigorously between the palms of both hands. ***Do not splash any of the medium up onto the tube closure.*** Return tube I to the water bath.
6. Agitate tube II to completely disperse the organisms and inoculate tube III with one loopful from tube II. Return tube II to the water bath.
7. Agitate tube III, flame its neck, and pour its contents into plate III.
8. Flame the necks of tubes I and II and pour their contents into their respective plates.
9. After the medium has completely solidified, incubate the *inverted* plates at 25°C for 24 to 48 hours.

Evaluation of the Two Methods

After 24 to 48 hours of incubation, examine all four petri plates. Look for colonies that are well isolated from the others. Note how crowded the colonies appear on plate I as compared with plates II and III. Plate I will be unusable. Either plate II or III will have the most favorable isolation of colonies. Can you pick out well-isolated colonies on your best pour plate that are distinct from one another?

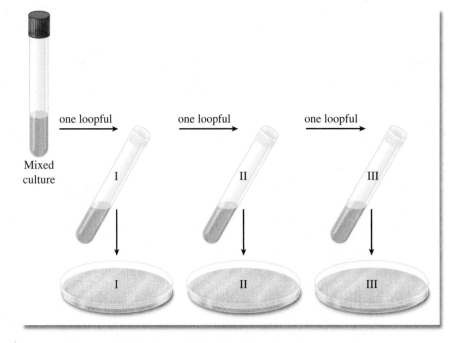

one loopful   one loopful   one loopful

Mixed culture   I   II   III

I   II   III

Figure 9.6   **Three steps in the loop dilution technique for separating out organisms.**

Draw the appearance of your streak plate and pour plates on the Laboratory Report.

Subculturing Techniques

The next step in the development of a pure culture is the transfer of an isolated colony from the petri plate to a tube of nutrient broth or a slant of nutrient agar. Use your loop to carefully pick an isolated colony and aseptically transfer a colony to the broth tube or slant. To insure that the broth is inoculated, rotate the loop in the broth several times before withdrawing it from the broth tube. For the slant, make an "S" motion by drawing the loop from the bottom of the tube up the surface of the slant. Use the following routine to subculture the different organisms that you have isolated.

Materials

- nutrient agar slants
- inoculating loops
- Bunsen burners

(1) Liquefy three nutrient agar pours, cool to 50°C, and let stand for 10 minutes.

(2) After shaking the culture to disperse the organisms, flame the loop and necks of the tubes.

(3) Transfer one loopful of the culture to tube I.

(4) Flame the loop and the necks of both tubes.

(5) Replace the caps on the tubes and return culture to the test-tube rack.

(6) Disperse the organisms in tube I by shaking the tube or rolling it between the palms.

continued

Figure 9.7　**Tube-handling procedure in making inoculations for pour plates.**

(7) Transfer one loopful from tube I to tube II. Return tube I to the water bath.

(8) After shaking tube II and transferring one loopful to tube III, flame the neck of each tube.

(9) Pour the inoculated pours into their respective petri plates.

Figure 9.7 *(continued)*

1. Label one tube *Micrococcus luteus* and a second *Serratia marcescens*.
2. Select a well-isolated red colony on either the streak plate or the pour plate. Use your inoculating loop to pick a well-isolated colony and transfer it to the tube labeled *S. marcescens*.
3. Repeat this procedure for a yellow/cream-colored colony and transfer the colony to the tube labeled *M. luteus*.
4. Incubate the tubes at 25°C for 24 to 48 hours.

Evaluation of Slants

After incubation, examine the slants. Is the *S. marcescens* culture red? If the culture was incubated at a temperature higher than 25°C it may not be red because higher temperatures inhibit the formation of the organism's red pigment. Draw the appearance of the slant. What color is the *M. luteus* culture? Draw the slant.

You cannot be sure that your cultures are pure until you have made a microscopic examination of the respective cultures. It is entirely possible that the *S. marcescens* culture harbors some contaminating *M. luteus* and vice versa. Prepare smears for each culture and Gram stain the smears (see Exercise 14). *S. marcescens* is a gram-negative short rod, whereas *M. luteus* is a gram-positive coccus. Draw the Gram-stained smears on the Laboratory Report.

Laboratory Report

Complete the Laboratory Report for this exercise.

9 Pure Culture Techniques

A. Results

1. **Evaluation of Streak Plate**

   Show within the circle the distribution of the colonies on your streak plate. To identify the colonies, use red for *Serratia marcescens*, yellow for *Micrococcus luteus,* and purple for *Chromobacterium violaceum.* If time permits, your instructor may inspect your plate and enter a grade where indicated.

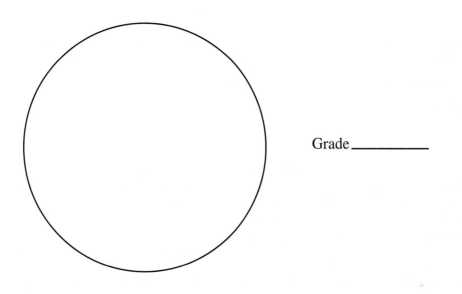

Grade _____

2. **Evaluation of Pour Plates**

   Show the distribution of colonies on plates II and III, using only the quadrant section for plate II. If plate III has too many colonies, follow the same procedure. Use colors.

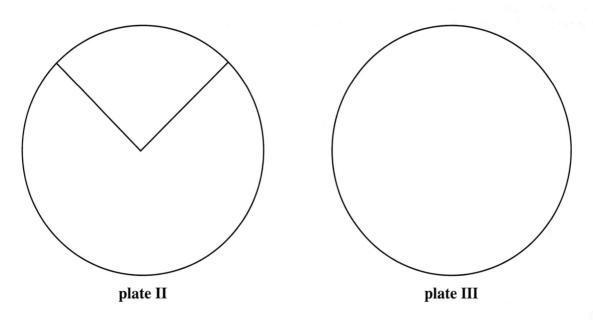

plate II   **plate III**

3. **Subculture Evaluation**

   With colored pencils, sketch the appearance of the growth on the slant diagrams below. Also, draw a few cells of each organism as revealed by Gram staining in the adjacent circle.

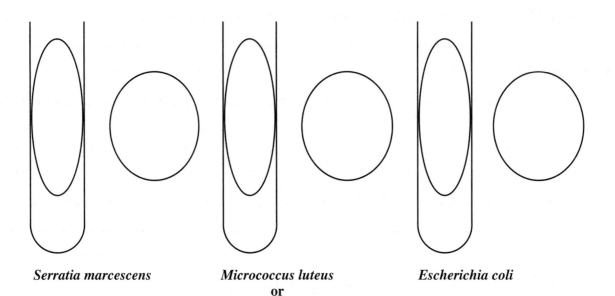

Serratia marcescens   *Micrococcus luteus*   *Escherichia coli*
                                                    or
                                    Chromobacterium violaceum

4. Compare the results of your streak and pour plates. Which method achieved the best separation of species?

   _____

   _____

5. Do your slants contain pure cultures? How would you confirm their purities?

   _____

   _____

B. **Short-Answer Questions**

   1. In regard to bacterial growth on solid media, define the term "colony."

      _____

      _____

   2. What colony characteristics can be used for differentiation of bacterial species? As an example, compare the properties of colonies of *Serratia marcescens* and *Micrococcus luteus* on your streak plate.

      _____

      _____

   3. Why is dilution a necessary part of pure culture preparation?

      _____

      _____

4. What advantage(s) does the streak-plate method have over the pour-plate method?

   _____

   _____

5. What advantage(s) does the pour-plate method have over the streak-plate method?

   _____

   _____

6. Why is the loop flamed before it is placed in a culture tube? Why is it flamed after completing the inoculation?

   _____

   _____

7. Before inoculating and pouring molten nutrient agar into a plate, why must the agar first be cooled to 50°C?

   _____

   _____

8. Explain why plates should be inverted during incubation.

   _____

   _____

Staining and Observation of Microorganisms

The eight exercises in this unit include the procedures for 10 slide techniques that one might employ in morphological studies of bacteria. A culture method in Exercise 17 is also included as a substitute for slide techniques when pathogens are encountered.

These exercises are intended to serve two equally important functions: (1) to help you to develop the necessary skills in making slides and (2) to introduce you to the morphology of bacteria. Although the title of each exercise pertains to a specific technique, the organisms chosen for each method have been carefully selected so that you can learn to recognize certain morphological features. For example, in the exercise on simple staining (Exercise 11), a single staining procedure applied to the selected organism can be used to demonstrate cell morphology, cell arrangement, and internal storage materials such as metachromatic granules. In Exercise 14 (Gram Staining),

Centers for Disease Control

you will learn how to perform an important differential stain that employs more than one stain. This procedure allows you to taxonomically differentiate between two different kinds of bacteria as well as distinguish their cell morphology—cocci or rods.

The importance of the mastery of these techniques cannot be overemphasized. Although one is seldom able to make species identification on the basis of morphological characteristics alone, it is a very significant starting point. This fact will become increasingly clear with subsequent experiments.

Although the steps in the various staining procedures may seem relatively simple, student success is often quite unpredictable. Unless your instructor suggests a variation in the procedure, try to follow the procedures exactly as stated, without improvisation. Photomicrographs in color have been provided for many of the techniques; use them as a guide to evaluate the slides you have prepared. Once you have mastered a specific technique, feel free to experiment.

MICROBIOLOGY

Smear Preparation

Learning Outcomes

After completing this exercise, you should be able to

1. Prepare a thin smear of bacteria on a microscope slide by transferring bacteria from both liquid and solid cultures.

2. Fix the bacterial cells using gentle heat to create a smear that can be stained.

The success of virtually all staining procedures depends upon the preparation of a good **smear.** The procedure outlined in this exercise should be followed for most stains unless you are instructed otherwise. Good smears are critical for discerning: (1) the morphology of cells, such as rods, cocci, and commas; (2) the arrangement of cells, such as single cells, chains, or bunches; and (3) internal structures, such as endospores and cell inclusions. Learning early how to prepare good smears will insure success in later exercises that you will perform, such as the staining exercises and the identification of an unknown bacterium.

There are several goals in preparing a smear. The first goal is to cause the cells to adhere to the microscope slide so that they are not washed off during subsequent staining and washing procedures. Second, it is important to insure that shrinkage of cells does not occur during staining, otherwise distortion and artifacts can result. A third goal is to prepare thin smears because the thickness of the smear will determine if you can visualize individual cells, their arrangement, or details regarding microstructures associated with cells. The arrangement of cells such as streptococci in chains or staphylococci in bunches is diagnostic for these groups of organisms. Also, internal structures such as polyphosphate granules (volutin or metachromatic granules) are important for identifying organisms such as *Corynebacterium diphtheriae.* Thick smears of cells with large clumps can obscure details about arrangement and the presence of internal structures. Furthermore, stain can become entrapped in the clumps of cells, preventing its removal by destaining and washing and leading to erroneous results for staining reactions. The procedure for making a smear is illustrated in figure 10.1.

The first step in preparing a bacteriological smear differs according to the source of the organisms. If the bacteria are growing in a liquid medium (broths, milk, saliva, urine, etc.), one starts by placing two or more loopfuls of the liquid medium directly on the slide.

From solid media such as nutrient agar, blood agar, or some part of the body, one starts by placing one or two loopfuls of water on the slide and then using an inoculating loop to disperse the organisms in the water. Bacteria growing on solid media tend to cling to each other and must be dispersed sufficiently by dilution in water; unless this is done, the smear will be too thick. *The most difficult concept for students to understand about making slides from solid media is that it takes only a very small amount of material to make a good smear.* When your instructor demonstrates this step, pay very careful attention to the amount of material that is placed on the slide.

The organisms to be used for your first slides may be from several different sources. If the plates from Exercise 6 were saved, some slides may be made from them. If they were discarded, the first slides may be made for Exercise 11, which pertains to simple staining. Your instructor will indicate which cultures to use.

From Broth Cultures

(Broths, saliva, milk, etc.)

If you are preparing a bacterial smear from liquid media, follow this routine, which is depicted on the left side of figure 10.1.

Materials

- microscope slides
- broth cultures of *Staphylococcus*, *Streptococcus*, and *Bacillus*
- Bunsen burner
- wire loop
- marking pen
- slide holder (clothespin)

1. Wash a slide with soap or Bon Ami and hot water, removing all dirt and grease. Handle the clean slide by its edges.

2. Write the initials of the organism or organisms on the left-hand side of the slide with a marking pen.
3. To provide a target on which to place the organisms, make a $\frac{1}{2}''$ circle on the *bottom* side of the slide, centrally located, with a marking pen. Later on, when you become more skilled, you may wish to omit the use of this "target circle."
4. Shake the culture vigorously and transfer two loopfuls of organisms to the center of the slide over the target circle. Follow the routine for inoculations shown in figure 10.2. *Be sure to flame the loop after it has touched the slide.*

Caution

Be sure to cool the loop completely before inserting it into a medium. A loop that is too hot will spatter the medium and move bacteria into the air.

5. Spread the organisms over the area of the target circle.
6. Allow the slide to dry by normal evaporation of the water. Don't apply heat.
7. After the smear has become completely air dried, place the slide in a clothespin and pass the slide several times through the Bunsen burner flame.

From Broth Cultures

"Target circle" on bottom of slide.

Two loopfuls of liquid containing organisms are placed in the center of the "target circle."

Organisms are dispersed over entire area of the "target circle."

From Plates and Slants

Two loopfuls of water are placed in center of "target circle."

A very small amount of organisms is dispersed with inoculating loop in water over entire area of "target circle."

The smear is allowed to dry at room temperature.

Slide is passed through flame several times to heat-kill and fix organisms to slide. Use of clothespin is suggested.

Figure 10.1 Procedure for making a bacterial smear.

Caution

Avoid prolonged heating of the slide as this can result in the slide shattering and injuring you. The underside of the slide should feel warm to the touch.

Note that in this step one has the option of using or not using a clothespin to hold the slide. *Use the option preferred by your instructor.*

When preparing a bacterial smear from solid media, such as nutrient agar or a part of the body, follow this routine, which is depicted on the right side of figure 10.1.

(1) Shake the culture tube from side to side to suspend organisms. Do not moisten the cap on the tube during shaking.

(2) Heat loop and wire to red-hot. Flame the handle slightly also.

(3) Remove the cap and flame the neck of the tube. Do not place the cap down on the table.

(4) After allowing the loop to cool for at least 5 seconds, aseptically remove a loopful of organisms from the tube. Avoid touching the sides of the tube.

(5) Flame the mouth of the culture tube again.

(6) Return the cap to the tube and place the tube in a test-tube rack.

continued

Figure 10.2 **Aseptic procedure for organism removal.**

89

(7) Place the loopful of organisms in the center of the target circle on the slide.

(8) Flame the loop again before removing another loopful from the culture or setting the inoculating loop aside.

Figure 10.2 *(continued)*

Materials

- microscope slides
- inoculating needle and loop
- marking pen
- slide holder (clothespin)
- Bunsen burner

1. Wash a slide with soap or Bon Ami and hot water, removing all dirt and grease. Handle the clean slide by its edges.
2. Write the initials of the organism or organisms on the left-hand side of the slide with a marking pen.
3. Mark a "target circle" on the bottom side of the slide with a marking pen. (See comments in step 3 on page 88.)
4. Flame an inoculating loop, let it cool, and transfer two loopfuls of water to the center of the target circle.

5. Flame an inoculating needle and then let it cool. Pick up *a very small amount of the organisms,* and mix it into the water on the slide. Disperse the mixture over the area of the target circle. Be certain that the organisms have been well emulsified in the liquid. *Be sure to flame the inoculating needle before placing it in its holder.*
6. Allow the slide to dry by normal evaporation of the water. Don't apply heat.
7. After the slide has become completely dry, place it in a clothespin and pass it several times through the flame of a Bunsen burner. Avoid prolonged heating of the slide as it can shatter from excessive exposure to heat.

Laboratory Report

Answer the questions on Laboratory Report 10 that relate to this exercise.

Laboratory Report

Student: _____

Date: _____ Section: _____

10 Smear Preparation

A. Short-Answer Questions

1. How does smear preparation of cells from a liquid medium differ from preparation of cells from a solid medium?

   _____

   _____

2. Why is it important to limit the quantity of cells used to prepare a smear?

   _____

   _____

3. For preparation of a smear on a slide, what is the purpose of heat fixation? What problems can arise when the slide is heated in a flame?

   _____

   _____

Simple Staining

Learning Outcomes

After completing this exercise, you should be able to

1. Prepare a thin smear of bacterial cells and stain them with a simple stain.

2. Understand why staining is necessary to observe bacteria with a brightfield microscope.

3. Observe the different morphologies of bacterial cells.

Because bacterial cells are composed of approximately 80% water, there is very little contrast and hence definition between the cell and the surrounding aqueous environment in which most cells occur. This lack of contrast makes it extremely difficult to visualize cells or their internal details in an aqueous suspension using a brightfield microscope. To enhance the contrast of bacterial cells so they can be visualized with a brightfield microscope, a smear of cells is prepared on a microscope slide which is heat-fixed to cause them to adhere to the slide and, importantly, to preserve the structural integrity of the cells. Smears are then stained using various dyes to enhance cell features and structures.

The use of a single stain to color a bacterial cell is referred to as **simple staining.** Commonly used dyes for performing simple staining are methylene blue, basic fuchsin, and crystal violet. These are referred to as **basic dyes** because they have color-bearing ionic groups (*chromophores*) that are positively charged (cationic). They work well with bacterial cells that have chemical groups on their surfaces that confer a net negative charge to the cell. Therefore, there is a pronounced electrostatic attraction between the cell and the cationic chromophore of the stain. In contrast to cationic chromophores, dyes that have anionic chromophores (negatively charged) are called **acidic dyes.** An example is eosin (sodium$^+$ eosinate$^-$) which does not stain bacterial cells because the anionic chromophore is electrostatically repelled by the negatively charged bacterial cell.

Simple stains can be used to determine the morphology of bacterial cells. Figure 11.1 illustrates most of the common shapes of bacterial cells. They can be grouped into three morphological types: **bacilli**

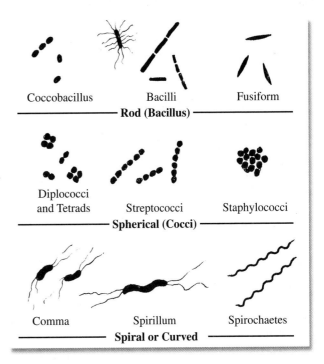

Figure 11.1 **Bacterial morphology.**

(rods), **cocci** (spherical), and **spirals** (corkscrew-shaped rods). Rods or bacilli can have rounded, flat, or tapered ends. The fusiform bacteria are rods with tapered ends and are prevalent in the human mouth. Cocci may occur singly, in chains, in tetrads (packets of four), or in irregular masses. Most streptococci occur in chains, whereas the staphylococci occur in bunches that resemble grapes. The spiral-shaped bacteria can exist as spirochaetes, as a spirillum, or as a comma-shaped, curved rod. *Treponema pallidum,* the causative agent of syphilis, is a spirochaete that is too thin to be observed by brightfield microscopy. *Vibrio cholerae,* the bacterium responsible for cholera, is a comma-shaped bacterium.

Cultures of *Staphylococcus aureus, Streptococcus lactis, Bacillus megaterium,* and *Corynebacterium xerosis* will be used in the exercise for the simple stain (figure 11.2). *C. xerosis* is related to *Corynebacterium diphtheriae,* the bacterium responsible for diphtheria in humans. The basic dye methylene blue (methylene$^+$ chloride$^-$) will be used in this exercise to stain these cells with their different morphologies and

(1) A bacterial smear is stained with methylene blue for 1 minute.

(2) Stain is briefly washed off slide with water.

(3) Water drops are carefully blotted off slide with bibulous paper.

Figure 11.2 **Procedure for simple staining.**

cell arrangements. It is informative to do the simple stain with corynebacteria because the stain demonstrates unique characteristics of these bacteria that are used in their identification. These characteristics are: pleomorphism, metachromatic granules, and palisade arrangement of cells.

Pleomorphism pertains to irregularity of form: that is, demonstrating several different shapes. While *C. diphtheriae* is basically rod-shaped, it also appears club-shaped, spermlike, or needle-shaped. *Bergey's Manual* uses the terms "pleomorphic" and "irregular" interchangeably (figure 11.3).

Metachromatic granules are distinct reddish-purple granules within cells that show up when the organisms are stained with methylene blue. These granules are masses of *volutin*, a polymetaphosphate.

Palisade arrangement pertains to parallel arrangement of rod-shaped cells. This characteristic, also called "picket fence" arrangement, is common to many corynebacteria.

Procedure

Prepare smears of *S. aureus*, *S. lactis*, *B. acillus megaterium*, and *C. xerosis* ATCC 373 as shown in figure 11.2. Refer to Exercise 10 for the smear preparation procedure.

Materials

- slant cultures of *S. aureus*, *S. lactis*, *B. megaterium*, and *C. xerosis*
- methylene blue (Loeffler's)
- wash bottle
- bibulous paper

After examining the slide, compare it with the photomicrograph in figure 11.4. Record your observations on Laboratory Report 11.

Figure 11.3 **Methylene blue stain of *Corynebacterium* showing club-shaped cells.**
Centers for Disease Control

Figure 11.4 **Simple stain of *Bacillus subtilis* and *Staphylococcus aureus*.**
© McGraw-Hill Education. Auburn University Research Instrumentation Facility/Michael Miller, photographer.

11   Simple Staining

A. Results

1. What three noteworthy physical characteristics of *Corynebacterium xerosis* are visible after performing a simple stain? Draw cells from your slide to demonstrate these characteristics.

Corynebacterium xerosis

2. What is the arrangement and cell morphology of the streptococci?

Streptococci

3. What is the arrangement and cell morphology of staphylococcus?

Staphlyococcus

4. What is the arrangement and cell morphology of bacillus?

Bacillus

B. Short-Answer Questions

1. What are chromophores?

   _____

2. Why do acidic dyes not stain bacterial cells?

   _____

3. Crystal violet is an example of what type of stain?

   _____

4. Volutin is composed of what?

   _____

5. What is meant by palisade arrangement of cells?

   _____

6. What is the difference between basic and acidic dyes?

   _____

7. What shape does *Vibrio cholerae* have?

   _____

8. Where are the fusiform bacteria usually found in humans?

   _____

Negative Staining

Learning Outcomes

After completing this exercise, you should be able to

1. Prepare a negative stain of bacterial cells using the slide-spreading or loop-spreading techniques.
2. Use the negative stain to visualize cells from your teeth and mouth.
3. Discern different morphological types of bacterial cells in a negative stain.

Negative stains can be useful in studying the morphology of bacterial cells and characterizing some of the external structures, such as capsules, that are associated with bacterial cells. Negative stains are acidic and thus have a negatively charged chromophore that does not penetrate the cell but rather is repelled by the similarly charged bacterial cell. The background surrounding the cell is colored by a negative stain, resulting in a negative or indirect staining of the cell. Usually cells appear as transparent objects against a dark background (see figure 12.1). Examples of negative stains are india ink and nigrosin. The negative stain procedure consists of mixing the organism with a small amount of stain and spreading a very thin film over the surface of a microscope slide. For negative stains, cells are not usually heat-fixed prior to the

application of the negative stain. Sometimes negative staining can be combined with positive staining to better demonstrate structures such as capsules. In this case, the capsule can be seen as a halo surrounding a positively stained cell against a dark background.

Negative staining can also be useful for accurately determining cell dimensions. Because heat fixation is not performed, no shrinkage of cells occurs and size determinations are more accurate than those determined on fixed material. Avoiding heat fixation is also important if the capsule surrounding the cell is to be observed because heat fixation will severely shrink this structure. The negative stain is also useful for observing spirochaetes, which tend to be very thin cells that do not readily stain with positive stains.

Two Methods

Negative staining can be done by two methods. Figure 12.2 illustrates the more commonly used method in which the organisms are mixed in a drop of nigrosin and spread over the slide with another slide. The goal is to produce a smear that is thick at one end and feather-thin at the other end. Somewhere between the too thick and too thin areas will be an ideal spot to study the organisms.

Figure 12.3 illustrates a second method, in which organisms are mixed in only a loopful of nigrosin instead of a full drop. In this method, the organisms are spread over a smaller area in the center of the slide with an inoculating needle. No spreader slide is used in this method.

Note in the procedure below that slides may be prepared with organisms from between your teeth or from specific bacterial cultures. Your instructor will indicate which method or methods you should use and demonstrate some basic aseptic techniques. Various options are provided here to ensure success.

Figure 12.1   Negative stain of Bacillus and Staphylococcus using nigrosin.
© McGraw-Hill Education. Lisa Burgess, photographer

Materials

- microscope slides (with polished edges)
- nigrosin solution or india ink
- slant cultures of *S. aureus* and *B. megaterium*
- inoculating straight wire and loop
- sterile toothpicks
- Bunsen burner
- marking pen

(1) Organisms are dispersed into a small drop of nigrosin or india ink. Drop should not exceed ⅛″ diameter and should be near the end of the slide.

(2) Spreader slide is moved toward drop of suspension until it contacts the drop, causing the liquid to be spread along its spreading edge.

(3) Once spreader slide contacts the drop on the bottom slide, the suspension will spread out along the spreading edge as shown.

(4) Spreader slide is pushed to the left, dragging the suspension over the bottom slide. After the slide has air-dried, it may be examined under oil immersion.

Figure 12.2   Negative staining technique, using a spreader slide.

(1) A loopful of nigrosin or india ink is placed in the center of a clean microscope slide.

(2) A sterile inoculating wire is used to transfer the organisms to the liquid and mix the organisms into the stain.

(3) Suspension of bacteria is spread evenly over an area of one or two centimeters with the straight wire.

(4) Once the preparation has completely air-dried, it can be examined under oil immersion. No heat should be used to hasten drying.

Figure 12.3   A second method for negative staining.

1. Swab down your tabletop with disinfectant in preparation for making slides.
2. Clean two or three microscope slides with Bon Ami to rid them of all dirt and grease.
3. By referring to figure 12.2 or 12.3, place the proper amount of stain on the slide.
4. **Oral Organisms:** Remove a small amount of material from between your teeth with a sterile straight toothpick and mix it into the stain on the slide. Be sure to break up any clumps of organisms with the toothpick or a sterile inoculating loop. When using a loop, *be sure to flame it first to make it sterile.*

Caution

If you use a toothpick, discard it into a beaker of disinfectant.

5. **From Cultures:** With a *sterile* straight wire, transfer a very small amount of bacteria from the slant to the center of the stain on the slide.
6. Spread the mixture over the slide according to the procedure used in figure 12.2 or 12.3.
7. Allow the slide to air-dry and examine with an oil immersion objective.

Laboratory Report

Draw a few representative types of organisms on Laboratory Report 12. If the slide is of oral organisms, look for yeasts and hyphae as well as bacteria. Spirochaetes may also be present.

12 Negative Staining

A. Results

1. Draw the different types of microorganisms that were found in the negative stain of the oral sample. How would you differentiate between oral streptococci, yeasts, and spirochaetes in your sample?

Oral organisms
(nigrosin)

B. Short-Answer Questions

1. What type of chromophore is associated with a negative stain?

   _____

   _____

2. What is an example of a negative stain?

   _____

   _____

3. What step normally associated with staining bacterial cells is omitted when the dimensions of cells are determined? Why?

   _____

   _____

4. What external bacterial cell structures can be demonstrated by a negative stain?

   _____

   _____

Capsular Staining

Learning Outcomes

After completing this exercise, you should be able to

1. Prepare a smear of an encapsulated bacterium and stain its capsule using the Anthony capsule stain.

2. Visualize the capsule and differentiate it from the cell body.

Many bacterial cells are surrounded by an extracellular gel-like layer that occurs outside of the cell wall. If the layer is distinct and gelatinous, it is referred to as a **capsule.** If the layer is diffuse and irregular, it is called a **slime layer.** The capsule or slime layer can vary in its chemical composition. If it is made up of polysaccharides, it is known as a **glycocalyx,** which literally means "sugar shell." However, the capsule found in *Bacillus anthracis* is composed of poly-D-glutamic acid, which forms a proteinaceous matrix.

Capsules or slime layers perform very important functions for a cell. In pathogens such as *Streptococcus pneumoniae,* they are protective structures because they prevent phagocytic white blood cells from engulfing and destroying the pathogen, enabling the organism to invade the lungs and cause pneumonia. Another function for capsules or slime layers is that of attachment of the bacterial cell to solid surfaces in the environment. For example, *Streptococcus mutans* produces a capsule that facilitates the attachment of the organism to the tooth surface, resulting in the formation of dental plaque. If not removed, plaque will contribute to the formation of dental caries. Evidence supports the view that probably all bacterial cells have some amount of slime layer, but in most cases the amount is not enough to be readily discerned.

Staining of the bacterial capsule cannot be accomplished by ordinary staining procedures. If smears are heat-fixed prior to staining, the capsule shrinks or is destroyed and therefore cannot be seen in stains. In the Anthony method (figure 13.1), smears are air-dried

(1) Two loopfuls of the organism are used to prepare a thin smear.

(2) The smear is allowed to only air-dry; do not heat-fix.

(3) Stain the smear with 1% crystal violet for 2 minutes.

(4) The crystal violet is washed off the slide with 20% copper sulfate into a proper waste container. Do not wash copper sulfate into the sink.

(5) Blot dry with bibulous paper.

Figure 13.1 **Procedure for demonstration of a capsule.**

and then stained with crystal violet for 2 minutes. The stain is washed off with an aqueous solution of 20% copper sulfate and blotted dry. Under oil immersion, the capsules will appear as halos around the cells and the cells will be dark purple (figure 13.2). You will use this procedure to stain the capsules of *Klebsiella pneumoniae*.

Materials

- 36–48 hour culture of *Klebsiella pneumoniae*
- 1% (wt/vol) crystal violet
- 20% (wt/vol) aqueous copper sulfate
- waste containers for the copper sulfate
- disinfectant solutions for used slides

Procedure

1. Prepare thin smears of *Klebsiella pneumoniae* on a microscope slide.
2. Allow the smear to only air-dry. **Do not heat-fix as this will cause the capsule to shrink or be destroyed.**

3. Apply 1% crystal violet and allow it to remain on the slide for 2 minutes.
4. With the slide over the proper waste container provided, gently wash off the crystal violet with 20% copper sulfate. **Caution: Do not wash the copper sulfate and stain directly into the sink.**
5. Blot the slide dry with bibulous paper.
6. Observe with the oil immersion lens and compare your stain with figure 13.2. Record your results on Laboratory Report 13.
7. Be sure to dispose of your used slide in the disinfectant container when you are finished.

Note: The capsule stain may also be done using the procedure for the negative stain in Exercise 12. Either nigrosin or india ink can be used as the stain. Cells will have a halo around them against a dark background.

Observation   Examine the slide under oil immersion and compare your slide with figure 13.2. Record your results on Laboratory Report 13.

(a)

(b)

Figure 13.2   **(a) Anthony capsule stain of *Klebsiella*; (b) negative stain of *Klebsiella* showing capsule.**
14.2 (a) © McGraw-Hill Education. Auburn University Research Instrumentation Facility/Michael Miller, photographer. (b) © Steven P. Lynch.

13 Capsular Staining

A. Results

1. Draw cells that display a capsule from your stained slide of *Klebsiella pneumoniae*. Explain how the capsule is visualized without the use of dyes that adhere to a capsule.

Klebsiella pneumoniae
(capsular stain)

B. Short-Answer Questions

1. What are two functions of the capsule or glycocalyx in bacterial cells?

   _____

   _____

2. What biological molecules can make up the bacterial capsule or glycocalyx?

   _____

   _____

3. What function does the capsule have for *Streptococcus mutans*?

   _____

   _____

4. A student heat-fixes his smear intended for capsule staining. What result might the student expect?

   _____

   _____

MICROBIOLOGY

Gram Staining

Learning Outcomes

After completing this exercise, you should be able to

1. Explain the importance of the Gram stain in microbiology.

2. Summarize the differences between gram-positive and gram-negative cell walls.

3. Explain how each step of the Gram stain procedure works to differentially stain the two types of bacteria.

4. Prepare thin smears of gram-positive and gram-negative cells, and stain them using the Gram stain method.

5. Differentiate between gram-positive and gram-negative cells under the microscope.

Reagent	Gram-positive	Gram-negative
None (Heat-fixed cells)		
Crystal Violet (30 seconds)		
Gram's Iodine (1 minute)		
Ethyl Alcohol (5–15 seconds)		
Safranin (1 minute)		

Figure 14.1 **Color changes that occur at each step in the Gram staining process.**

In 1884, the Danish physician Hans Christian Gram was trying to develop a staining technique that would differentiate bacterial cells from eukaryotic nuclei in diseased lung tissue. He discovered that certain stains were retained by some types of bacterial cells but removed from others during the staining process. His published work served as the foundation of what would become the most important stain in bacteriology, the Gram stain.

Gram staining is a valuable diagnostic tool used in the clinical and research setting. Although newer molecular techniques have been developed, the Gram stain is still a widely used method for the identification of unknown bacteria. It is often the first test conducted on an unknown species in the laboratory, and in some cases, it can provide presumptive identification of the organism. For physicians, a Gram stain of a clinical specimen may be used to determine an appropriate treatment for a bacterial infection.

The Gram stain is an example of a differential stain. Differential staining reactions take advantage of the fact that cells or structures within cells display dissimilar staining reactions that can be distinguished by the use of different dyes. In the Gram stain, two kinds of cells, gram-positive and gram-negative, are differentiated based on their cell wall structure and composition. These types of cells can be identified by their respective colors, purple and pink or red, after performing the staining method. The procedure is based on the fact that gram-positive bacteria retain a purple dye complex, whereas gram-negative bacteria are decolorized and must be counterstained with a red dye in order to be visualized by microscopy.

Figure 14.1 illustrates the appearance of cells after each step in the Gram stain procedure. Initially, both gram-positive and gram-negative cells are stained by the **primary stain,** crystal violet. In the second step of the procedure, Gram's iodine is added to the smear. Iodine is a **mordant** that combines with the crystal violet and forms an insoluble complex in gram-positive cells. At this point, both types of cells will still appear as purple. During **decolorization** with alcohol and/or acetone, gram-positive cells retain the crystal violet-iodine complex, and therefore these cells will appear purple under the microscope. Alternatively, the dye-mordant complex is removed from gram-negative cells, leaving them colorless. Safranin is applied as a counterstain, coloring the gram-negative cells pink or red. The safranin also sticks to the gram-positive cells, but their appearance is unchanged because the crystal violet is a much more intense stain than safranin.

The mechanism for how the Gram stain works is not completely understood, but it is known to be related to structural and chemical differences in the cell walls of gram-positive and gram-negative bacteria (figure 14.2). When viewed by electron microscopy, gram-positive cells have a thick layer of **peptidoglycan** that comprises the cell wall of these organisms. In

Figure 14.2
Comparison of gram-positive and gram-negative cell walls.

Gram-positive cell

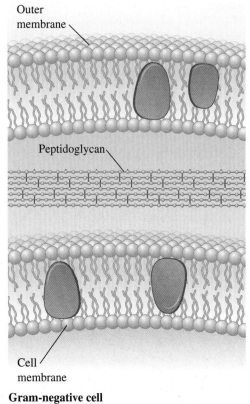

Gram-negative cell

contrast, the cell wall in gram-negative cells consists of an outer membrane that covers a much thinner layer of peptidoglycan. It is believed that the thick, tightly linked peptidoglycan molecules of gram-positive cells trap the crystal violet-iodine complexes, preventing their removal when the smear is correctly decolorized. In contrast, the decolorizer dissolves the lipids in the outer membrane of gram-negative bacteria, allowing the dye-mordant complexes to escape through the thin peptidoglycan layer.

Some bacteria are considered gram-variable because some cells will retain the crystal violet stain, while others will not and appear red from the counterstain. Other bacteria, called acid-fast bacteria, have a unique cell wall made of "waxy" lipids. Mycobacteria, the causative agents of tuberculosis and Hansen's disease (leprosy), are acid-fast bacteria. These cells may appear as either nonreactive or gram-positive after the Gram stain technique, but a special acid-fast staining technique (see Exercise 16) can be used to identify bacteria with this type of cell wall.

Although the Gram stain technique may seem quite simple, performing it with a high degree of reliability requires some practice and experience. Several factors can affect the outcome of the procedure:

1. It is important to use cultures that are 16–18 hours old. Gram-positive cultures older than this can convert to gram-variable or gram-negative and give erroneous results. (It is important to note that gram-negative bacteria never convert to gram-positive.)

2. It is critical to prepare thin smears. Thin smears allow the observation of individual cells and any arrangement in which the cells occur. However, thick smears can entrap the primary stain, preventing decolorization. Cells that occur in the entrapped stain may falsely appear gram-positive.

3. Decolorization is the most critical step in the Gram stain procedure. If the decolorization is overapplied, the dye-mordant complex can eventually be removed from gram-positive cells, causing them to incorrectly appear as gram-negative cells.

During this laboratory period, you will be provided an opportunity to stain several different kinds of bacteria to see if you can achieve the degree of success that is required. Remember, if you don't master this technique now, you may have difficulty with future lab exercises, so take time to repeat stains that yield incorrect results.

Materials

- broth cultures of *Staphylococcus aureus, Pseudomonas aeruginosa, Escherichia coli,* and *Moraxella catarrhalis*
- nutrient agar slant culture of *Bacillus megaterium*
- Gram staining kit and wash bottle
- bibulous paper

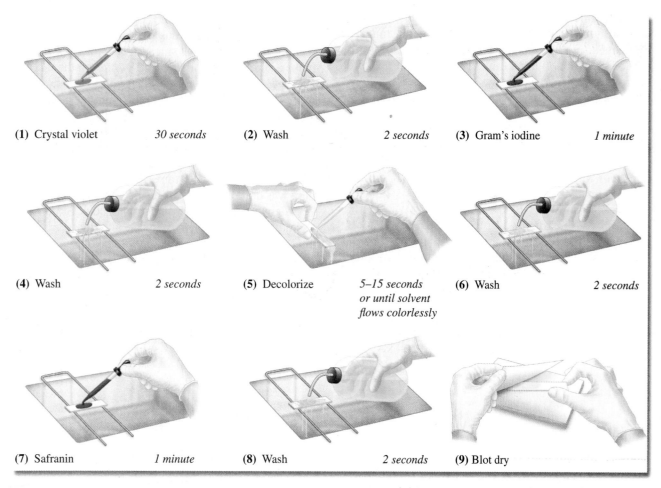

(1) Crystal violet   *30 seconds*   **(2)** Wash   *2 seconds*   **(3)** Gram's iodine   *1 minute*

(4) Wash   *2 seconds*   **(5)** Decolorize   *5–15 seconds or until solvent flows colorlessly*   **(6)** Wash   *2 seconds*

(7) Safranin   *1 minute*   **(8)** Wash   *2 seconds*   **(9)** Blot dry

Figure 14.3   **The Gram staining procedure.**

Gram Stain Procedure

1. Cover a heat-fixed smear with **crystal violet** and let stand for *30 seconds* (see figure 14.3).
2. Briefly wash off the stain, using a wash bottle of distilled water. Drain off excess water.
3. Cover the smear with **Gram's iodine** solution and let it stand for *1 minute.* (Your instructor may prefer only 30 seconds for this step.) Wash off the Gram's iodine.
4. Hold the slide at a 45-degree angle and apply the **decolorizer,** allowing it to flow down the surface of the slide. Do this until the decolorizer is colorless as it flows from the smear down the surface of the slide. *This should take no more the 15 seconds for properly prepared smears.* **Note:** Thick smears can take longer for decolorization.
5. Stop decolorization by washing the slide with a gentle stream of water.
6. Cover the smear with **safranin** for 1 minute.
7. Wash gently for a few seconds, blot dry with bibulous paper, and air-dry.
8. Examine the slide under oil immersion.

Gram Staining Exercises

The organisms that will be used here for Gram staining represent a diversity of form and staining characteristics. Some of the rods and cocci are gram-positive; others are gram-negative. Once you practice the technique on a known gram-positive and a known gram-negative organism, you will use these bacteria as controls for determining the Gram reaction of other bacterial species. You will also observe the Gram stain reaction of a spore-forming bacterium.

Gram Stain Practice Slides Prepare two slides with three smears on each slide. On the left portion of each slide, make a thin smear of *Staphylococcus aureus*. On the right portion of each slide, make a thin smear of *Pseudomonas aeruginosa*. In the middle of the slide, make a thin smear that is a mixture of both organisms, using two loopfuls of each organism. ***Be sure to flame the loop sufficiently to avoid contaminating cultures.***

Using the Gram stain procedure shown in figure 14.3, stain one slide first, saving the other one if you need to repeat the technique. If done properly, *Staphylococcus aureus* should appear as purple cocci, and *Pseudomonas aeruginosa* should appear as pink or red rods, as shown in figure 14.4a.

Call your instructor over to evaluate your slide. If the slide is improperly stained, the instructor may be able to tell what went wrong by examining all three smears. He or she will inform you how to correct your technique and if time allows, you can repeat the exercise with the other practice slide.

Record your results on Laboratory Report 14 by drawing a few cells in the appropriate circle.

Slides to Determine Gram Reaction Using a similar procedure to the previous exercise, you will again be making slides with triple smears. As before, you should make a smear of *Staphylococcus aureus* on the left and *Pseudomonas aeruginosa* on the right. In this exercise, make one slide with a mixture of *Bacillus megaterium* and *Escherichia coli* in the center. On another slide, make a smear of *Moraxella catarrhalis* in the center. Because you already know the Gram reaction of *S. aureus* and *P. aeruginosa*, these bacterial smears can be used as controls to determine if the Gram stain procedure was done correctly for the bacteria on each slide.

Stain both of these slides using the Gram stain technique, and then observe under oil immersion. Draw a few cells in the appropriate circles on your Laboratory Report sheet, and identify the Gram reaction for all three types of bacteria. As you examine the slide with *Bacillus megaterium*, look for clear areas in the rods which represent endospores. Since endospores are impermeable to the dyes used in this technique, they will appear as transparent holes in the cells as seen in figure 14.4d.

S. aureus

P. aeruginosa

(a) Gram stain: *Staphylococcus aureus* and *Pseudomonas aeruginosa*

E. coli

B. megaterium

(b) Gram stain: *Bacillus megaterium* and *Escherichia coli*

(c) Gram stain: Gram-negative diplococci in patient specimen

Endospore

(d) Gram stain: Endospore-forming rods

Figure 14.4 Photomicrographs of Gram stained bacteria.
(a), (b) © McGraw-Hill Education. Auburn University Photographic Services; (c) CDC/Dr. Norman Jacobs; (d) CDC/Dr. Gilda Jones

14   Gram Staining

A. Results

Draw cells from the Gram stained slides, and label each bacterial species. Based on your results, note the Gram reaction, cell shape, and cell arrangement of each bacterium in the given table. Verify these results to determine if you are correct.

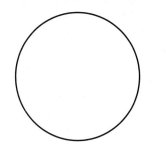

P. aeruginosa and **S. aureus**

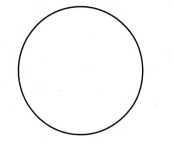

B. megaterium and **E. coli**

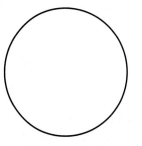

M. catarrhalis

BACTERIAL SPECIES	GRAM REACTION	CELL SHAPE	CELL ARRANGEMENT
Staphylococcus aureus			
Pseudomonas aeruginosa			
Bacillus megaterium			
Escherichia coli			
Moraxella catarrhalis			

B. Short-Answer Questions

1. Why is the Gram stain considered a differential stain?

   It takes advantage of the fact ther cells or structures within cells display chemical staining reactions ther can be distinguished by use of different dyes. Allow determination of size cell, morphology, arraymens bout informanon gbout other features as well.

2. How do gram-positive and gram-negative bacteria differ in cellular structure, and how does this contribute to their differential staining properties?   Based on cell wall structured componens, Gram+ve have a thick layer of peptidoglycan that compose cell of organisms. Gram -ve consist of thinner layer of peptidoglycan.

3. How does the age of a culture affect the Gram stain reaction? What is an optimum culture age for a valid Gram reaction?

   _____

   _____

4. Which step in the Gram stain procedure is most prone to error? If done incorrectly, how might that step affect the end result?

   _____

   _____

5. What is the function of a mordant, and which reagent serves this purpose in the Gram stain procedure?

   _____

   _____

6. List the reagents of the Gram stain technique in order and their general role in the staining process.

   _____

   _____

7. In what type of cell, gram-positive or gram-negative, would you find lipopolysaccharide in its cell wall?

   _____

   _____

C. Matching Questions

   Match the expected result (purple, pink, or colorless) to the following descriptions of Gram stained cells. Consult your chart at the beginning of this lab report if you need help remembering the correct Gram reaction for each species. Choices may be used more than once.

   1. *Staphylococcus aureus* before the primary stain _____ NO colur _____

   2. *Pseudomonas aeruginosa* after the primary stain _____ purple _____

   3. *Bacillus megaterium* after the addition of the mordant _____ purple _____

   4. *Staphylococcus aureus* after decolorization _____ purple _____

   5. *Moraxella catarrhalis* after decolorization _____ NO color _____

   6. *Pseudomonas aeruginosa* after decolorization _____ colorless _____

   7. *Bacillus megaterium* after adding the counterstain _____ purple _____

   8. *Escherichia coli* under the microscope if you forgot to apply safranin _____ colorless _____

   9. *Escherichia coli* under the microscope if you forgot to apply decolorizer _____ purple _____

   10. *Bacillus megaterium* under the microscope if you forgot to apply iodine _____ pink _____

Spore Staining:
Two Methods

Learning Outcomes

After completing this exercise, you should be able to

1. Prepare an endospore stain of bacterial cells and demonstrate endospores in the stained preparation.
2. Differentiate between vegetative cells and endospores.

When species of bacteria belonging to the genera *Bacillus* and *Clostridia* exhaust essential nutrients, they undergo a complex developmental cycle that produces resting stages called **endospores.** Endospores allow these bacteria to survive environmental conditions that are not favorable for growth. If nutrients once again become available, the endospore can go through the process of germination to form a new vegetative cell, and growth will resume. Endospores are very dehydrated structures that are not actively metabolizing. Furthermore, they are resistant to heat, radiation, acids, and many chemicals, such as disinfectants, that normally harm or kill vegetative cells. Their resistance is due in part to the fact that they have a protein coat, or **exosporium,** that forms a protective barrier around the spore. Heat resistance is associated with the water content of endospores. The higher the water content of an endospore, the less heat resistant the endospore will be. During sporulation, the water content of the endospore is reduced to 10–30% of the vegetative cell. During endospore formation, calcium dipicolinate and spore-specific proteins form a cytoplasmic gel that reduces the protoplasmic volume of the endospore to a minimum. In addition, a thick cortex forms around the endospore, and contraction of the cortex results in a smaller dehydrated structure. Calcium dipicolinate is not present in vegetative cells. The gel formed by this chemical and the spore-specific proteins controls the amount of water that can enter the endospore, thus maintaining its dehydrated state.

Since endospores are not easily destroyed by heat or chemicals, they define the conditions necessary to establish sterility. For example, to destroy endospores by heating, they must be exposed for 15–20 minutes to steam under pressure, which generates temperatures of 121°C. Such conditions are produced in an **autoclave.**

The resistant properties of endospores also mean that they are not easily penetrated by stains. For example in Exercise 14, you observed that endospores did not readily Gram stain. If endospore-containing cells are stained by basic stains such as crystal violet, the spores appear as unstained areas in the vegetative cell. However, if heat is applied while staining with malachite green, the stain penetrates the endospore and becomes entrapped in the endospore. The malachite green is not removed by subsequent washing with decolorizing agents or water. In this instance, heat is acting as a mordant to facilitate the uptake of the stain.

Schaeffer-Fulton Method

The Schaeffer-Fulton method, which is depicted in figure 15.1, utilizes malachite green to stain the endospore and safranin to stain the vegetative portion of the cell. Utilizing this technique, a properly stained spore-former will have a green endospore contained in a pink sporangium. Figure 15.1 reveals what *Bacillus* and *Clostridium* look like after staining for endospores.

Prepare a smear of *Bacillus megaterium* and allow the smear to air-dry. Heat-fix the dried smear and follow the steps for staining outlined in figure 15.2.

Materials

- 24–36 hour nutrient agar slant culture of *Bacillus megaterium*
- electric hot plate and small beaker (25 ml)
- spore-staining kit consisting of a bottle each of 5% malachite green and safranin

Dorner Method

The Dorner method for staining endospores produces a red spore within a colorless sporangium. Nigrosin is used to provide a dark background for contrast. The six steps involved in this technique are shown in figure 15.3. Although both the sporangium and endospore are stained during boiling in step 3, the sporangium is decolorized by the diffusion of carbolfuchsin molecules into the nigrosin.

Nigrosin is negative (acidic) stains gram -ve organisms only

111

(a)

(b)

Figure 15.1 Spore stain of (a) *Bacillus* and (b) *Clostridium*.
15.1(a) CDC/Courtesy of Larry Stauffer, Oregon State Public Health Laboratory 15.1 (b) © McGraw-Hill Education. Auburn University Research Instrumentation Facility/Michael Miller, photographer.

(1) Cover smear with small piece of paper toweling and saturate it with malachite green. Steam over boiling water for *5 minutes.* Add additional stain if stain boils off.

(2) After the slide has cooled sufficiently, remove the paper toweling and rinse with water for *30 seconds.*

(3) Counterstain with safranin for about *20 seconds.*

(4) Rinse briefly with water to remove safranin.

(5) Blot dry with bibulous paper, and examine slide under oil immersion.

Figure 15.2 The Schaeffer-Fulton spore stain method.

Prepare a slide of *Bacillus megaterium* that utilizes the Dorner method. Follow the steps in figure 15.3.

Materials

- carbolfuchsin
- nigrosin
- electric hot plate and small beaker (25 ml)
- small test tube (10 × 75 mm size)
- test-tube holder
- 24–36 hour nutrient agar slant culture of *Bacillus megaterium*

Quick Spore Stain

A variation on the Schaeffer-Fulton method is a quick method that uses the same stains.

Materials

- *Bacillus megaterium* slant cultures, older than 36 hours
- malachite green stain
- safranin stain
- staining racks
- clothespins

Procedure

1. Prepare a smear of the organism and allow it to air-dry.
2. Grasp the slide with the air-dried smear with a clothespin and pass it through a Bunsen burner flame 10 times. Be careful not to overdo the heating as the slide can break.
3. Immediately flood the smear with malachite green and allow to stand for 5 minutes.
4. Wash the smear with a gentle stream of water.
5. Stain with safranin for 45 seconds. Spores will be green and the vegetative cell will be red.

Laboratory Report

After examining the organisms under oil immersion, draw a few cells in the appropriate circles in Laboratory Report 15.

(1) Make a heavy suspension of bacteria by dispersing several loopfuls of bacteria in 5 drops of sterile water.

(2) Add 5 drops of carbolfuchsin to the bacterial suspension.

(3) Heat the carbolfuchsin suspension of bacteria in a beaker of boiling water for *10 minutes*.

(4) Mix several loopfuls of bacteria in a drop of nigrosin on the slide.

(5) Spread the nigrosin-bacteria mixture on the slide in the same manner as in Exercise 13 (Negative Staining).

(6) Allow the smear to air-dry. Examine the slide under oil immersion.

Figure 15.3 **The Dorner spore stain method.**

15 Spore Staining: Two Methods

A. Results

Draw cells from the spore slides. Differentiate endospores from vegetative cells.

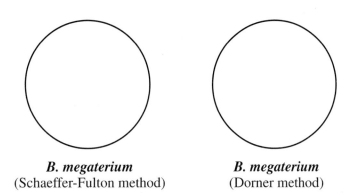

B. megaterium
(Schaeffer-Fulton method)

B. megaterium
(Dorner method)

B. Short-Answer Questions

1. What are the functions of endospores in bacteria?

   _____

   _____

2. What external structure on the endospore acts as a protective barrier? What is its composition?

   _____

   _____

3. Compared to a vegetative cell, how much less water is present in an endospore?

   _____

   _____

4. What is the mordant in the spore stain?

   _____

   _____

5. What unique chemical compound is important in the heat resistance of endospores?

   _____

   _____

6. What conditions are necessary to destroy endospores? In what device are these conditions achieved?

_____

_____

7. What is the color of endospores after Gram staining? After spore staining?

_____

_____

8. What is the secondary stain in the spore stain?

_____

_____

9. What is the color of the vegetative cell after the spore stain?

_____

_____

10. Of the three genera of bacteria, which does not produce endospores: A. *Clostridium*; B. *Mycobacterium*; C. *Bacillus*? _____

Acid-Fast Staining:
Kinyoun Method

Learning Outcomes

After completing this exercise, you should be able to

1. Prepare an acid-fast stain of bacterial cells.

2. Differentiate between acid-fast and non-acid-fast cells in a mixed stain.

3. Explain the basis for the stain and why the stain is important in clinical microbiology.

Bacteria such as *Mycobacterium* and some *Nocardia* have cell walls with a high lipid content. One of the cell wall lipids is a waxy material called **mycolic acid.** This material is a complex lipid that is composed of fatty acids and fatty alcohols that have hydrocarbon chains up to 80 carbons in length. It significantly affects the staining properties of these bacteria and prevents them from being stained by many of the stains routinely used in microbiology. The acid-fast stain is an important diagnostic tool in the identification of *Mycobacterium tuberculosis,* the causative agent of tuberculosis, and *Mycobacterium leprae,* the bacterium that causes leprosy in humans.

To facilitate the staining of acid-fast bacteria, it is necessary to use methods that make the cells more permeable to stains because the mycolic acid in their cell walls prevents the penetration of most stains. In the Ziehl-Neelsen method the primary stain, carbolfuchsin, contains phenol, and the cells are heated for 5 minutes during the staining procedure. Phenol and heat facilitate the penetration of the carbolfuchsin into the cell. Heat is acting as a mordant to make the stain complex more permeable to the mycolic acid and cell wall lipids. Subsequent treatment of the cells with acid-alcohol, a decolorizer, does not remove the entrapped stain from the cells. Hence, these bacteria are termed **acid-fast.** In order for non-acid-fast bacteria to be visualized in the acid-fast procedure, they must be counterstained with methylene blue, as the primary stain is removed from these bacteria by the acid-alcohol. In the Ziehl-Neelsen method, the application of heat to cells during staining with carbolfuchsin and phenol is not without safety concerns. Phenol can vaporize when heated, giving rise to noxious fumes that are toxic to the eyes and mucus membranes. The **Kinyoun acid-fast method**

is a modification in which the concentrations of primary stain, basic fuchsin (substituted for carbolfuchsin), and phenol are increased, making it unnecessary to heat the cells during the staining procedure. The increased concentrations of basic fuchsin and phenol are sufficient to allow the penetration of the stain into cells, and the basic fuchsin is not removed during destaining with acid-alcohol. This procedure is safer because phenol fumes are not generated during staining of the cells.

In the acid-fast staining method, acid-fast bacteria such as *Mycobacterium* are not decolorized by acid-alcohol and are therefore stained pink to red by the basic fuchsin. Because non-acid-fast bacteria such as *Staphylococcus* are decolorized by the acid-alcohol, a secondary stain, methylene blue, must be applied to visualize these cells in stained preparations. These appear blue after staining is completed (figure 16.1).

In the following exercise, you will prepare an acid-fast stain of a mixture of *Mycobacterium smegmatis* and *Staphylococcus aureus* using the Kinyoun method for acid-fast staining. *M. smegmatis* is a nonpathogenic, acid-fast rod that occurs in soil and on the external genitalia of humans. *S. aureus* is a non-acid-fast coccus that is also part of the normal flora of humans but can also be a serious opportunistic pathogen.

Figure 16.1 **Acid-fast stain of *Mycobacterium smegmatis* and *Staphylococcus aureus*.**
© McGraw-Hill Education. Auburn University Research Instrumentation Facility/Michael Miller, photographer.

(1) Cover the heat-fixed smear with basic fuchsin and stain for *5 minutes*.

(2) Wash with water and shake off any excess water.

(3) Decolorize with acid-alcohol for approximately *1 minute*. Check by tilting the slide and adding more destain to make sure that no more stain runs off of the smear.

(4) Stop decolorization by the acid-alcohol by rinsing *briefly* with water.

(5) Counterstain with methylene blue for *30 seconds*.

(6) Rinse briefly with water to remove excess methylene blue.

(7) Blot dry with bibulous paper. Examine under oil immersion.

Figure 16.2   **Kinyoun acid-fast staining procedure.**

Materials

- nutrient agar slant culture of *Mycobacterium smegmatis* (48-hour culture)
- nutrient broth culture of *S. aureus*
- acid-fast staining kit (basic fuchsin, acid-alcohol, and methylene blue)

Smear Preparation   Prepare a mixed culture smear by placing two loopfuls of *S. aureus* on a slide and transferring a small amount of *M. smegmatis* to the broth on the slide with an inoculating needle. Since the Mycobacteria are waxy and tend to cling to each other in clumps, break up the masses of organisms with the inoculating needle. After air-drying the smear, heat-fix it.

Staining   Follow the staining procedure outlined in figure 16.2.

Examination   Examine under oil immersion and compare your slide with figure 16.1.

Laboratory Report

Record your results in Laboratory Report 16.

Laboratory Report

Student: _____

Date: _____ Section: _____

16 Acid-Fast Staining: Kinyoun Method

A. Results

Draw cells from the acid-fast slide. Differentiate acid-fast from non-acid-fast cells.

M. smegmatis and **S. aureus**
(Kinyoun method)

B. Short-Answer Questions

1. What makes *Mycobacterium* resistant to staining?

   _____

   _____

2. What other bacterial genus is acid-fast?

   _____

   _____

3. What is the primary stain in the acid-fast stain?

   _____

   _____

4. What is the secondary stain in the acid-fast stain?

   _____

   _____

5. In the Ziehl-Neelsen acid-fast stain, what is the mordant?

   _____

   _____

6. What is the color of *Mycobacterium* after the addition of the secondary stain?

   _____

   _____

7. Which stain would be most useful in staining the leprosy bacillus: the Gram stain, spore stain, or acid-fast stain? _____

8. What is the advantage of the Kinyoun staining procedure over the Ziehl-Neelsen method?

   _____

   _____

Motility Determination

Learning Outcomes

After completing this exercise, you should be able to

1. Prepare a wet mount of bacterial cells.

2. Differentiate between motile and nonmotile bacteria in the wet mount.

3. Explain the difference between true motility, Brownian motion, and motion resulting from other forces on the slide.

Motility occurs in rod-shaped bacteria and spirochetes and is almost never found in cocci. The major organelles of motility in bacteria are **flagella.** Flagella allow cells to move toward nutrients in the environment or move away from harmful substances, such as acids, in a complicated process called **chemotaxis.** The flagellum is a rigid helical structure that extends as much as 10 microns out from the cell. However, flagella are very thin structures, less than 0.2 microns, and, therefore, they are below the resolution of the light microscope. For flagella to be observed by light microscopy, they must be stained by special techniques. An individual bacterial flagellum is composed of a rigid filament that occurs in the form of a helix. This constitutes the main body of the flagellum structure. The filament is connected to a hook that is attached to a shaft that is inserted into a series of rings whose number differs for gram-positive and gram-negative cells (figure 17.1). Gram-positive cells contain the S and M rings that are situated in the area of the cell membrane. Gram-negative cells also possess the S and M rings and two additional rings, the L and P rings, that are associated with the outer membrane and peptidoglycan of the cell. The shaft, rings, and accessory proteins make up the basal body of the flagellum. The basal body is situated in the cell membrane/cell wall area of the bacterial cell. Rotation of the flagellum is powered by a proton motive force (pmf) that is established when proteins associated with the basal body transport protons across the cell membrane, creating a charge differential across the membrane. The pmf induces the S and M rings to rotate, which results in the rotation of the shaft, hook, and filament. Other proteins in the basal body can reverse the direction of rotation of the flagellum. The movement of the rigid and helical filament is analogous to the rotation of a propeller on a boat engine. Hence,

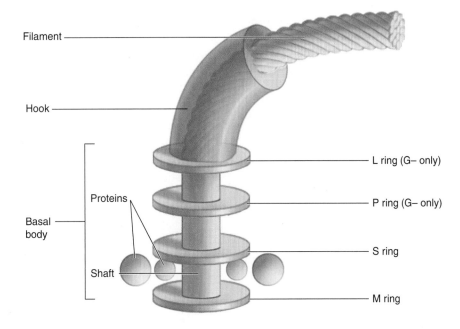

Filament

Hook

L ring (G− only)

P ring (G− only)

Proteins

Basal body

S ring

Shaft

M ring

Figure 17.1 **Structure of the gram-negative bacterial flagellum.**

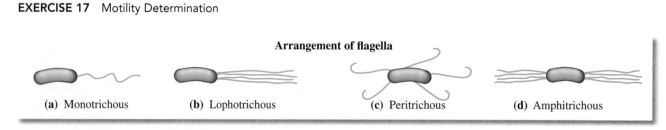

Arrangement of flagella

(a) Monotrichous   **(b)** Lophotrichous   **(c)** Peritrichous   **(d)** Amphitrichous

Figure 17.2   **Arrangement of flagella**

the movement of the filament propels the bacterial cell in much the same way that a boat is moved through the water by its engine and propeller. This is in contrast to a eukaryotic flagellum, which causes the cell to move because the flagellum beats like a whip.

Motility and the arrangement of flagella around the cell (figure 17.2) are important taxonomic characteristics that are useful in characterizing bacteria. Motility can be determined by several methods: 1. wet mounts; 2. hanging drop slide; 3. semisolid media. It can be determined microscopically by observing cells in a **wet mount.** In this procedure, a drop of viable cells is placed on a microscope slide and covered with a cover glass. The slide is then observed with a phase-contrast microscope. The rapid swimming movement of cells in the microscopic field confirms motility. However, wet mounts can easily dry out by evaporation, which is especially troublesome if observations need to made for prolonged periods of time. Drying can be delayed by using the **hanging drop technique,** shown in figure 17.3. In this procedure, a drop of cells is placed on a cover glass, which is then placed over a special slide that has a concave depression in its center. The coverslip is held in place with petroleum jelly, thus forming an enclosed glass chamber that prevents drying.

For the beginner, true swimming motility under the microscope must be differentiated from **Brownian motion** of cells or movement caused by currents under the cover glass. Brownian motion is movement due to molecular bombardment of cells causing cells to shake or "jiggle about" but not move in any vectorial way. Cells can also appear to move because currents can be created under the cover glass when pressure is exerted by focusing the oil immersion lens or by the wet mount drying out. This causes cells to "sweep" across the field.

Another method for determining motility involves inoculating semisoft agar medium. This medium has an agar concentration of 0.4%, which does not inhibit bacteria from "swimming" through the medium. In this procedure, the organisms are inoculated by stabbing the semisolid medium with an inoculating needle. If the organisms are motile, they will swim away from the line of inoculation into the uninoculated surrounding medium, causing the medium to be turbid. Nonmotile bacteria will be found only along the line of inoculation. For pathogenic bacteria, such as the typhoid bacillus, the use of semisoft agar

medium to determine motility is often preferred over microscope techniques because of the potential for infection posed by pathogens in making wet mounts.

In the following exercise, you will use both microscopic and culture media procedures to determine the motility of bacterial cultures.

🕐 First Period

During the first period, you will make wet mounts and hanging drop slides of two organisms: *Proteus vulgaris* and *Micrococcus luteus.* Tube media (semisolid medium or SIM medium) and a soft agar plate will also be inoculated. The media inoculations will have to be incubated to be studied in the next period. Proceed as follows:

Materials

- microscope slides and cover glasses
- depression slide
- 2 tubes of semisolid or SIM medium
- 1 petri plate of soft nutrient agar (20–25 ml of soft agar per plate)
- nutrient broth cultures of *Micrococcus luteus* and *Proteus vulgaris* (young cultures)
- inoculating loop and needle
- Bunsen burner or incinerator

Wet Mounts   Prepare wet mount slides of each of the organisms, using several loopfuls of the organism on the slides. Examine under an oil immersion objective. Observe the following guidelines:

- Use only scratch-free clean slides and cover glasses. This is particularly important when using phase-contrast optics.
- Label each slide with the name of the organism.
- By manipulating the diaphragm and voltage control, reduce the lighting sufficiently to make the organisms visible. Unstained bacteria are very transparent and difficult to see.
- For proof of true motility, look for directional movement that is several times the long dimension of the bacterium. The movement will also occur in different directions in the same field.
- Ignore Brownian movement. *Brownian movement* is vibrational movement caused by invisible

(1) A small amount of Vaseline is placed near each corner of the cover glass with a toothpick.

(2) Two loopfuls of organisms are placed in the cover glass.

(3) Depression slide is pressed against Vaseline on cover glass and quickly inverted.

Cover glass

Vaseline

Organisms

(4) The completed preparation can be examined under oil immersion.

Figure 17.3 **The hanging drop slide.**

molecules bombarding bacterial cells. If the only movement you see is vibrational and not directional, the organism is nonmotile.

- If you see only a few cells exhibiting motility, consider the organism to be motile. Characteristically, only a few of the cells will be motile at a given moment.
- Don't confuse water current movements with true motility. Water currents are due to capillary action caused by temperature changes and drying out. All objects move in a straight line in one direction.
- And, finally, always *examine a wet mount immediately,* once it has been prepared, because motility decreases with time after preparation.

Hanging Drop Slides By referring to figure 17.3, prepare hanging drop slides of each organism. Be sure to use clean cover glasses and label each slide with a china marking pen. When placing loopfuls of organisms on the cover glass, be sure to flame the loop between applications. Once the slide is placed on the microscope stage, do as follows:

1. Examine the slide first with the low-power objective. If your microscope is equipped with an automatic stop, avoid using the stop; instead, use the coarse adjustment knob for bringing the image into focus. The greater thickness of the depression

slide prevents one from being able to focus at the stop point.
2. Once the image is visible under low power, swing the high-dry objective into position and readjust the lighting. Since most bacteria are drawn to the edge of the drop by surface tension, **focus near the edge of the drop.**
3. If your microscope has phase-contrast optics, switch to high-dry phase. Although a hanging drop does not provide the shallow field desired for phase contrast, you may find that it works fairly well.
4. If you wish to use oil immersion, simply rotate the high-dry objective out of position, add immersion oil to the cover glass, and swing the oil immersion lens into position.
5. Avoid delay in using this setup. Water condensation may develop to decrease clarity, and the organisms become less motile with time.
6. Review all the characteristics of bacterial motility that are stated on page 122 under wet mounts.

Tube Method Inoculate tubes of semisolid or SIM media with each organism according to the following instructions:

1. Label the tubes of semisolid (or SIM) media with the names of the organisms. Place your initials on the tubes also.

(1) Wire with organisms is brought into tube without touching walls of tube.

(2) Wire penetrates medium to two-thirds of its depth.

(3) Wire is withdrawn from medium and tube. Neck of tube is flamed and plugged.

Figure 17.4   **Stab technique for motility test.**

2. Flame and cool the inoculating needle, and insert it into the culture after flaming the neck of the tube.
3. Remove the cap from the tube of medium, flame the neck, and stab it two-thirds of the way down to the bottom, as shown in figure 17.4. Flame the neck of the tube again before returning the cap to the tube.
4. Repeat steps 2 and 3 for the other culture.
5. Incubate the tubes at room temperature for 24 to 48 hours.

Plate Method   Mark the bottom of a plate of soft agar with two 1/2" circles about 1" apart. Label one circle ML and the other PV. These circles will be targets for your culture stabs. Put your initials on the plate also.

Using proper aseptic techniques, stab the medium in the center of the ML circle with *M. luteus* and the center of the other circle with *P. vulgaris*. Incubate the plate for 24 to 48 hours at room temperature.

⏱ Second Period

Assemble the following materials that were inoculated during the last period and incubated.

Materials

- culture tubes of motility medium that have been incubated
- inoculated petri plate that has been incubated

Compare the two tubes that were inoculated with *M. luteus* and *P. vulgaris*. Look for cloudiness as evidence of motility. *Proteus* should exhibit motility. Does it? Record your results on the Laboratory Report.

Compare the appearance of the two stabs in the soft agar. Describe the differences that exist in the two stabs.

Does the plate method provide any better differentiation of results than the tube method?

Laboratory Report

Complete Laboratory Report 17 for this exercise.

17 Motility Determination

A. Results

1. Which bacterial species exhibited true motility on the slides?

   _____

   _____

2. Draw the appearance of the inoculated SIM tubes. Did the semisolid medium inoculations concur with the slide results?

   _____

   _____

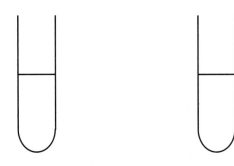

Micrococcus luteus   *Proteus vulgaris*

B. Short-Answer Questions

1. Describe the structure of a flagellum. How do flagella generate cell motility?

   _____

   _____

2. If you compared two motile bacterial species and determined one was considerably more motile than the other, which arrangement of flagella would you expect to be associated with the highly motile species? How would you confirm this supposition?

   _____

   _____

3. Differentiate between the following types of movement observed in a wet mount or hanging drop slide:

   a. directional motility.

   _____

   b. Brownian movement.

   _____

   c. water current movement.

   _____

4. Between wet mount and hanging drop slide preparations, which is more resistant to evaporation? Which works best with phase-contrast microscopy?

_____

_____

5. What concentration of agar is used in a semisolid medium for motility determination? How does that compare to a typical solid medium (*see Exercise 18*)? Explain.

_____

_____

6. Why are semisolid media sometimes preferred over slide techniques for evaluating bacterial motility?

_____

_____

7. If SIM medium was used for motility determination for *Proteus vulgaris,* what noticeable change to the medium was observed? *(Hint: see Exercise 38 to find out what the letters "SIM" represent.)*

_____

_____

Culture Methods

All nutritional types are represented among microorganisms. This metabolic diversity requires a multiplicity of culture methods. This unit presents those techniques that have proven most successful for the culture of bacteria, molds, and slime molds.

The three exercises (18, 19, and 20) pertain to basic techniques applicable to the cultivation of bacteria.

This unit culminates the basic techniques phase of this course. A thorough understanding of microscopy, slide techniques, and culture methods provides a substantial foundation for the remainder of the exercises in this manual. If independent study projects are to be pursued as a part of this course, the completion of this unit will round out your background knowledge and skills for such work.

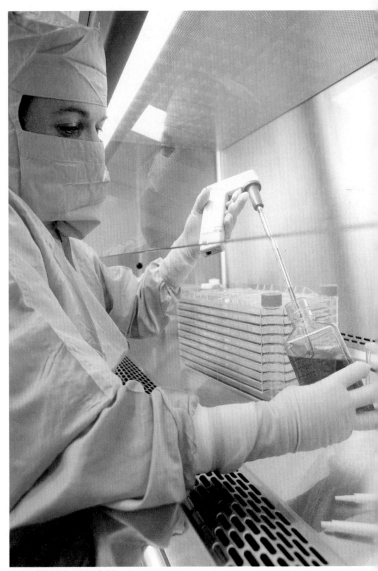

© Pixtal/AGE Fotostock RF

Culture Media Preparation

Learning Outcomes

After completing this exercise, you should be able to

1. State the six basic components that are found in all bacteriological media.
2. Explain the difference between a defined and complex medium.
3. Explain the difference between differential and selective media.
4. Prepare and sterilize a complex agar medium.

The cultivation of microorganisms on an artificial growth medium requires that the medium supply all the nutritional and energy requirements necessary for growth. However, in some cases, we may not know what the specific nutrient requirements are for a certain organism to grow. In order to cultivate such an organism, we construct a medium using rich extracts of meat or plants that would supply all the amino acids, nucleotide bases, vitamins, and other growth factors required by our organism. Such a medium is called a **complex medium** because the exact composition and amounts of the individual amino acids, vitamins, growth factors, and other components that make up the medium are not exactly known. Many of the media used in microbiology are complex, such as nutrient agar, which is used to cultivate a variety of bacteria, especially those used in exercises in this manual. For some organisms we know what the specific nutritional requirements are for growth. For example, for *Escherichia coli,* we can prepare a medium composed of specific components and amounts, such as glucose and various salts. This medium is called a **defined medium** because the specific chemical composition is known and the individual components are weighed out exactly to make up the medium.

Nutritional Requirements of Bacteria

Any medium, be it complex or defined, must supply certain basic nutritional requirements that are necessary for all cells to grow. These include a carbon source, energy source, nitrogen, minerals, vitamins and growth factors, and water.

Carbon Source Organisms can be divided into two groups based on their carbon requirements. **Heterotrophs** obtain their carbon from organic compounds such as polysaccharides, carbohydrates, amino acids, peptides, and proteins. Meat and plant extracts are added to complex media to supply these nutrients. In contrast, **autotrophs** derive their carbon requirements from fixing carbon dioxide. From the latter, they must synthesize all the complex molecules that comprise the bacterial cell.

Energy Source Bacterial cells require energy to carry out biosynthetic processes that lead to growth. These include synthesizing nucleic acids, proteins, and structural elements such as cell walls. **Chemoorganotrophs** derive their energy needs from the breakdown of organic molecules by fermentation or respiration. Most bacteria belong to this metabolic group. **Chemolithotrophs** oxidize inorganic ions such as nitrate or iron to obtain energy to fix carbon dioxide. Examples of the chemolithotrophs are the nitrifying and iron bacteria. **Photoautotrophs** contain photosynthetic pigments such as chlorophyll or bacteriochlorophyll that convert solar energy into chemical energy by the process of photosynthesis. Energy derived from photosynthesis can then be used by the cell to fix carbon dioxide and synthesize the various cellular materials necessary for growth. For these organisms, no energy source is supplied in the medium but rather energy is supplied in the form of illumination. The cyanobacteria and the green and purple sulfur bacteria are examples of photoautotrophs that carry out photosynthesis. A few photosynthetic bacteria are **photoheterotrophs.** These organisms also derive their energy requirements from photosynthesis, but their carbon needs come from growth on organic molecules such as succinate or glutamate. Some of the purple nonsulfur bacteria are found in this category. Cells also require phosphorus, which is important in the synthesis of ATP. The breakdown of energy sources by chemoorganotrophs as well as photosynthesis by photoautotrophs leads directly to the synthesis of ATP, the source of energy for cell biosynthesis.

Nitrogen Nitrogen is an essential element in biological molecules such as amino acids, nucleotide bases, and vitamins. Some bacteria can synthesize these compounds using carbon intermediates and inorganic forms of nitrogen (e.g., ammonia and nitrate). Others lack this capability and must gain their nitrogen from organic molecules such as proteins, peptides, or amino acids. Beef extract and peptones are incorporated into complex media to provide a source of nitrogen for these bacteria. Some bacteria are even capable of fixing atmospheric nitrogen into inorganic nitrogen, which can then be used for biosynthesis of amino acids. Bacteria such as *Rhizobium* and *Azotobacter* are examples of nitrogen-fixing bacteria.

Minerals Metals are essential in bacterial metabolism because they are cofactors in enzymatic reactions and are integral parts of molecules such as cytochromes, bacteriochlorophyll, and vitamins. Metals required for growth include sodium, potassium, calcium, magnesium, manganese, iron, zinc, copper, and cobalt. Most are required in catalytic or very small amounts.

Vitamins and Growth Factors Vitamins serve as coenzymes in metabolism. For example, the vitamin niacin is a part of the coenzyme NAD, and flavin is a component of FAD. Some bacteria, like the streptococci and lactobacilli, require vitamins because they are unable to synthesize them. Other bacteria (e.g., *Escherichia coli*) can synthesize vitamins and hence do not require them in media in order to grow. However, sometimes, even in addition to supplying all the normal components, it is necessary to add growth factors for ample growth of certain bacteria. Many pathogens are fastidious and grow better if blood or serum components are incorporated into their media. Blood and serum may provide additional metabolic factors not found in the normal components.

Water The cell consists of 70–80% water. Cells require an aqueous environment because enzymatic reactions and transport only occur in its presence. Furthermore, water maintains the various components of the cytoplasm in solution. When preparing media, it is essential to always use either **deionized** or **distilled water.** Tap water can contain minerals such as calcium, phosphorus, and magnesium ions that could react with peptones and meat extracts to cause unwanted precipitates and cloudiness.

In addition to having the right components, it is important to make sure that the pH of the medium is adjusted to optimal values so that growth is not inhibited. Most bacteria grow best at a neutral pH value around 7. Fungi prefer pH values around 5 for best growth. Most commercial media do not require adjusting the pH, but it would probably be necessary to adjust the pH of a defined synthetic medium. This can be done with acids such as HCl or bases such as sodium hydroxide.

Differential and Selective Media

Media can be made with components that will allow certain bacteria to grow but will inhibit others from growing. Such a medium is a **selective medium.** Antibiotics, dyes, and various inhibitory compounds are often incorporated into media to create selective conditions for growing specific organisms. For example, the dyes eosin and methylene blue when incorporated into EMB media do not affect the growth of gram-negative bacteria but do inhibit the growth of gram-positive bacteria. Incorporation of sodium chloride into mannitol-salt agar selects for the staphylococci but inhibits the growth of other bacteria that cannot tolerate the salt concentration.

A **differential medium** contains substances that cause some bacteria to take on an appearance that distinguishes them from other bacteria. When *Staphylococcus aureus* grows on mannitol-salt agar, it ferments mannitol, changing a pH indicator from red to yellow around colonies. Other staphylococci cannot ferment mannitol, and their growth on this medium results in no change in the indicator. EMB is also a differential medium. Gram-negative bacteria that ferment lactose in this medium form colonies with a metallic-green sheen. Non-lactose-fermenting bacteria do not form colonies with the characteristic metallic sheen.

Media can be prepared in liquid or solid form depending on their application. Liquid broth cultures are used to grow large volumes of bacteria. Fermentation studies, indole utilization, and the methyl red and Voges-Proskauer tests are done in broth cultures. Streaking of bacteria and selection of isolated colonies are done on solid media. **Agar,** a complex polysaccharide isolated from seaweed, is added in a concentration of 1.5% to solidify liquid media. The use of agar in bacteriological media was first introduced in Robert Koch's laboratory. Agar has unique properties that make it ideal for use in microbiology. First, it melts at 100°C but does not solidify until it cools to 45°C. Bacteria can be inoculated into agar media at this temperature (e.g., pour plates) without killing the cells. Second, agar is not a nutrient for most bacteria (the exceptions are a few bacteria found in marine environments). Sometimes, agar is added at lower concentrations (e.g., 0.4%) to make semisoft media. This type of medium is used in motility studies.

Prior to 1930, it was necessary for laboratory workers to prepare media using various raw materials. This often involved boiling plant material or meat to prepare extracts that would be used in the preparation of media. Today, commercial companies prepare and sell media components, which are used for most routine bacteriological media. It is only necessary to weigh out a measured amount of a specific medium, such as nutrient agar, and dissolve it in water. In some cases, it may be necessary to adjust the pH prior to sterilizing the medium.

Before any medium can be used to grow bacteria or other microorganisms, it must be **sterilized,** that is, any contaminating bacteria introduced during preparation must be killed or removed. Most media can be **autoclaved** to achieve sterilization. This involves heating the media to **121°C** for at least **15 minutes** at **15 psi** of steam pressure. These conditions are sufficient to kill cells and any endospores present. Sometimes a medium may require a component that is heat sensitive such as a vitamin, and it cannot be subjected to autoclave temperatures. The component can be filter sterilized by passing a solution through a bacteriological filter of 0.45 microns. This filter will retain any cells and endospores that may be present. After filtration, the component can be added to the sterilized medium.

Media Preparation Assignment

In this laboratory period, you will work with your laboratory partner to prepare tubes of media that will be used in future laboratory experiments (figure 18.1). Your instructor will indicate which media you are to prepare. Record in the space below the number of tubes of specific media that have been assigned to you and your partner.

nutrient broth	_____
nutrient agar pours	_____
nutrient agar slants	_____
other	_____

Several different sizes of test tubes are used for media, but the two sizes most generally used are either 16 mm or 20 mm diameter by 15 cm long. Select the correct size tubes first, according to these guidelines:

Large tubes (20 mm dia): Use these test tubes for *all pours* (nutrient agar, Sabouraud's agar, EMB agar, etc.). Pours are used for filling petri plates.

Small tubes (16 mm dia): Use these tubes for all *broths, deeps,* and *slants*.

If the tubes are clean and have been protected from dust or other contamination, they can be used without cleaning. If they need cleaning, scrub out the insides with warm water and detergent, using a test-tube brush. Rinse twice, first with tap water and finally with distilled water to rid them of all traces of detergent. Place them in a wire basket or rack, inverted, so that they can drain. Do not dry with a towel.

Measurement and Mixing

The amount of medium you make for a batch should be determined as precisely as possible to avoid shortage or excess.

Materials

- graduated cylinder, beaker, glass stirring rod
- bottles of dehydrated media
- Bunsen burner and tripod, or hot plate

Figure 18.1 **Basic supplies for making up a batch of medium.**
© McGraw-Hill Education. Auburn University Photographic Services

1. Measure the correct amount of water needed to make up your batch. The following volumes required per tube must be taken into consideration:

pours . . . . 12 ml
deeps . . . . 6 ml
slants . . . . 4 ml
broths . . . . 5 ml
broths with fermentation tubes . . . . 5–7 ml

2. Consult the label on the bottle to determine how much powder is needed for 1000 ml and then determine by proportionate methods how much you need for the amount of water you are using. Weigh this amount on a balance and add it to the beaker of water (figure 18.2). If the medium does not contain agar, the mixture usually goes into solution without heating (figure 18.3).

3. **If the medium contains agar,** heat the mixture on a stirring hot plate (figure 18.4) or on an electric hot plate until it comes to a boil. To safeguard against water loss, *before heating, mark the level of the top of the medium on the side of the beaker*

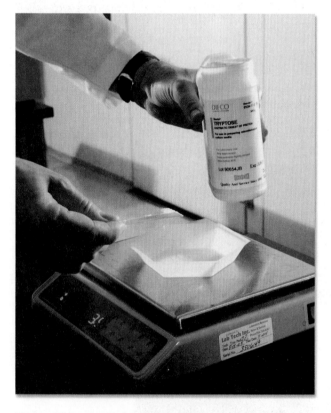

Figure 18.2 Correct amount of dehydrated medium is carefully weighed on a balance.
© McGraw-Hill Education. Auburn University Photographic Services

with a china marking pencil. As soon as it "froths up," turn off the heat. If an electric hot plate is used, the medium must be removed from the hot plate or it will boil over the sides of the container.

Caution

Be sure to keep stirring the medium so that it does not char on the bottom of the beaker.

4. Check the level of the medium with the mark on the beaker to note if any water has been lost. Add sufficient distilled water as indicated. Keep the temperature of the medium at about 60°C to avoid solidification. The medium will solidify at around 40°C.

Adjusting the pH

Although dehydrated media contain buffering agents to keep the pH of the medium in a desired range, the pH of a batch of medium may differ from that stated on the label of the bottle. Before the medium is tubed, therefore, one should check the pH and make any necessary adjustments.

If a pH meter (figure 18.5) is available and already standardized, use it to check the pH of your medium. If the medium needs adjustment, use the bottles of HCl and NaOH to correct the pH. If no meter is available, pH papers will work about as well. Make pH adjustment as follows:

Materials

- beaker of medium
- acid and base kits (dropping bottles of 1N and 0.1N HCl and NaOH)
- glass stirring rod
- pH papers
- pH meter (optional)

1. Dip a piece of pH test paper into the medium to determine the pH of the medium.
2. **If the pH is too high,** add a drop or two of HCl to lower the pH. For large batches use 1N HCl. If the pH difference is slight, use the 0.1N HCl. Use a glass stirring rod to mix the solution as the drops are added.
3. **If the pH is too low,** add NaOH, one drop at a time, to raise the pH. For slight pH differences, use 0.1N NaOH; for large differences use 1N NaOH. Use a glass stirring rod to mix the solution as the drops are added.

Figure 18.3   Dehydrated medium is dissolved in a measured amount of water.
© McGraw-Hill Education. Auburn University Photographic Services

Figure 18.4   If the medium contains agar, it must be heated to dissolve the agar.
© McGraw-Hill Education. Auburn University Photographic Services

Figure 18.5   The pH of the medium is adjusted by adding acid or base as per recommendations.
© McGraw-Hill Education. Auburn University Photographic Services

Figure 18.6 An automatic pipetting machine will deliver precise amounts of media to test tubes.
© McGraw-Hill Education. Auburn University Photographic Services

Filling the Test Tubes

Once the pH of the medium is adjusted, it must be dispensed into test tubes. If an automatic pipetting machine is to be used, as shown in figure 18.6, it will have to be set up for you by your instructor. These machines can be adjusted to deliver any amount of medium at any desired speed. When large numbers of tubes are to be filled, the automatic pipetting machine should be used.

Materials

- automatic pipetters

1. Follow the instructions provided by your instructor for setting up the automatic pipetter. This will involve adjusting the desired amount of medium to be delivered to each test tube and possibly other settings. If you are using an automatic pipette aid, you will need to repeatedly draw up medium in a pipette and deliver the desired amount by pressing the release button on the pipette aid (figure 18.7).
2. Place the supply tube into the medium and proceed to fill each test tube according to the type of delivery system you are using. Your instructor will help you with this step.

Figure 18.7 Small batches of media can be delivered with hand-held automatic pipetters.
© McGraw-Hill Education. Auburn University Photographic Services

3. If you are delivering agar medium, keep the beaker of medium on a stirring hot plate to maintain the agar in solution. A magnetic stirring bar placed in the medium will aid in constantly stirring the solution.

4. If the medium is to be used for fermentation, add a Durham tube to each tube before filling the test tube. This should be placed in the tube *with the open end of the Durham tube down.* When medium is placed in the test tube, the Durham tube may float on the top of the medium, but it will submerge during autoclaving.

Capping the Tubes

The last step before sterilization is to provide a closure for each tube. Plastic (polypropylene) caps are suitable in most cases. All caps that slip over the tube end have inside ridges that grip the side of the tube and provide an air gap to allow steam to escape during sterilization (figure 18.8). If you are using tubes with plastic screw-caps, *the caps should not be screwed tightly before sterilization; instead, each one must be left partly unscrewed about a quarter of a turn.*

If no slip-on caps of the correct size are available, it may be necessary to make up some cotton plugs. A properly made cotton plug should hold firmly in the tube so that it is not easily dislodged.

Sterilization

As soon as the tubes of media have been stoppered, they must be sterilized. Organisms on the walls of the tubes, in the distilled water, and in the dehydrated medium will begin to grow within a short period of time at room temperature, degrading the medium.

Prior to sterilization, the tubes of media should be placed in a test tube rack with a label taped on the outside of the test tube rack. The label should indicate the type of medium, the date, and your name.

Sterilization must be done in an autoclave (figure 18.9). The following considerations are important in using an autoclave:

• Check with your instructor on the procedure to be used with your particular type of autoclave. Complete sterilization occurs at 250°F (121.6°C). To achieve this temperature the autoclave has to develop 15 pounds per square inch (psi) of steam pressure. To reach the correct temperature, there must be some provision in the chamber for the escape of air. On some of the older units it is necessary to allow the steam to force air out through the door before closing it.

• *Don't overload the chamber.* One should not attempt to see how much media can be packed into it. Provide ample space between racks of media to allow for circulation of steam.

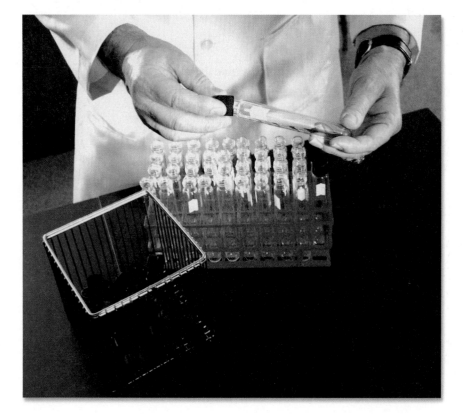

Figure 18.8 **Once the medium has been dispensed, tubes are capped prior to autoclaving.**
© McGraw-Hill Education. Auburn University Photographic Services

Figure 18.9 Media is sterilized in an autoclave for 15–20 minutes at 15 psi steam pressure.
© McGraw-Hill Education. Auburn University Photographic Services

- *Adjust the time of sterilization to the size of load.* Small loads may take only 10–15 minutes. An autoclave full of media may require 30 minutes for complete sterilization.

After Sterilization

Slants   If you have a basket of tubes that are to be converted to slants, it is necessary to lay the tubes down in a near-horizontal manner as soon as they are removed from the autoclave. The easiest way to do this is to use a piece of rubber tubing ($\frac{1}{2}''$ dia) to support the capped end of the tube as it rests on the countertop. Solidification should occur in about 30–60 minutes.

Other Media   Tubes of broth, agar deeps, nutrient gelatin, etc., should be allowed to cool to room temperature after removal from the autoclave. Once they have cooled down, place them in a refrigerator or cold-storage room.

Storage   If tubes of media are not to be used immediately, they should be stored in a cool place. When stored for long periods of time at room temperature, media tend to lose moisture. At refrigerated temperatures media will keep for months.

Laboratory Report

Complete the Laboratory Report for Exercise 18.

Student: _____

Date: _____   Section: _____

18   Culture Media Preparation

A. Short-Answer Questions

1. Differentiate between complex and defined media.

_____

_____

2. Name six basic nutritional requirements supplied in all culture media.

_____

_____

3. What growth factor is often supplied for cultivation of fastidious bacterial pathogens?

_____

_____

4. A powdered complex medium can be stored for months in the laboratory. However, after preparing the medium, it must be sterilized almost immediately. Why?

_____

_____

5. An autoclave is typically used for sterilization of media.

    a. Define sterilization.

_____

    b. Under what conditions are media typically sterilized in an autoclave?

_____

    c. What type of media components cannot be sterilized in an autoclave?

_____

    d. What is an alternative to autoclaving for sterilizing heat sensitive materials?

_____

6. Mannitol-salt agar (MSA) is a selective and differential medium.

    a. What is a selective medium? What component(s) of MSA make it selective?

_____

b. What is a differential medium? What component(s) of MSA make it differential?

_____

c. This medium is useful for the isolation and characterization of which microorganisms?

_____

7. Agar is a solidifying agent used in media preparation.

   a. What is its origin?

_____

   b. What makes it ideal for cultivation of microbes?

_____

   c. How and why does the agar concentration in semisolid media differ from conventional solid media?

_____

B. Multiple Choice

Select the answer that best completes the following questions.

1. Most bacteria derive their carbon and energy needs from organic molecules and are classified as
   a. chemolithotrophs.
   b. chemoorganotrophs.
   c. photoautotrophs.
   d. photoheterotrophs.

2. The cyanobacteria use solar energy to fix carbon dioxide. They are classified as
   a. chemolithotrophs.
   b. chemoorganotrophs.
   c. photoautotrophs.
   d. photoheterotrophs.

3. *Rhizobium* and *Azotobacter* are examples of nitrogen-fixing soil bacteria that are classified as
   a. chemolithotrophs.
   b. chemoorganotrophs.
   c. photoautotrophs.
   d. photoheterotrophs.

4. Some purple nonsulfur bacteria utilize solar energy but require an organic carbon source. They are classified as
   a. chemolithotrophs.
   b. chemoorganotrophs.
   c. photoautotrophs.
   d. photoheterotrophs.

ANSWERS

Multiple Choice

1. _____

2. _____

3. _____

4. _____

Enumeration of Bacteria:
The Standard Plate Count

EXERCISE
19

Learning Outcomes

After completing this exercise, you should be able to

1. Explain the various methods used to count bacteria.
2. Perform a standard plate count to determine the number of bacteria in a sample.
3. Use a spectrophotometer to measure the increase in culture turbidity by determining optical density.
4. Correlate turbidity determinations with standard plate counts of a bacterial culture.

It is often essential to determine the number of bacteria in a sample. For example, the grading of milk is based on the numbers of bacteria present. Whether a patient has a bladder infection is dependent on a certain threshold level of bacteria present in a urine sample. Sometimes it is just important to know how many bacteria are present in food or water to ensure safety. Several different methods can be used to determine the number of bacterial cells, and each method has its own advantages and disadvantages. The use of one method over another will be dictated by the purpose of the study. The following are some of the direct methods for determining bacterial numbers:

Microscopic Counts A sample can be diluted and the cells in the sample can be counted with the aid of a microscope. Special slides, such as the Petroff-Hauser chamber, facilitate counting because the slide has a grid pattern and the amount of sample delivered to the grid is known. Milk samples can be counted by microscopic means with a great deal of reliability and confidence.

Most Probable Number (MPN) The number of bacteria in a sample can be determined by the relationship of some growth parameter to statistical probability. The safety of drinking water is dependent on there being no sewage contamination of potable water, and this is tested using the MPN method. Indicator bacteria called __coliforms,__ which are found in the intestines of humans and warm-blooded animals, ferment lactose to produce acid and gas. The presence of these bacteria in a water sample suggests the potential for disease. A series of

tubes with lactose is inoculated with water samples, and the pattern of tubes showing acid and gas is compared to statistical tables that give the probable numbers of coliforms present. You will use this procedure in Exercise 59 to test water for the presence of coliforms. The MPN method is limited to testing where statistical tables have been set up for a particular growth parameter.

Standard Plate Count (SPC) The standard plate, or viable count, is one of the most common methods for determining bacterial numbers in a sample. A sample is diluted in a series of dilution blanks, as shown in figure 19.1. Aliquots of the dilutions are then plated onto media and the numbers of colonies are counted after incubation for 24 to 48 hours. It is assumed that the bacterial cells are diluted to an end point where a single cell divides, giving rise to a visible colony on a plate. The number of bacteria in the original sample is determined by multiplying the number of colonies by the dilution factor. However, the assumption that a colony represents a single cell is not always correct because cells in a chain, such as *Streptococcus,* will also give rise to a colony on a plate. Because of the uncertainty in how many actual cells form a colony, counts by the SPC are reported as **colony forming units (CFUs).** Only numbers between 30 and 300 CFUs are considered statistically valid. If the CFUs are greater than 300, there is a probability that overcrowding on the

Figure 19.1 **Quantitative plating procedure.**

plate could have inhibited some cells from growing. Less than 30 CFUs could involve a sampling error and an underestimate of numbers. The SPC method determines only viable cells, whereas a microscopic count determines both living and dead cells. Also, the SPC method is biased because specific conditions and media are used, and these factors may exclude certain bacteria in the counts. For example, the SPC would severely underestimate the numbers of bacteria in a soil sample because the conditions and the medium used for the count probably favor heterotrophs that grow aerobically at neutral pH values. These conditions do not allow for the growth of anaerobes, chemolithotrophs, or bacteria that may grow at extremes of pH.

Indirect Methods Sometimes one only wants to know if cells are growing and, therefore, increasing in number. Growth can be related to some parameter that increases with cell division. Growing cells increase their protein, nucleic acid content, and mass because cells are dividing. Thus, measurements of protein, DNA, and dry weight can be used to monitor growth. Likewise, a culture will become more turbid as cells divide, and the **turbidity** of a culture can be determined and related to growth. Cell turbidity can be measured in a spectrophotometer, which measures the **absorbance** or **optical density** of a culture. Oftentimes, a standard curve is constructed that relates optical density to actual numbers of bacteria determined by an SPC. However, one must bear in mind that both living and dead cells will contribute to the culture turbidity, which is also a disadvantage of this method.

In the following exercise, you will use the SPC to determine the numbers of bacteria in a culture. You will also measure the turbidity of a culture and plot the optical density values of diluted samples.

Quantitative Plating Method

(Standard Plate Count)

In determining the number of organisms present in water, milk, and food, the **standard plate count (SPC)** procedure is universally used. It is relatively easy to perform and gives excellent results. We can also use this basic technique to calculate the number of organisms in a bacterial culture. It is in this respect that this assignment is set up.

The procedure consists of diluting the organisms with a series of sterile water blanks, as illustrated in figure 19.1. Generally, only three bottles are needed, but more could be used if necessary. By using the dilution procedure indicated here, a final dilution of 1:1,000,000 occurs in blank C. From blanks B and C, measured amounts of the diluted organisms are transferred into empty petri plates. Nutrient agar,

cooled to 50°C, is then poured into each plate. After the nutrient agar has solidified, the plates are incubated for 24 to 48 hours and examined. A plate that has between 30 and 300 colonies is selected for counting. From the count it is a simple matter to calculate the number of organisms per milliliter of the original culture. It should be pointed out that greater accuracy can be achieved by pouring two plates for each dilution and averaging the counts. Your instructor will advise if duplicate plating is to be performed.

Pipette Handling

Success in this experiment depends considerably on proper pipetting techniques. Pipettes may be available to you in metal canisters or in individual envelopes; they may be disposable or reusable. Because of potential hazards, no mouth pipetting is allowed, and all pipetting is therefore done using pipette aids. Your instructor will indicate the techniques that will prevail in this laboratory. If this is the first time that you have used sterile pipettes, consult figure 19.2, keeping the following points in mind:

- When removing a sterile pipette from a canister, do so without contaminating the ends of the other pipettes with your fingers. This can be accomplished by *gently* moving the canister from side to side in an attempt to isolate one pipette from the rest.
- After removing your pipette, replace the cover on the canister to maintain sterility of the remaining pipettes.
- Don't touch the body of the pipette with your fingers or lay the pipette down on the table before or after you use it. **Keep that pipette sterile** until you have used it, and don't contaminate the table or yourself with it after you have used it.
- Always use a mechanical pipetting device such as the one in illustration 3, figure 19.2.
- Remove and use only one pipette at a time; if you need three pipettes for the whole experiment and remove all three of them at once, there is no way that you will be able to keep two of them sterile while you are using the first one.
- When finished with a pipette, place it in the *discard canister*. The discard canister will have a disinfectant in it. At the end of the period, reusable pipettes will be washed and sterilized by the laboratory assistant. Disposable pipettes will be discarded. Students have been known to absent-mindedly return used pipettes to the original sterile canister, and, occasionally, even toss them into the wastebasket. We are certain that no one in this laboratory would *ever* do that!

Diluting and Plating Procedure

Proceed as follows to dilute out a culture of *E. coli* and pour four plates, as illustrated in figure 19.1.

(1) Reusable pipettes may be available in disposable envelopes or metal canisters. When using pipettes from canisters, be sure to cap them after removing a pipette.

(2) Never touch the tip or barrel of a pipette with your fingers. Contaminating the pipette will contaminate your work.

(3) Use a pipette aid for all pipetting in this laboratory. Pipetting by mouth is too hazardous.

(4) After using a pipette, place it in the discard canister. Even "disposable" pipettes must be placed here.

Figure 19.2   **Pipette-handling techniques.**

Materials

per 4 students:
- 1 bottle (40 ml) broth culture of *E. coli*

per student:
- 1 bottle (80 ml) nutrient agar
- 4 petri plates
- 1.1 ml pipettes
- 3 sterile 99 ml water blanks
- canister for discarded pipettes

1. Liquefy a bottle of nutrient agar. While it is being heated, label three 99 ml sterile water blanks **A, B,** and **C.** Also, label the four petri plates **1:10,000, 1:100,000, 1:1,000,000,** and **1:10,000,000.** In addition, indicate with labels the amount to be pipetted into each plate (**0.1 ml** or **1.0 ml**).

2. Shake the culture of *E. coli* and transfer 1 ml of the organisms to blank A, using a sterile 1.1 ml pipette. After using the pipette, place it in the discard canister.

3. Shake blank A 25 times in an arc of 1 foot for 7 seconds with your elbow on the table, as shown in figure 19.3. Forceful shaking not only brings about good distribution, but it also breaks up clumps of bacteria.

4. With a different 1.1 ml pipette, transfer 1 ml from blank A to blank B.

5. Shake water blank B 25 times in the same manner.

6. With another sterile pipette, transfer 0.1 ml from blank B to the 1:100,000 plate and 1.0 ml to the 1:10,000 plate. With the same pipette, transfer 1.0 ml to blank C.

7. Shake blank C 25 times.

Figure 19.3 **Standard procedure for shaking water blanks.**

Figure 19.4 **Serial dilution plates of *E. coli*.**
© McGraw-Hill Education. Lisa Burgess, photographer

Figure 19.5 **Colony counts are determined on a colony counter.**
© McGraw-Hill Education. Lisa Burgess, photographer

8. With another sterile pipette, transfer from blank C 0.1 ml to the 1:10,000,000 plate and 1.0 ml to the 1:1,000,000 plate.

9. After the bottle of nutrient agar has boiled for 8 minutes, cool it down in a water bath at 50°C for **at least 10 minutes.**

10. Pour one-fourth of the nutrient agar (20 ml) into each of four plates. Rotate the plates **gently** to get adequate mixing of medium and organisms. **This step is critical!** Too little action will result in poor dispersion and too much action may slop inoculated medium over the edge.

11. After the medium has cooled completely, incubate at 35°C for 48 hours, inverted.

Counting and Calculations

Materials

- 4 culture plates
- colony counter
- mechanical hand counter
- felt-tip pen (optional)

1. Lay out the plates on the table in order of dilution and compare them. *Select the plates that have no fewer than 30 nor more than 300 colonies for your count.* Plates with less than 30 or more than 300 colonies are statistically unreliable (figure 19.4).

2. Place the plate on the colony counter with the lid removed. See figure 19.5. Start counting at the top of the plate, using the grid lines to prevent counting the same colony twice. Use a mechanical hand counter. Count every colony, regardless of how small or insignificant. Record counts on the table in section A of Laboratory Report 19.

   Alternative Counting Method: Another way to do the count is to remove the lid and place the plate upside down on the colony counter. Instead of using the grid to keep track, use a felt-tip pen to mark off each colony as you do the count.

3. Calculate the number of bacteria per ml of undiluted culture using the data recorded in section A of Laboratory Report 19. Multiply the number of colonies counted by the dilution factor (the reciprocal of the dilution).

   Example: If you counted 220 colonies on the plate that received 1.0 ml of the 1:1,000,000 dilution: 220 × 1,000,000 (or 2.2×10^8) bacteria per ml. If 220 colonies were counted on the plate that received 0.1 ml of the 1:1,000,000 dilution, then the above results would be multiplied by 10 to convert from number of bacteria per 0.1 ml to number of bacteria per 1.0 ml (2,200,000,000, or 2.2×10^9).

   Use only two significant figures. If the number of bacteria per ml was calculated to be 227,000,000, it should be recorded as 230,000,000, or 2.3×10^8.

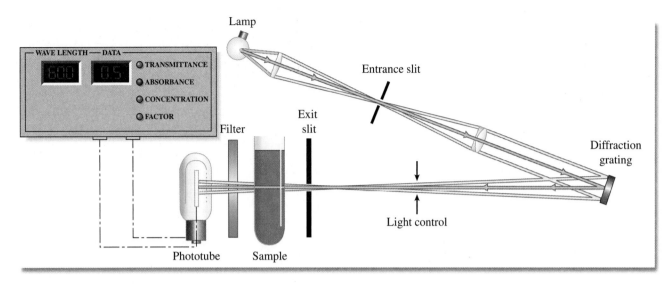

Figure 19.6 Schematic of a spectrophotometer.

Determination of Growth by Absorbance (Optical Density)

As the number of cells in a culture increases with time, the turbidity of the culture will also increase. The turbidity of a bacterial culture can be measured using an instrument called a **spectrophotometer.** In turbidity measurements, the culture is acting like a colloidal suspension. As light at a defined wavelength passes through the culture, it will be absorbed by the bacterial cells, and the light emerging from the culture will be proportionately decreased by the number of cells present. This is given by the following equation:

$$\%T = (l/l_0) \times 100$$ where: %T is the percent of transmitted light

l_0 is the intensity of light striking the sample

l is the intensity of light after passing through the sample

Spectrophotometers determine the %T and also determine the **absorbance** or **optical density (O.D.)** of a suspension of cells. The absorbance (A) or O.D. is a logarithmic function of T and is expressed as follows:

$$A = \log (1/T) = \log (l_0/l)$$
Note: a. at 100% transmittance; A = 0

b. absorbance has no units

c. the highest calibrated unit of A on an instrument is 2.0

Therefore, within certain defined limits, the amount of light absorbed is proportional to the number of cells present.

Figure 19.6 illustrates the path of light through a spectrophotometer. A beam of white light passes through a series of lenses and a slit and onto a diffraction grating, where the light is separated into different wavelengths of the visible spectrum. A specific wavelength of monochromatic light can be selected from the diffraction grating by an exit slit. Adjusting the wavelength control on the instrument will reorient the diffraction grating so that a different wavelength can be selected by the exit slit. The monochromatic light then passes through a sample and activates a photomultiplier tube, which measures the percent transmittance (%T) and the absorbance (A) values. This is provided as a digital readout depending on which parameter is selected. (Note: On analog instruments, %T and A can be read simultaneously on the meter.) The higher the absorbance, the greater the concentration of bacterial cells.

Absorbance determinations do not give specific bacterial numbers like those obtained with the standard plate count. However, standard plate counts and absorbance determinations for a bacterium growing under defined conditions and on a specific medium can be graphed to generate a standard curve that can be used to determine the

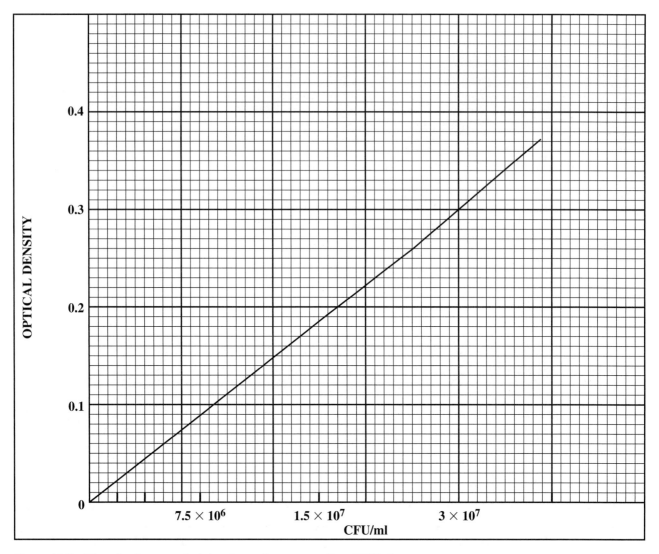

Figure 19.7   Plot of culture optical density and corresponding CFU/ml.

number of cells in a suspension by only measuring the absorbance of the culture. This is achieved by determining the increase in bacterial cell numbers in a culture over time using the standard plate count method and also determining the absorbance values for each of the samples that were plated. The absorbance values are then graphed versus the plate counts (CFU/ml) to generate a standard curve. For the bacterium growing under the same defined conditions, new absorbance values are determined and used to ascertain the corresponding number of cells in the sample from the standard curve (figure 19.7). For example, in figure 19.7, an absorbance of 0.3 is equal to 3×10^7 CFU/ml. It is important to understand that this method is only valid when the medium and growth conditions are defined for a particular organism. One cannot use numbers generated for an *Escherichia coli* culture to determine the number of cells in a *Staphylococcus aureus* culture

grown under different conditions. Also, the wave length used to determine absorbance will vary for each type of bacterium.

In the following exercise, you will demonstrate the relationship between absorbance and cell turbidity by measuring the absorbance values for various dilutions of a culture. There should be a proportional relationship between the concentration of bacterial cells and the absorbance or optical density of a culture. To demonstrate this relationship, you will measure the absorbance values of various dilutions provided to you. These values will be plotted on a graph as a function of culture dilution. For lower values of absorbance there may be a linear relationship between the concentration of cells and absorbance. At higher values, however, the relationship may not be linear. That is, for a doubling in cell concentration, there may be less than a doubling of absorbance.

(a) Turn on the instrument with the on/off switch (left front knob). Allow the spectrophotometer to warm up for 15 minutes. Set the wavelength to 550 nm and position the filter to correspond to this wavelength. Select the Transmittance mode and make sure the cuvette holder cover is closed. Adjust the digital readout to zero % transmittance using the left front (on/off) knob.

(b) Set the mode to Absorbance. Insert a cuvette containing sterile nutrient broth into the sample holder and close the cover. Adjust the Absorbance to 0 (zero) using the right front knob. This may require several turns of the knob.

(c) Remove the blank and insert a cuvette with one of the bacterial cell samples. Close the cover and read the absorbance/O.D.

(d) Occasionally blank the instrument to zero with the sterile nutrient broth during the course of reading the cell samples.

Figure 19.8 **Calibration procedure for the B & L Spectronic 200 digital spectrophotometer.**
© McGraw-Hill Education. Auburn University Photographic Services

Materials

- broth culture of *E. coli* (same one as used for plate count)
- spectrophotometer cuvettes (2 per student)
- 4 small test tubes and test-tube rack
- 5 ml pipettes
- bottle of sterile nutrient broth (20 ml per student)

1. Calibrate the spectrophotometer using the procedure described in figure 19.8. These instructions apply specifically to the Bausch and Lomb Spectronic 200 digital spectrophotometer. It is important to blank the spectrophotometer by adjusting the instrument to an absorbance of 0 (zero) using uninoculated nutrient broth. The

medium contains components that cause it to be slightly colored and, hence, it will absorb some light, adding to the light absorbance of the bacterial culture tubes. Blanking the instrument using the uninoculated medium will subtract the absorbance resulting from the medium. In handling the cuvettes, keep the following in mind:

a. Rinse the cuvette with distilled or deionized water to clean it before using.
b. Keep the lower part of the cuvette free of liquids, smudges, and fingerprints by carefully wiping the surface only with Kim wipes or lint-free tissue provided. Do not use paper towels or handkerchiefs for this purpose. If smudges,

Bacterial culture

4 ml   4 ml   4 ml   4 ml   4 ml

1:1
(undiluted)   1:2   1:4   1:8   1:16

4 ml of sterile nutrient broth in each of these tubes

Figure 19.9   Dilution procedure for cuvettes.

liquids, or fingerprints occur on the cuvette surface, they can contribute to light absorbance and erroneous readings.

c. Insert the cuvette into the sample holder with the index line aligned with the index line on the cuvette holder. Properly seat the cuvette by exactly aligning the lines on the cuvette and holder.

d. Handle cuvettes with care as they are of optical quality and expensive.

2. Label a cuvette 1:1 (near top of tube) and four test tubes 1:2, 1:4, 1:8, and 1:16. These tubes will be used for the serial dilutions shown in figure 19.9.
3. With a 5 ml pipette, dispense 4 ml of sterile nutrient broth into tubes 1:2, 1:4, 1:8, and 1:16.
4. Shake the culture of *E. coli* vigorously to suspend the organisms, and with the same 5 ml pipette, transfer 4 ml to the 1:1 cuvette and 4 ml to the 1:2 test tube.
5. Mix the contents in the 1:2 tube by drawing the mixture up into the pipette and discharging it into the tube three times.
6. Transfer 4 ml from the 1:2 tube to the 1:4 tube, mix three times, and go on to the other tubes in a similar manner. Tube 1:16 will have 8 ml of diluted organisms.
7. Measure the optical density of each of the five tubes, starting with the 1:16 tube first. The contents of each of the test tubes must be transferred to a cuvette for measurement. Be sure to close the lid on the sample holder when making measurements. A single cuvette can be used for all the measurements.
8. Record the O.D. values in the table of Laboratory Report 19.
9. Plot the O.D. values on the graph of Laboratory Report 19.

Laboratory Report

Student: _____

Date: _____ Section: _____

19 Enumeration of Bacteria: The Standard Plate Count

A. Results

1. Quantitative Plating Method

   a. Record your plate counts in this table:

DILUTION BOTTLE	ml PLATED	DILUTION	DILUTION FACTOR	NUMBER OF COLONIES
b (1:10,000)	1.0	1:10,000	10^4	
b (1:10,000)	0.1	1:100,000	10^5	
c (1:1,000,000)	1.0	1:1,000,000	10^6	
c (1:1,000,000)	0.1	1:10,000,000	10^7	

   b. How many cells per milliliter were in the undiluted culture?_____

2. Optical Density Determination

   a. Record the optical density values for your dilutions in the following table.

DILUTION	OPTICAL DENSITY
1:1	
1:2	
1:4	
1:8	
1:16	

b. Plot the optical densities versus the concentration of organisms. Complete the graph by drawing a line between plot points.

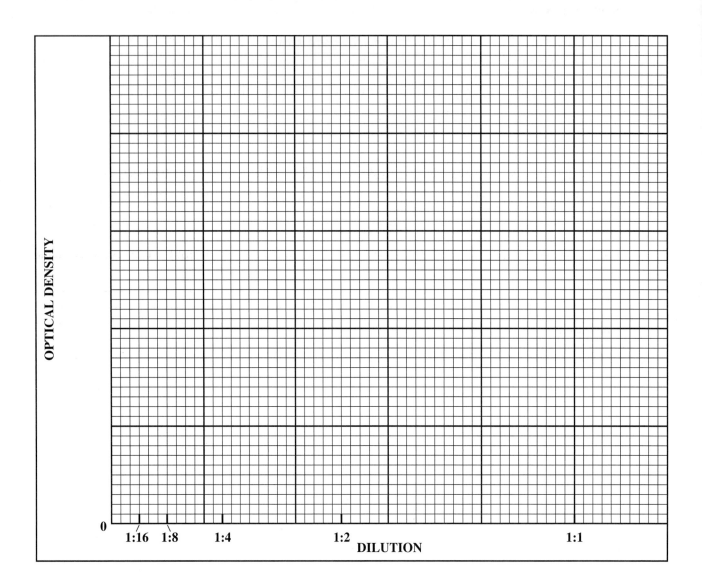

c. What is the maximum O.D. that is within the linear portion of the curve? _____

d. What is the corrected or true O.D. of the undiluted culture? (Hint: If the O.D. for the 1:2 dilution but not the 1:1 dilution is within the linear portion of the curve, then the O.D. of the 1:1 dilution should not be considered correct. The correct or true O.D. of the undiluted culture in this example could be estimated by multiplying the O.D. of the 1:2 dilution by 2.) _____

e. What is the correlation between corrected O.D. and cell number for your culture? _____

_____

_____

B. Short-Answer Questions

1. When performing a standard plate count, why are the counts reported as colony forming units (CFUs) rather than cells?

_____

_____

2. How would you inoculate a plate to get a 1:10 dilution? A 1:100 dilution?

_____

_____

3. Why is it necessary to perform a plate count in conjunction with the turbidimetry procedure?

_____

_____

4. For the following methods of bacterial enumeration, does the method determine total count or viable count?
    a. MPN_____
    b. Microscopic count_____
    c. Standard plate count_____
    d. Turbidity_____

Slide Culture:
Fungi

Learning Outcomes

After completing this exercise, you should be able to

1. Prepare a slide culture for cultivating fungal colonies.
2. Use the slide culture to observe fruiting structures, mycelium, and other structures associated with a fungal culture.

The isolation, culture, and microscopic examination of fungi require the use of suitable selective media and special microscopic slide techniques. Simple wet mounts prepared from fungal cultures usually do not reveal the arrangement of spores on fruiting bodies because the manipulation of the culture disrupts the fruiting structures and the hyphae of the culture. The type of fruiting structure and spore arrangement and morphology are important in the identification and taxonomy of these microorganisms. One way to preserve the integrity of the fruiting structure is to prepare a slide culture that can then be stained. This allows the observation of the fruiting structure *in situ* and does not disrupt the arrangement of the spores. In this exercise, a slide culture method will be used to prepare stained slides of molds. The method is superior to wet mounts in that the hyphae, sporangiophores, and spores remain more or less intact when stained.

When fungi are collected from the environment, as in Exercise 7, Sabouraud's agar is most frequently used. It is a simple medium consisting of 1% peptone, 4% glucose, and 2% agar-agar. The pH of the medium is adjusted to 5.6, which favors the growth of fungi but inhibits most bacterial growth.

Unfortunately, for some fungi the pH of Sabouraud's agar is too low and the glucose content is too high. A better medium for these organisms is one suggested by C. W. Emmons that contains only 2% glucose, with 1% neopeptone, and an adjusted pH of 6.8–7.0. To inhibit bacterial growth, 40 mg of chloramphenicol is added to 1 liter of the medium.

In addition to the above two media, cornmeal agar, Czapek solution agar, and others are available for special applications in culturing molds.

Figure 20.1 illustrates the procedure that will be used to produce a fungal culture on a slide that can be stained directly on the slide. Note that a sterile cube of

Sabouraud's agar is inoculated on two sides with spores from a mold colony. Figure 20.2 illustrates how the cube is held with a scalpel blade as inoculation takes place. The cube is placed in the center of a microscope slide with one of the inoculated surfaces placed against the slide. On the other inoculated surface of the cube is placed a cover glass. The assembled slide is incubated at room temperature for 48 hours in a moist chamber (petri dish with a small amount of water). After incubation, the cube of medium is carefully separated from the slide and discarded.

During incubation the fungal culture will grow over the glass surfaces of the slide and cover glass. By adding a little stain to the slide, a semipermanent slide can be made by placing a cover glass over it. The cover glass can also be used to make another slide by placing it on another clean slide with a drop of stain on it. Before the stain (lactophenol cotton blue) is used, it is desirable to add to the hyphae a drop of alcohol, which acts as a wetting agent.

🕐 First Period

(Slide Culture Preparation)

Proceed as follows to make slide cultures of one or more mold colonies.

Materials

- petri dishes, glass, sterile
- filter paper (9 cm dia, sterile)
- glass U-shaped rods
- fungal culture plate (mixture)
- 1 petri plate of Sabouraud's agar or Emmons' medium per 4 students
- scalpels
- inoculating loop
- sterile water
- microscope slides and cover glasses (sterile)
- forceps

1. Aseptically, with a pair of forceps, place a sheet of sterile filter paper in a petri dish.
2. Place a sterile U-shaped glass rod on the filter paper. (Rod can be sterilized by flaming, if held by forceps.)
3. Pour enough sterile water (about 4 ml) on filter paper to completely moisten it.

Figure 20.1 Procedure for making two stained slides from slide culture.

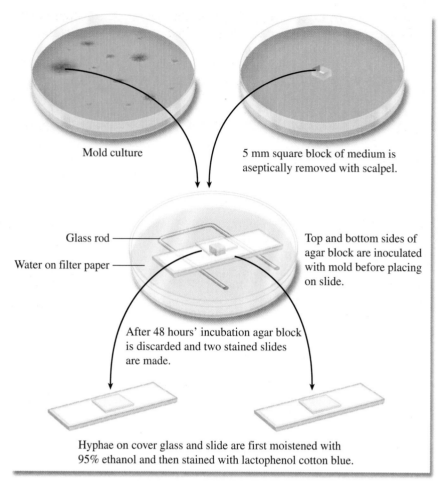

Mold culture

5 mm square block of medium is aseptically removed with scalpel.

Glass rod

Water on filter paper

Top and bottom sides of agar block are inoculated with mold before placing on slide.

After 48 hours' incubation agar block is discarded and two stained slides are made.

Hyphae on cover glass and slide are first moistened with 95% ethanol and then stained with lactophenol cotton blue.

4. With forceps, place a sterile slide on the U-shaped rod.
5. *Gently* flame a scalpel to sterilize, and cut a 5 mm square block of the medium from the plate of Sabouraud's agar or Emmons' medium.

6. Pick up the block of agar by inserting the scalpel into one side, as illustrated in figure 20.2. Inoculate both top and bottom surfaces of the cube with spores from the mold colony. Be sure to flame and cool the loop prior to picking up spores.

Figure 20.2 Inoculation technique.

7. Place the inoculated block of agar in the center of a microscope slide. Be sure to place one of the inoculated surfaces down.
8. Aseptically, place a sterile cover glass on the upper inoculated surface of the agar cube.
9. Place the cover on the petri dish and incubate at room temperature for 48 hours.
10. After 48 hours, examine the slide under low power. If growth has occurred, you should see hyphae and spores. If growth is inadequate and spores are not evident, allow the fungus to grow another 24 to 48 hours before making the stained slides.

Second Period

(Application of Stain)

As soon as there is evidence of spores on the slide, prepare two stained slides from the slide culture, using the following procedure:

Materials

- microscope slides and cover glasses
- 95% ethanol
- lactophenol cotton blue stain
- forceps

1. Place a drop of lactophenol cotton blue stain on a clean microscope slide.

2. Remove the cover glass from the slide culture and discard the block of agar.
3. Add a drop of 95% ethanol to the hyphae on the cover glass. As soon as most of the alcohol has evaporated, place the cover glass, mold side down, on the drop of lactophenol cotton blue stain on the slide. This slide is ready for examination.
4. Remove the slide from the petri dish, add a drop of 95% ethanol to the hyphae, and follow this up with a drop of lactophenol cotton blue stain. Cover the entire preparation with a clean cover glass.
5. Compare both stained slides under the microscope; one slide may be better than the other one.

Laboratory Report

There is no Laboratory Report for this exercise.

Questions

1. Why is a block of agar inoculated rather than streaking the fungal culture on a plate?
2. Why is chloramphenicol added to the fungal culture?

Bacterial Viruses

Viruses differ from bacteria in being much smaller and therefore below the resolution of the light microscope. The smallest virus is one million times smaller than a typical eukaryotic cell. Viruses are obligate intracellular parasites that require a host cell in order to replicate and reproduce, and hence they cannot be grown on laboratory media. Despite these obstacles, we can detect their presence by the effects that they have on their host cells.

Viruses infect all types of cells, eukaryotic and prokaryotic. They are composed of RNA or DNA but never both, and a protein coat, or capsid, that surrounds the nucleic acid. Their dependence on cells is due to their lack of metabolic machinery necessary for the synthesis of viral components. By invading a host cell, they can utilize the metabolic systems of the host cells to achieve their replication.

The study of viruses that parasitize plant and animal cells is time-consuming and requires special tissue culture techniques. Bacterial viruses are relatively simple to study, utilizing ordinary bacteriological techniques. It is for this reason that bacterial viruses will be studied here. However, the principles learned from studying the viruses that infect bacteria apply to viruses of eukaryotic cells.

Viruses that infect bacterial cells are called bacteriophages, or phages. They are diverse in their morphology and size. Some of the simplest ones have single-stranded DNA. Most phages are composed of head, sheath, and tail fibers as seen in figure 21.1. The capsid (head) may be round, oval, or polyhedral and is composed of individual protein subunits called

© MedicalRF.com RF

capsomeres. It forms a protective covering around the viral genome. The tail structure or sheath is composed of a contractile protein that surrounds a hollow core, which is a conduit for the delivery of viral nucleic acid into the host cell. At the end of the tail is a base plate with tail fibers and spikes attached to it. The tail fibers bind to chemical groups on the surface of the bacterial cell and are responsible for recognition. Lysozyme associated with the tail portion of the virus erodes and weakens the cell wall of the host cell. This facilitates the injection of the viral nucleic acid by the sheath contracting and forcing the hollow core through the weakened area in the cell wall.

Infections by viruses can have two outcomes. The lytic cycle involves virulent phages that cause lysis and death of the host cell. The lysogenic cycle involves temperate phages, which can either lyse the host or integrate their DNA into host cell DNA and alter the genetics of the host cell.

In the lytic cycle, the virus assumes control of cell metabolism and uses the cell's metabolic machinery to manufacture phage components (i.e., nucleic acid, capsid, sheath, tail fibers, spikes, and base plates). Mature phage particles are assembled and released from the cell where they can in turn invade new host cells. The result of a lytic infection for the host cell is almost always death (figure VI.1).

In the lysogenic cycle, the viral DNA of the temperate virus is integrated into host DNA, and no mature phages are made. Cells grow normally and are immune to further infections by the same phage. There is no visible evidence to indicate that a virus is even present in the cell. In some cases, the virus can carry genes that confer new genetic capabilities on the virally infected cell, or lysogen. For example, when *Corynebacterium diphtheriae* is infected with a certain lysogenic phage, because the phage carries a toxin gene in its genome, the host cells begin to produce a potent toxin responsible for the symptoms of diphtheria. This phenomenon is known as lysogenic conversion and is responsible for some of the toxins produced by various pathogens. Periodically, the lysogenic phage DNA can excise from the host DNA and initiate the lytic cycle and the production of mature phages. This results in the lysis of the host cell.

Visual evidence for lysis can be demonstrated by mixing phages with host cells and plating them onto media. The bacteria form a confluent lawn of growth, and where the phages cause lysis of the bacterial cells, there will be seen clear areas called plaques.

Some of the most studied bacteriophages are those that infect *Escherichia coli,* such as the T-even phages and lambda phage. They are known as the **coliphages.** Because *E. coli* is an intestinal bacterium, the coliphages can readily be isolated from raw sewage and coprophagous (dung-eating) insects such as flies. The exercises in this section will demonstrate some of the techniques for isolating, assaying, and determining the burst size of bacteriophages. It is recommended that you thoroughly understand the various stages in the lytic cycle before you begin the experiments in this section.

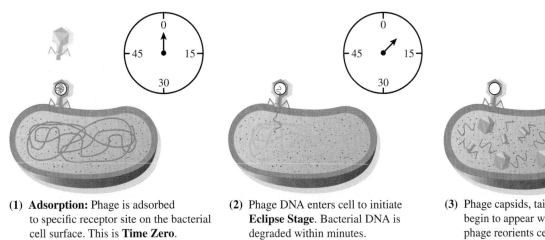

(1) **Adsorption:** Phage is adsorbed to specific receptor site on the bacterial cell surface. This is **Time Zero**.

(2) Phage DNA enters cell to initiate **Eclipse Stage**. Bacterial DNA is degraded within minutes.

(3) Phage capsids, tails, and phage DNA begin to appear within 12 minutes as phage reorients cell metabolism to its own fabrication processes.

(4) Components of phage are assembled into mature infective virions.

(5) Cell wall opens up due to enzymatic action to release mature virions. **Burst size** is the number of units released by cell. Total time: 40 minutes.

Figure VI.1   **The lytic cycle of a virulent bacteriophage.**

Determination of a Bacteriophage Titer

Learning Outcomes

After completing this exercise, you should be able to

1. Define the structures associated with a bacteriophage.
2. Define the steps in the infection of a bacterial cell by a bacteriophage.
3. Determine the number of infective phage particles in a sample using the plaque assay procedure.

Bacteriophages are viruses that infect bacterial cells. They were first described by Twort and d'Herelle in 1915 when they both noted that bacterial cultures spontaneously cleared and the bacteria-free liquid that remained could cause new cultures of bacteria to also clear. Because it appeared that the cultures were being "eaten" by some unknown agent, d'Herelle coined the term *bacteriophage,* which means "bacterial eater." Like all viruses, bacteriophages, or phages, for short, are **obligate intracellular parasites,** that is, they must invade a host cell in order to replicate and reproduce. This is due to the fact that viruses are composed primarily of only a single kind of nucleic acid molecule encased in a protein coat, or **capsid,** that protects the nucleic acid. All viruses lack metabolic machinery, such as energy systems, and protein synthesis components necessary for independent replication. In order to replicate and reproduce, they must use the host cell's metabolic machinery to synthesize their various component parts.

Viruses also exhibit specificity for their hosts. For example, a certain bacteriophage may only infect a specific strain of a bacterium. Examples are the T-even bacteriophages that infect *Escherichia coli* B, whereas other phages infect *E. coli* K12, a different strain of the organism. A phage that infects *Staphylococcus aureus* does not infect *E. coli* and vice versa. This specificity can be used in phage typing of pathogens (Exercise 23).

The structure of a T4 bacteriophage is shown in figure 21.1. A phage consists of a **nucleocapsid,** which is the nucleic acid and protein capsid. The nucleocapsid is attached to a protein **sheath** that is contractile and contains a hollow tube in its center. The sheath sits on a **base plate** to which **tail fibers**

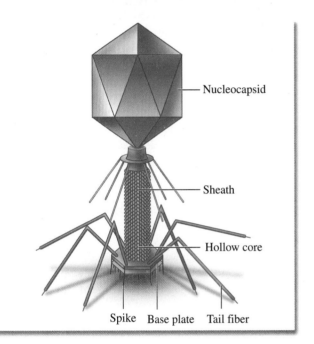

Nucleocapsid

Sheath

Hollow core

Spike **Base plate** **Tail fiber**

Figure 21.1 **T4 Bacteriophage.**

and **spikes** are attached. Most of the phage structure is necessary for delivery of the phage nucleic acid into its host. A single virus or phage particle is also called a **virion.**

The steps in a lytic phage infection of a bacterial cell are basically as follows:

Adsorption A bacteriophage recognizes its host by its tail fibers binding to chemical groups associated with receptors on the surface of the host cell. These receptors perform other functions for the bacterial cell. Examples of receptors are pili and lipopolysaccharide in gram-negative bacteria and teichoic acids in gram-positive bacteria. The specificity in phage infections resides in differences in the receptors. If the groups that are recognized by tail fibers are not present on a bacterial cell, a phage cannot bind to that cell and cause an infection.

Entry of Viral Genome The phage particle settles onto the surface of the host bacterial cell, and **lysozyme** that is associated with the phage begins to erode a localized area of the cell wall, thus weakening the wall. Then the sheath contracts, forcing a hollow core

159

that is connected to the phage capsid through the weakened area of the cell wall. This results in the viral genome being injected into the bacterial cytoplasm.

Synthesis of Phage Components   Only the phage nucleic acid enters the host cell, while the capsid and remainder of the phage structure remain on the outside of the bacterial cell. Phages such as the T phages (T2, T4, T6) produce a **lytic infection** in *E. coli*. These phages cause a **productive infection** in which new phages are made and released from the cell. To incapacitate the host cell and ensure that phage components are synthesized, a viral nuclease injected with the viral DNA begins to degrade *E. coli* DNA. As a result, the host cell cannot carry out any of its own metabolic functions. However, the virus leaves intact host cell metabolic machinery for producing energy and for synthesizing nucleic acids and proteins. The virus uses these systems to replicate its component parts such as capsid, sheath, and tail fibers. During this time no mature virus can be detected by assay procedures, and it is thus referred to as the **eclipse period**. Once all the parts are synthesized, they come together by the process of **self-assembly** to form mature phage particles. Genes on the viral genome also encode for the synthesis of lysozyme that begins to degrade the cell wall and weaken it from the inside. This facilitates the release of the virus from the cell.

In contrast to lytic phages, when **lysogenic** or **temperate** bacteriophages infect a bacterial cell, the phage genome can integrate into host DNA to become a stable genetic element in the host cell genome. In this instance, the genes for the synthesis of viral components are not usually expressed, and, hence no mature phages are produced in these cells. However, each time the bacterial genome replicates during cell division, phage DNA is also replicated, and all bacterial progeny contain phage DNA as part of their genetic makeup. Bacterial cells that have temperate phage DNA integrated into their DNA are called **lysogens**, and the integrated phage genome is known as a **prophage** or **provirus**. An example of a temperate bacteriophage that causes this type of infection is λ (lambda) phage that infects *Escherichia coli*.

Occasionally, the viral DNA in a lysogenic infection can excise from the host DNA, which results in the expression of viral genes responsible for the synthesis of phage components. Mature phages are made by the same steps that are responsible for the production of lytic phages and the host cell is lysed.

Release of Virus   The combination of the weakened cell wall resulting from the action of lysozyme plus the pressure exerted by the phage particles in the cell causes the cell to burst, releasing mature phages into the environment where they can infect susceptible cells.

Lytic phages will produce new lytic infections, whereas lysogenic phages will produce new lysogenic infections. One phage particle infecting a single host cell can produce as many as 200 virions. This number is called the **burst size** and will vary for each virus.

If bacterial cells are mixed with bacteriophage in soft agar, the bacteria will first grow to produce a **confluent lawn** of cells. Phages will infect the cells, causing them to undergo lysis and form clear areas in the confluent lawn called **plaques.** Each plaque is formed by the progeny of a single virion that has replicated and lysed the bacterial cells. Like colony-forming units, **plaque-forming units** (PFUs) can be counted to determine the number of viral particles in a suspension of phage.

In the following exercise, you will work in pairs and determine the number of phage particles or PFUs in a suspension of T4 bacteriophage. You will use *E. coli* B as the host for this experiment.

Materials

- 18- to 24-hour broth culture of *Escherichia coli* B
- 2 ml suspension of T4 bacteriophages with a titer of at least 10,000 phages/ml
- 5 trypticase soy agar (TSA) plates. These should be warmed to 37°C before use.
- 5 tubes of soft agar (0.7% agar). Prior to use, melt and hold at 50°C in a water bath.
- 5 tubes of 9.0 ml trypticase soy (TSB) broth
- 1 ml sterile pipettes
- pipette aids

1. Label the 5 TSA plates with your name and dilutions from 1:10 to 1:100,000.
2. Label 5 TSB tubes with the dilutions 1:10 to 1:100,000 (figure 21.2).
3. Prepare serial tenfold dilutions of the phage stock suspensions by transferring 1 ml of the phage suspension to the first dilution blank. Mix well and transfer 1 ml of the first dilution to the second dilution blank (1:100). Repeat this same procedure until the original phage stock has been diluted 1:100,000 (figure 21.2).
4. Aseptically transfer 2 drops of *E. coli* B broth culture to each of the 5 soft agar overlay tubes.
5. Transfer 1 ml of the first (1:10) phage dilution tube to a soft agar overlay and mix thoroughly but gently. After mixing, pour the contents of the soft agar tube onto the respective TSA plate. Make sure that the soft agar completely covers the surface of the TSA plate. This can be accomplished by gently swirling the plate several times after pouring and while the soft agar is still liquid.
6. Repeat this procedure for each dilution of the phage suspension.

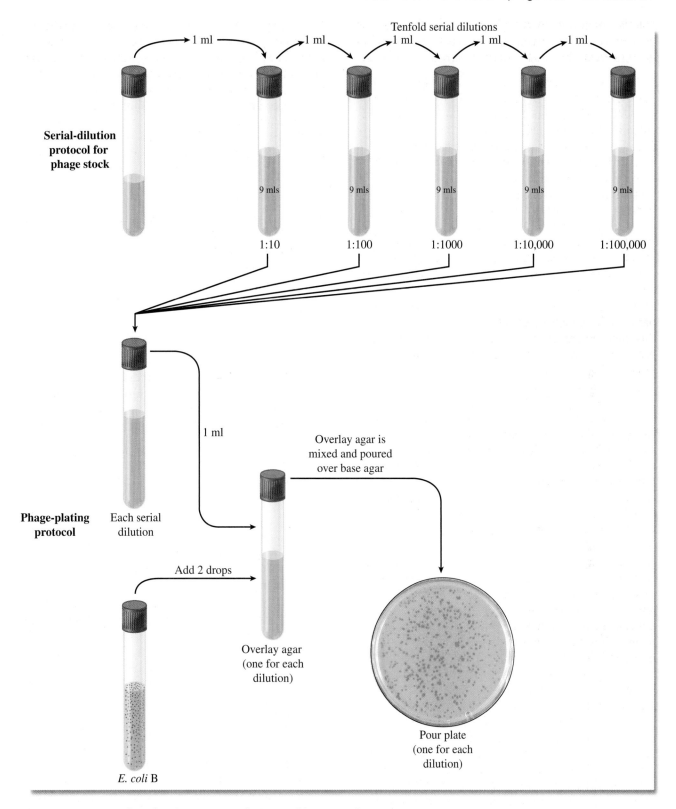

Figure 21.2 **Procedure for determining the titer of bacteriophage.**

7. Incubate the plates at 37°C for 24 hours. If the exercise cannot be completed at this time, refrigerate the plates until the next laboratory period.

8. Observe the plates. Plaques will appear as clear areas in the bacterial lawn. Count the plaques on the plates. Only include counts between 25 and 250 plaques. This can be facilitated with a bacterial colony counter. Multiply the number of plaques times the dilution factor to determine the number of phage particles in the original suspension of phages.

9. Record the phage titer in Laboratory Report 21.

Student: _____

Date: _____ Section: _____

21   Determination of a Bacteriophage Titer

A. Results

DILUTION	PLAQUE NUMBER	PFU/ML
1:10		
1:100		
1:1000		
1:10,000		
1:100,000		

For the PFU/ml calculation, only plates giving 25–250 plaques should be used. For the plaque number, indicate when the bacterial cells are cleared due to too many phage particles in the infection.

What is the number of phage particles in the original phage stock? _____

B. Short-Answer Questions

1. On gram-negative bacteria, to what chemical groups might bacteriophages attach?

   _____

   _____

2. Why are viruses called obligate intracellular parasites?

   _____

   _____

3. What type of viral infection can lead to altered host cell genetics?

   _____

   _____

4. What is lysogenic conversion?

   _____

   _____

5. Compare and contrast bacteriophage plaques with bacterial colonies.

   _____

   _____

6. Describe the four steps in a lytic phage infection.

   _____

   _____

7. What part of the host cell is degraded after phage nucleic acid has entered a host cell?

_____

_____

8. Name two stages in a lytic infection where lysozyme is used. Describe the role of lysozyme during each of these stages.

_____

_____

Isolation of Phages from Flies

Learning Outcomes

After completing this exercise, you should be able to

1. Understand where bacteriophages that infect *E. coli* naturally occur in the environment.
2. Isolate bacteriophages that infect *E. coli* from flies.

As stated earlier, coprophagous insects (insects that feed on fecal material and dung, as well as raw sewage) contain various kinds of bacterial viruses. Houseflies are coprophagous because they deposit their eggs in fecal material where the young larvae feed, grow, pupate, and emerge as adult flies. This type of environment is heavily populated by *E. coli* and the various bacteriophages that infect this bacterium.

Fly Collection

To increase the probability of success in isolating phages, it is desirable that one use 20 to 24 houseflies. A smaller number might be sufficient; the larger number, however, increases the probability of initial success. Houseflies should not be confused with the smaller blackfly or the larger blowfly. An ideal spot for collecting these insects is a barnyard or riding stable. One should not use a cyanide killing bottle or any other chemical means. Flies should be kept alive until just prior to crushing and placing them in the growth medium. There are many ways that one might use to capture them—use your ingenuity!

Enrichment

Within the flies' digestive tracts are several different strains of *E. coli* and bacteriophages. Our first concern is to enhance the growth of both organisms to ensure an adequate supply of phages. To accomplish this the flies must be ground up with a mortar and pestle and then incubated in a special growth medium for a total of 48 hours. During the last 6 hours of incubation, a lysing agent, sodium cyanide, is included in the growth medium to augment the lysing properties of the phage.

Figures 22.1 and 22.2 illustrate the procedure.

Materials

- bottle of phage growth medium* (50 ml)
- bottle of phage lysing medium* (50 ml)
- Erlenmeyer flask (125 ml capacity) with cap
- mortar and pestle (glass)
  *see appendix C for composition

1. Into a clean, nonsterile mortar place 24 freshly killed houseflies. Pour half of the growth medium into the mortar and grind the flies to a fine pulp with the pestle.
2. Transfer this fly-broth mixture to an empty flask. Use the remainder of the growth medium to rinse out the mortar and pestle, pouring all the medium into the flask.
3. Wash the mortar and pestle with soap and hot water before returning them to the cabinet.
4. Incubate the fly-broth mixture for 42 hours at 37°C.
5. At the end of the 42-hour incubation period, add 50 ml of lysing medium to the fly-broth mixture. Incubate this mixture for another 6 hours.

Centrifugation

Before attempting filtration, you will find it necessary to separate the fly fragments and miscellaneous bacteria from the culture medium. If centrifugation is incomplete, the membrane filter will clog quickly and filtration will progress slowly. To minimize filter clogging, a triple centrifugation procedure will be used. To save time in the event filter clogging does occur, an extra filter assembly and an adequate supply of membrane filters should be available. These filters have a maximum pore size of 0.45 μm, which holds back all bacteria, allowing only the phage virions to pass through.

Materials

- centrifuge
- 6–12 centrifuge tubes
- 2 sterile membrane filter assemblies (funnel, glass base, clamp, and vacuum flask)
- package of sterile membrane filters (0.45 μm)
- sterile Erlenmeyer flask with cap (125 ml size)
- vacuum pump and rubber hose

(1) Twenty to 24 flies are ground up in phage growth medium with a mortar and pestle.

(2) Crushed flies are incubated in growth medium for 42 hours at 37°C. After adding lysing medium it is incubated for another 6 hours.

(3) Fly-broth culture is triple-centrifuged at 2500 rpm.

(4) Membrane filter assembly is set up for filtration. This step must be done aseptically.

(5) Centrifuged supernatant is filtered to produce bacteria-free phage filtrate.

(6) Phage filtrate is dispensed to a sterile Erlenmeyer flask from which layered plates will be made (figure 22.2).

Figure 22.1   **Procedure for preparation of bacteriophage filtrate from houseflies.**

1. Into 6 or 8 centrifuge tubes, dispense the enrichment mixture, filling each tube to within $\frac{1}{2}''$ of the top. Place the tubes in the centrifuge so that the load is balanced. Centrifuge the tubes at 2500 rpm for 10 minutes.
2. Without disturbing the material in the bottom of the tubes, decant all material from the tubes to within $1''$ of the bottom into another set of tubes.
3. Centrifuge this second set of tubes at 2500 rpm for another 10 minutes. While centrifugation is taking place, rinse out the first set of tubes.

4. When the second centrifugation is complete, pour off the top two-thirds of each tube into the clean set of tubes and centrifuge again in the same manner.

Filtration

While the third centrifugation is taking place, aseptically place a membrane filter on the glass base of a sterile filter assembly (illustration 4, figure 22.1). Use flamed forceps. Note that the filter is a thin sheet with grid lines on it. Place the glass funnel over the filter

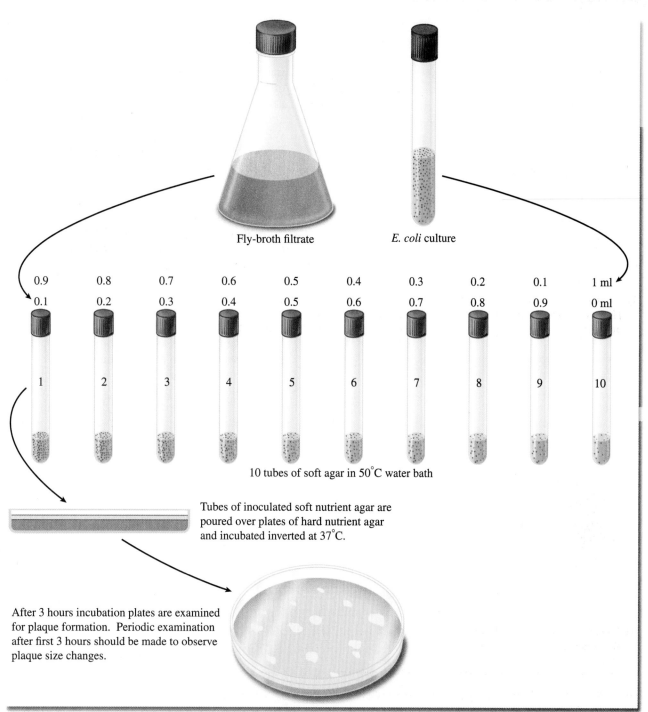

Fly-broth filtrate

E. coli culture

| 0.9 | 0.8 | 0.7 | 0.6 | 0.5 | 0.4 | 0.3 | 0.2 | 0.1 | 1 ml |
| 0.1 | 0.2 | 0.3 | 0.4 | 0.5 | 0.6 | 0.7 | 0.8 | 0.9 | 0 ml |

1   2   3   4   5   6   7   8   9   10

10 tubes of soft agar in 50°C water bath

Tubes of inoculated soft nutrient agar are poured over plates of hard nutrient agar and incubated inverted at 37°C.

After 3 hours incubation plates are examined for plaque formation. Periodic examination after first 3 hours should be made to observe plaque size changes.

Figure 22.2   Inoculation of *Escherichia coli* with bacteriophage from fly-broth filtrate.

and fix the clamp in place. Hook up a rubber hose between the vacuum flask and pump.

Now, carefully decant the top three-fourths of each tube into the filter funnel. Take care not to disturb the material in the bottom of the tube. Turn on the vacuum pump. If centrifugation and decanting have been performed properly, filtration will occur almost instantly. If the filter clogs before you have enough filtrate, recentrifuge all material and pass it through the spare filter assembly.

Aseptically, transfer the final filtrate from the vacuum flask to a sterile 125 ml Erlenmeyer flask that has a sterile closure. Putting the filtrate in a small flask is necessary to facilitate pipetting. Be sure to flame the necks of both flasks while pouring from one to the other.

Inoculation and Incubation

To demonstrate the presence of bacteriophages in the fly-broth filtrate, a strain of phage-susceptible *E. coli* will be used. To achieve an ideal proportion of phages to bacteria, a proportional dilution method will be used. The phages and bacteria will be added to tubes of soft nutrient agar that will be layered over plates of hard nutrient agar. Soft nutrient agar contains only half as much agar as ordinary nutrient agar. (This medium and *E. coli* provide an ideal "lawn" for phage growth.) Its jellylike consistency allows for better diffusion of phage particles; thus, more even development of plaques occurs.

Figure 22.2 illustrates the overall procedure. It is best to perform this inoculation procedure in the morning so that the plates can be examined in late afternoon. As plaques develop, one can watch them increase in size with the multiplication of phages and simultaneous destruction of *E. coli.*

Materials

- nutrient broth cultures of *Escherichia coli* (ATCC #8677 phage host)
- flask of fly-broth filtrate
- 10 tubes of soft nutrient agar (5 ml per tube) with metal caps
- 10 plates of nutrient agar (15 ml per plate, and prewarmed at 37°C)
- 1 ml serological pipettes, sterile

1. Liquefy 10 tubes of soft nutrient agar and cool to 50°C. Keep tubes in water bath to prevent solidification.

2. With a marking pen, number the tubes of soft nutrient agar 1 through 10. Keep the tubes sequentially arranged in the test-tube rack.

3. Label 10 plates of prewarmed nutrient agar 1 through 10. Also, label plate 10 "negative control." Prewarming these plates will allow the soft agar to solidify more evenly.

4. With a 1 ml serological pipette, deliver 0.1 ml of fly-broth filtrate to tube 1, 0.2 ml to tube 2, etc., until 0.9 ml has been delivered to tube 9. Refer to figure 22.2 for sequence. **Note that no fly-broth filtrate is added to tube 10.** This tube will be your negative control.

5. With a fresh 1 ml pipette, deliver 0.9 ml of *E. coli* to tube 1, 0.8 ml to tube 2, etc., as shown in figure 22.2. **Note that tube 10 receives 1.0 ml of *E. coli.*** Make sure to gently but thoroughly mix all the tubes.

6. After flaming the necks of each of the tubes, pour them into similarly numbered plates.

7. When the agar has cooled completely, put the plates, inverted, into a 37°C incubator.

8. **After about 3 hours** of incubation, examine the plates, looking for plaques. If some are visible, measure them and record their diameters on Laboratory Report 22.

9. If no plaques are visible, check the plates again in another **2 hours.**

10. Check the plaque size again at **12 hours,** if possible, recording your results. Incubate a total of 24 hours.

11. Complete Laboratory Report 22.

22 Isolation of Phages from Flies

A. Results

1. **Plaque Size Increase**
   With a china marking pencil, circle and label three plaques on one of the plates and record their sizes in millimeters at 1-hour intervals.

TIME	PLAQUE SIZE (millimeters)		
	Plaque No. 1	**Plaque No. 2**	**Plaque No. 3**
3 hours			
5 hours			
12 hours			
24 hours			

   a. Were any plaques seen on the negative control plate? _____

   b. Do the plates show a progressive increase in number of plaques with increased amount of

   fly-broth filtrate? _____

   c. Did the phage completely "wipe out" all bacterial growth on any of the plates? _____
   If so, which plates? _____

2. **Observations**
   Count all the plaques on each plate and record the counts in the following table. If the plaques are very numerous, use a colony counter and hand counting device. If this exercise was performed as a class project with individual students doing only one or two plates from a common fly-broth filtrate, record all counts on the chalkboard on a table similar to the one below.

Plate Number	1	2	3	4	5	6	7	8	9	10
E. coli (ml)	0.9	0.8	0.7	0.6	0.5	0.4	0.3	0.2	0.1	1.0
Filtrate (ml)	0.1	0.2	0.3	0.4	0.5	0.6	0.7	0.8	0.9	0
Number of plaques										

B. **Short-Answer Questions**

1. How does the life cycle of houseflies contribute to the presence of *E. coli* bacteriophages in their guts?

2. From what other environments might *E. coli* bacteriophages be readily isolated?

3. What is the purpose of including a tube in the phage assay in which only the *E. coli* culture is inoculated? Explain.

4. Why would phages pass through the filters used to separate fragments of bacterial cells after the centrifugation step in the exercise?

Phage Typing

Learning Outcomes

After completing this exercise, you should be able to

1. Differentiate strains of *S. aureus* based on their susceptibility to different bacteriophage types.

2. Understand how phage typing is used in epidemiological investigations.

The host specificity of bacteriophages is such that it is possible to differentiate strains of individual species of bacteria based on their susceptibility to various kinds of bacteriophages. In epidemiological studies, where it is important to discover the source of a specific infection, determining the phage type of the causative organism can be an important tool in solving the riddle. For example, if it can be shown that the phage type of *S. typhi* in a patient with typhoid fever is the same as the phage type of an isolate from a suspected carrier, chances are excellent that the two cases are epidemiologically related. Since most bacteria are probably infected by bacteriophages, it is theoretically possible to classify each species into strains based on their phage susceptibility. Such phage-typing groups have been determined for *Staphylococcus aureus*, *Salmonella typhi*, and several other pathogens. The following table illustrates the lytic phage groups for *S. aureus*.

LYTIC GROUP	PHAGES IN GROUP
I	29, 52, 52A, 79, 80
II	3A, 3B, 3C, 55, 71
III	6, 7, 42E, 47, 53, 54, 75, 77, 83A
IV	42D
not allotted	81, 187

In bacteriophage typing, a suspension of the organism to be typed is uniformly swabbed over an agar surface. The bottom of the plate is marked off into squares and each square labeled to indicate which phage type is applied to the square. A small drop of each bacteriophage type is added to its respective square. After incubation, the plate is examined to determine which

Agar is uniformly swabbed with organism to be typed.

Different phage types are added to swabbed surface of medium.

37°C 24 hr

Bacteriophages that cause plaque formation determine the phage type of the unknown.

Figure 23.1 **Bacteriophage typing.**

phage caused lysis of the test organism. In this exercise, you will determine which phage causes the lysis of *S. aureus* (see figure 23.1).

Materials

- 1 petri plate of tryptone yeast extract agar or trypticase soy agar
- bacteriophage cultures (available types)
- nutrient broth cultures of *S. aureus* with cotton swabs

1. Mark the bottom of a plate of tryptone yeast extract agar with as many squares as there are phage types to be used. Label each square with the phage type numbers.
2. Uniformly swab the entire surface of the agar with the organisms.
3. Deposit 1 drop of each phage in its respective square.
4. Incubate the plate at 37°C for 24 hours and record the lytic group and phage type of the culture.
5. Record your results in Laboratory Report 23.

23 Phage Typing

A. Results

1. To which phage types was this strain of *S. aureus* susceptible?

   _____

2. To what lytic group does this strain of staphylococcus belong?

   _____

B. Short-Answer Questions

1. What factors are responsible for bacteriophage host specificity?

   _____

   _____

2. Why is phage typing an important clinical tool?

   _____

   _____

PART 7

Environmental Influences and Control of Microbial Growth

The exercises in the following section are concerned with the effects that factors such as oxygen, temperature, pH, water activity, UV light, antibiotics, disinfectants, antiseptics, and hand washing have on the growth of bacteria. The microbiologist is concerned with providing the optimum conditions for the growth of microorganisms. In contrast, the medical practitioner is concerned with limiting microbial growth to prevent disease. Understanding both of these will enhance the job of each of these individuals.

In Part 5, the primary concern was in formulating a medium that contained all the essential nutrients to support the growth of a microorganism. However, very little emphasis was placed on other limit-

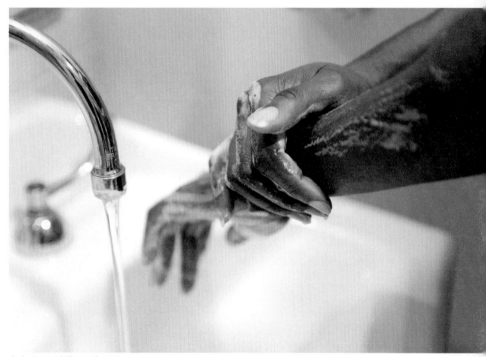

© O. Dimier/ PhotoAlto RF

ing factors such as temperature, oxygen, or pH. Even though all its nutritional needs are provided, an organism may fail to grow if these other factors are not considered. The total environment must be considered to achieve the desired growth of microorganisms.

Effects of Oxygen on Growth

Learning Outcomes

After completing this exercise, you should be able to

1. Define the groups of bacteria based on their metabolic need for oxygen.

2. Explain the different methods of growing anaerobic bacteria.

3. Differentiate the different growth patterns for aerobes, facultative aerobes, microaerophiles, aerotolerant anaerobes, and anaerobes in fluid thioglycollate medium.

Bacteria can be classified as either aerobes or anaerobes based upon their metabolic need for oxygen, which comprises approximately 20% of the atmospheric gases. After they have been classified as aerobes or anaerobes, they can be further separated into categories based on their sensitivity to oxygen.

Obligate (Strict) Aerobes: These bacteria must grow in oxygen because their metabolism requires oxygen. They carry out respiration in which oxygen is utilized as the terminal electron acceptor in the electron transport chain. Examples are *Pseudomonas, Micrococcus,* and many *Bacillus.*

Microaerophiles: These aerobic bacteria prefer to grow in oxygen concentrations of 2–10% rather than the 20% found in the atmosphere. The lower concentration of oxygen is necessary for their respiratory metabolism. Their sensitivity to the higher concentrations of oxygen is not, however, completely understood. *Helicobacter pylori* is a microaerophile that causes stomach ulcers in humans. The oxygen concentration in the stomach is less than the 20% that occurs in the atmosphere.

Facultative Anaerobes: These bacteria grow very well aerobically but also have the capacity to grow anaerobically if oxygen is not present. Their metabolism is flexible because under aerobic conditions they can carry out respiration to produce energy, but if oxygen is absent they can switch to fermentation that does not require oxygen for energy production. *Escherichia coli* is a facultative anaerobe.

Aerotolerant Anaerobes: These anaerobes can tolerate oxygen and even grow in its presence, but they do not require oxygen for energy production. Because they produce their energy strictly by fermentation and not by respiratory means, they are also called **obligate fermenters**. Examples are the streptococci that produce many food products by fermentation such as cheese, yogurt, and sour cream. Other examples are *Enterococcus faecalis* found in the human intestinal tract and *Streptococcus pyogenes,* a pathogen that causes several diseases in humans, such as strep throat and heart and kidney infections.

Obligate (Strict) Anaerobes: Obligate anaerobes cannot tolerate oxygen and must be cultured under conditions in which oxygen is completely eliminated, otherwise they are harmed or killed by its presence. These organisms carry out fermentation or **anaerobic respiration**, in which inorganic compounds, such as nitrates and sulfate, replace oxygen in electron transport as the terminal electron acceptor. Obligate anaerobes are only found among the prokaryotes and in some protozoa. Strict anaerobes occur in environments such as the soil, the rumen (stomach) of cattle, and in anaerobic sewage digesters. *Clostridium, Methanococcus,* and *Bacteroides*, a bacterium found in the human intestine, are examples of obligate anaerobes.

The reason for the sensitivity of strict anaerobes to oxygen is not completely understood. Toxic forms of oxygen such as hydrogen peroxide and superoxide are generated by various chemical mechanisms, and these toxic compounds are abundant in most environments. Toxic forms of oxygen are highly reactive compounds that can damage biological molecules such as nucleic acids, proteins, and small molecules such as coenzymes. Most aerobes possess enzyme systems that will convert the toxic forms of oxygen to less harmful compounds that are not damaging to the cell. For example, **catalase** will degrade hydrogen peroxide into oxygen and water. **Peroxidase** will also degrade hydrogen peroxide. **Superoxide dismutase** will act on the superoxide anion and convert it to oxygen and hydrogen peroxide; the latter can be degraded by catalase. Most strict anaerobes do not possess catalase or superoxide dismutase. However, some strict anaerobes contain other detoxifying systems to deal with superoxide. Therefore, it is not solely the lack of catalase and superoxide dismutase in these bacteria that is responsible for oxygen sensitivity, but it may involve other factors. Interestingly, aerotolerant anaerobes such as the streptococci possess superoxide dismutase, but they lack catalase.

The cultivation of obligate anaerobes requires specialized conditions that eliminate oxygen and therefore its toxic forms. This can be achieved in anaerobic incubators or in anaerobic jars that employ chemical catalysts to eliminate oxygen. These bacteria can also be cultivated in specialized media that contain chemicals such as thioglycollate, which reacts with oxygen to create anaerobic conditions.

The growth of some bacteria such as the streptococci can be enhanced by cultivation in a candle jar, where the concentration of oxygen is less than that in the atmosphere. Cultures are set up in a jar in which a lighted candle is placed. A lid is placed on the jar and tightened, and the candle is extinguished because the oxygen is partially consumed by combustion. The oxygen concentration decreases and the carbon dioxide increases to about 3.5% in the jar.

Figure 24.1 illustrates where the various classes of bacteria grow in a tube in relation to the level of oxygen tension in the medium. In this experiment, you will inoculate various media with several organisms that have different oxygen requirements. The media you will use are fluid thioglycollate (FTM), tryptone glucose yeast extract agar (TGYA), and Brewer's anaerobic agar. Each medium will serve a different purpose. A description for each medium follows:

FTM Fluid thioglycollate medium is a rich liquid medium that supports the growth of both aerobic and anaerobic bacteria. It contains glucose, cystine, and sodium thioglycollate to reduce its oxidation/reduction (O/R) potential. It also contains the dye resazurin, which is an indicator for the presence of oxygen. In the presence of oxygen the dye becomes pink. Since the oxygen tension is always higher near the surface of the medium, the medium will be pink at the top and colorless in the middle and bottom. The medium also contains a small amount of agar, which

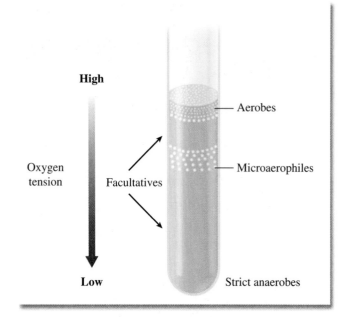

Figure 24.1 Oxygen needs of microorganisms.

helps to localize the organisms and favors anaerobiasis in the bottom of the tube.

TGYA Shake This solid medium will be used to prepare "shake tubes." The medium is not primarily for the cultivation of anaerobes but will be used to determine the oxygen requirements of different bacteria. It will be inoculated in the liquefied state, shaken to mix the organisms throughout the medium, and allowed to solidify. After incubation one determines the oxygen requirements on the basis of where growth occurs in the shake tube: top, middle, or bottom.

Brewer's Anaerobic Agar This solid medium is excellent for culturing anaerobic bacteria in petri dishes. It contains thioglycollate, a reducing agent, and resazurin, an oxidation/reduction (O/R) indicator. For strict anaerobic growth, it is essential that plates be incubated in an oxygen-free environment.

To provide an oxygen-free incubation environment for the petri plates of anaerobic agar, we will use the **GasPak anaerobic jar.** Note in figure 24.2 that hydrogen is generated in the jar, which removes the oxygen by forming water. Palladium pellets catalyze the reaction at room temperature. The generation of hydrogen is achieved by adding water to a plastic envelope of chemicals. Note also that CO_2 is produced, which is a requirement for the growth of many fastidious bacteria. To make certain that anaerobic conditions actually exist in the jar, an indicator strip of methylene blue becomes colorless in the total absence

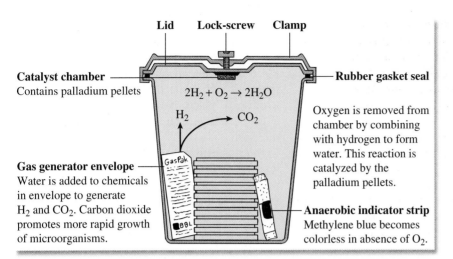

Figure 24.2 **The GasPak anaerobic jar.**

Catalyst chamber
Contains palladium pellets

Lid **Lock-screw** **Clamp**

$2H_2 + O_2 \rightarrow 2H_2O$

$H_2 \longrightarrow CO_2$

Rubber gasket seal

Oxygen is removed from chamber by combining with hydrogen to form water. This reaction is catalyzed by the palladium pellets.

Gas generator envelope
Water is added to chemicals in envelope to generate H_2 and CO_2. Carbon dioxide promotes more rapid growth of microorganisms.

Anaerobic indicator strip
Methylene blue becomes colorless in absence of O_2.

of oxygen. If the strip is not reduced (decolorized) within 2 hours, the jar has not been sealed properly, or the chemical reaction has failed to occur.

In addition to doing a study of the oxygen requirements of six organisms in this experiment, you will have an opportunity during the second period to do a microscopic study of the types of endospores formed by three spore-formers used in the inoculations. Proceed as follows:

First Period

(Inoculations and Incubation)

Since six microorganisms and three kinds of media are involved in this experiment, it will be necessary for economy of time and materials to have each student work with only three organisms. The materials list for this period indicates how the organisms will be distributed.

During this period, each student will inoculate three tubes of medium and only one petri plate of Brewer's anaerobic agar. All of the plates will be placed in a GasPak. Both the GasPak containing the plates and the FTM and TGYA tubes that you inoculated will be incubated at 37°C. Students will share results.

Materials

per student:
- 3 tubes of fluid thioglycollate medium
- 3 TGYA shake tubes (liquefied)
- 1 petri plate of Brewer's anaerobic agar

broth cultures for **odd-numbered students:**
- *Staphylococcus aureus, Enterococcus faecalis,* and *Clostridium sporogenes*

broth cultures for **even-numbered students:**
- *Bacillus subtilis, Escherichia coli,* and *Clostridium beijerinckii (ATCC 14950)*

Figure 24.3 **Organisms are dispersed in medium by rolling tube gently between palms.**

- GasPak anaerobic jar, 3 GasPak generator envelopes, 1 GasPak anaerobic generator strip, scissors, one 10 ml pipette, and water baths at student stations (electric hot plate, beaker of water, and thermometer)

1. Set up a 45°C water bath at your station in which you can keep your tubes of TGYA shakes from solidifying. One water bath for you and your laboratory partner will suffice. (Note in the materials list that the agar shakes have been liquefied for you prior to lab time.)

2. Label the six tubes with the organisms assigned to you (one organism per tube), your initials, and assignment number.

   Note: *Handle the tubes gently to avoid taking on any unwanted oxygen into the media. If the tubes of FTM are pink in the upper 30%, they must be boiled a few minutes to drive off the oxygen, then cooled to inoculate.*

3. Heavily inoculate each of the TGYA shake tubes with several loopfuls of the appropriate organism for that tube. To get good dispersion of the organisms in the medium, roll each tube gently between the palms as shown in figure 24.3. To prevent oxygen

179

Figure 24.4 Three organisms are streaked on agar plate as straight-line streaks.

uptake, do not overly agitate the medium. Allow these tubes to solidify at room temperature.

4. Inoculate each of the FTM tubes with the appropriate organisms.
5. Streak your three organisms on the plate of anaerobic agar in the manner shown in figure 24.4. Note that only three straight-line streaks, well separated, are made. Place the petri plate (inverted) in a canister with the plates of other students that is to go into the GasPak jar.
6. Once all the students' plates are in canisters, place the canisters into the jar.
7. To activate and seal the GasPak jar, proceed as follows:
   a. Peel apart the foil at one end of a GasPak indicator strip and pull it halfway down. The indicator will turn blue on exposure to the air. Place the indicator strip in the jar so that the wick is visible.
   b. Cut off the corner of each of three GasPak gas generator envelopes with a pair of scissors. Place them in the jar in an upright position.
   c. Pipette 10 ml of tap or distilled water into the open corner of each envelope. Avoid forcing the pipette into the envelope.
   d. Place the inner section of the lid on the jar, making certain it is centered on top of the jar. Do not use grease or other sealant on the rim of the jar since the O-ring gasket provides an effective seal when pressed down on a clean surface.
   e. Unscrew the thumbscrew of the outer lid until the exposed end is completely withdrawn into

the threaded hole. Unless this is done, it will be impossible to engage the lugs of the jar with the outer lid.
   f. Place the outer lid on the jar directly over the inner lid and rotate the lid slightly to allow it to drop in place. Now rotate the lid firmly to engage the lugs. The lid may be rotated in either direction.
   g. Tighten the thumbscrew by turning it clockwise. If the outer lid raises up, the lugs are not properly engaged.
8. Place the jar (containing the plates of anaerobic agar) and the tubes in a 37°C incubator. After 2 or 3 hours, check the jar to determine if the indicator strip has lost its blue color. If decolorization has not occurred, replace the palladium pellets and reinoculate new plates.
9. Incubate the tubes and plates for 24 to 48 hours.

🕐 Second Period

(Culture Evaluations and Spore Staining)

Remove the lid from the GasPak jar. If vacuum holds the inner lid firmly in place, break the vacuum by sliding the lid to the edge. When transporting the plates and tubes to your desk *take care not to agitate the FTM tubes*. The position of growth in the medium can be easily changed if handled carelessly.

Materials

- tubes of FTM
- shake tubes of TGYA
- 2 Brewer's anaerobic agar plates
- spore-staining kits and slides

1. Compare the six FTM and TGYA shake tubes that you and your laboratory partner share with figures 24.5, 24.6, and 24.7 to evaluate the oxygen needs of the six organisms.
2. Compare the growths (or lack of growth) on your petri plate and the plate of your laboratory partner.
3. Record your results on Laboratory Report 24.
4. If time permits, make a combined slide with three separate smears of the three spore-formers, using the spore-staining methods in Exercise 15. Draw the organisms in the circles provided in Laboratory Report 24.

Laboratory Report

Complete Laboratory Report 24.

Figure 24.5   **Thioglycollate tubes: (a) control, (b) microaerophile, (c) strict aerobe, (d) anaerobe, and (e) facultative anaerobe.**
© McGraw-Hill Education. Lisa Burgess, photographer

(a)   (b)   (c)   (d)   (e)

(a)   (b)   (c)

Figure 24.6   **Closeup photos showing relative positions of (a) facultative anaerobe, (b) aerobe, and (c) anaerobe. Reazurin (red dye) indicates presence of oxygen.**
© McGraw-Hill Education. Lisa Burgess, photographer

Figure 24.7   **Closeup of microaerophile growth in thioglycollate broth.**
© McGraw-Hill Education. Lisa Burgess, photographer

Student: _____

Date: _____ Section: _____

24 Effects of Oxygen on Growth

A. Results

1. **Tube Inoculations**

   After carefully comparing the appearance of the six cultures belonging to you and your laboratory partner, select the best tube for each organism and sketch its appearance in the tubes below. Indicate under each name the type of medium (FTM or TGYA).

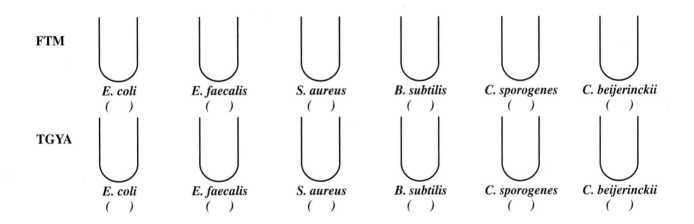

FTM

| *E. coli* | *E. faecalis* | *S. aureus* | *B. subtilis* | *C. sporogenes* | *C. beijerinckii* |
| ( ) | ( ) | ( ) | ( ) | ( ) | ( ) |

TGYA

| *E. coli* | *E. faecalis* | *S. aureus* | *B. subtilis* | *C. sporogenes* | *C. beijerinckii* |
| ( ) | ( ) | ( ) | ( ) | ( ) | ( ) |

2. **Plate Inoculations**

   After comparing the growths on the two plates of Brewer's anaerobic agar with the growths in the six tubes, classify each organism as to its oxygen requirements:

   Escherichia coli: _____   *Bacillus subtilis:* _____

   Enterococcus faecalis: _____   *Clostridium sporogenes:* _____

   Staphylococcus aureus: _____   *Clostridium beijerinckii:* _____

3. **Spore Study**

   If a spore-stained slide is made of the three spore-formers, draw a few cells of each organism in the spaces provided.

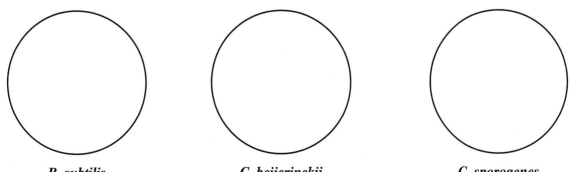

| *B. subtilis* | *C. beijerinckii* | *C. sporogenes* |

B. Short-Answer Questions

1. What is the role of oxygen for cellular respiration?

   _____

   _____

2. What type of metabolism occurs in the absence of oxygen?

   _____

   _____

3. Name two enzymes that are present in obligate aerobes but lacking in obligate anaerobes. What is the function of each enzyme?

   _____

   _____

4. Differentiate between a microaerophile and an aerotolerant organism.

   _____

   _____

5. Why is resazurin a useful media additive for the study of anaerobes?

   _____

   _____

6. Why is a GasPak anaerobic jar necessary for the culture of anaerobes on plates of Brewer's anaerobic agar but not in tubes of fluid thioglycollate medium (FTM)?

   _____

   _____

Temperature:
Effects on Growth

Learning Outcomes

After completing this exercise, you should be able to

1. Define the temperature groups that characterize bacteria.

2. Understand how temperature affects certain metabolic functions such as pigment production in *Serratia marcescens*.

3. Understand how temperature affects the growth rate of bacterial cells.

Microorganisms grow over a broad temperature range that extends from below 0°C to greater than 100°C. Based on their temperature requirements, they can be divided into four groups that define their optimal growth:

psychrophiles: optimal growth between −5°C and 20°C; these bacteria can be found in the supercooled waters of the Arctic and Antarctic.

mesophiles: optimal growth between 20°C and 50°C; most bacteria fall into this class, for example most pathogens grow between 35°C and 40°C.

thermophiles: optimal growth between 50°C and 80°C; bacteria in this group occur in soils where the midday temperature can reach greater than 50°C or in compost piles where fermentation activity can cause temperatures to exceed 60–65°C.

hyperthermophiles: growth optimum above 80°C; many of the Archaea occupy environments that are heated by volcanic activity where water is superheated to above 100°C. These organisms have been isolated from thermal vents deep within the ocean floor and from volcanic heated hot springs.

It should be noted that one single organism is not capable of growth over the entire range but would be restricted to one of the temperature classes. However, some bacteria within the classes are capable of growth at temperatures lower or higher than their optima. For example, some mesophilic bacteria such as *Proteus, Pseudomonas, Campylobacter,* and *Leuconostoc* can grow at 4°C, refrigerator temperatures, and cause food spoilage. These bacteria are referred to as **psychrotrophs.**

Temperature can affect several metabolic factors in the cell. Enzymes are directly affected by temperature, and any one enzyme will have a minimum, optimum, and maximum temperature for activity. Maximal enzyme activity will occur at the optimum temperature. At temperatures above the maximum, enzymes will begin to denature and lose activity. Below the minimum temperature, chemical activity slows down and some denaturation can also occur. In addition to the effects on enzyme activity, temperature can also greatly affect cell membranes and transport. As temperature decreases, transport of nutrients into the cell also decreases due to fluidity changes in the membrane. If the temperature increases above the maximum of an organism, membrane lipids can be destroyed, resulting in serious damage to the membrane and death of the organism. Last, ribosomes can be directly affected by temperature, and if extremes of temperature occur, they will cease to function adequately.

In this experiment, we will attempt to measure the effects of various temperatures on two physiological reactions: pigment production and growth rate. Nutrient broth and nutrient agar slants will be inoculated with three different organisms that have different optimum growth temperatures. One organism, *Serratia marcescens,* produces a red pigment called *prodigiosin.* This compound is an antibiotic and has also been tested in preclinical trials for the treatment of human pancreatic cancer. *S. marcescens* only produces prodigiosin in a certain temperature range. It is our goal here to determine the optimum temperature for prodigiosin production and the approximate optimum growth temperatures for all three microorganisms. To determine optimum growth temperatures, we will be incubating cultures at five different temperatures. A spectrophotometer will be used to measure turbidity densities in the broth cultures after incubation.

🕑 First Period

(Inoculations)

To economize on time and media, it will be necessary for each student to work with only two organisms and seven tubes of media. Refer to table 25.1 to determine your assignment. Figure 25.1 illustrates the procedure.

Table 25.1 Inoculation Assignments

STUDENT NUMBER	S. MARCESCENS	G. STEAROTHERMOPHILUS	E. COLI
1, 4, 7, 10, 13, 16, 19, 22, 25	2 slants and 5 broths		
2, 5, 8, 11, 14, 17, 20, 23, 26	2 slants	5 broths	
3, 6, 9, 12, 15, 18, 21, 24, 27	2 slants		5 broths

Materials

- nutrient broth cultures of *Serratia marcescens*, *Geobacillus stearothermophilus*, and *Escherichia coli*

per student:
- 2 nutrient agar slants
- 5 tubes of nutrient broth

1. Label the tubes as follows:
   Slants: Label both of them *S. marcescens;* label one tube 25°C and the other tube 38°C.
   Broths: Label each tube of nutrient broth with your other organism and one of the following five temperatures: 5°C, 25°C, 38°C, 42°C, or 55°C.
2. Using a wire loop, inoculate each of the tubes with the appropriate organisms.
3. Place each tube in one of the five baskets that is labeled according to incubation temperature.
   Note: The instructor will see that the 5°C basket is placed in the refrigerator and the other four are placed in incubators that are set at the proper temperatures.

⊘ Second Period

(Tabulation of Results)

Materials

- slants and broth cultures that have been incubated at various temperatures
- spectrophotometer and cuvettes
- tube of sterile nutrient broth

1. Compare the nutrient agar slants of *S. marcescens*. Using colored pencils, draw the appearance of the growths on Laboratory Report 25.
2. Shake the broth cultures and compare them, noting the differences in turbidity. Those tubes that appear to have no growth should be compared with a tube of sterile nutrient broth.
3. If a spectrophotometer is available, determine the turbidity of each tube following the instructions in Exercise 19.
4. If no spectrophotometer is available, record turbidity by visual observation. Laboratory Report 25 indicates how to do this.
5. Exchange results with other students to complete data collection for the experiment.

Laboratory Report

After recording all data, answer the questions in Laboratory Report 25.

Two nutrient agar slants are streaked with *S. marcescens* and incubated at different temperatures for pigment production.

Five nutrient broths are inoculated with one of three organisms and incubated at five different temperatures to determine optimum growth temperatures for each organism.

Figure 25.1 **Inoculation procedure.**

Laboratory Report

25

Student: _____
Date: _____ Section: _____

25 Temperature: Effects on Growth

A. Results

1. Pigment Formation and Temperature

   a. Draw the appearance of the growth of *Serratia marcescens* on the nutrient agar slants using colored pencils.

   b. Which temperature seems to be closest to the optimum temperature for pigment formation?

   _____

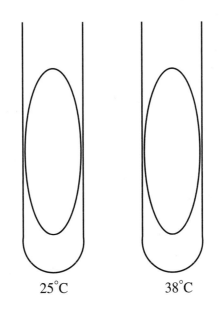

25°C   38°C

2. Growth Rate and Temperature

   a. If a spectrophotometer is available, dispense the cultures into labeled cuvettes and determine the optical density of each culture.

   If no spectrophotometer is available, record only the visual readings as +, + +, + + +, and none.

Temp. °C	NAME OF ORGANISM		NAME OF ORGANISM		NAME OF ORGANISM	
	Visual Reading	O.D.	Visual Reading	O.D.	Visual Reading	O.D.
5						
25						
38						
42						
55						

b. Growth curves of *Serratia marcescens, Escherichia coli,* and *Geobacillus stearothermophilus* as related to temperature.

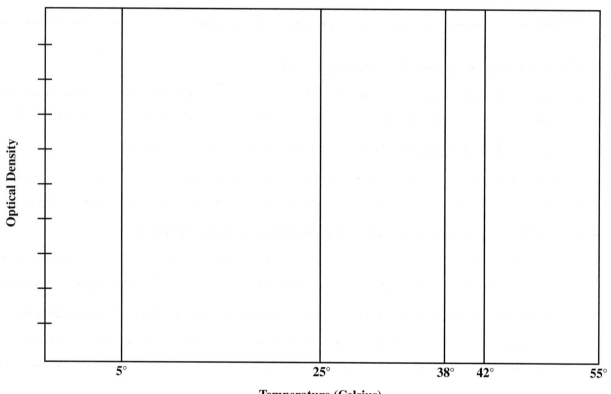

c. On the basis of the preceding graph, estimate the optimum growth temperature of the three organisms.

Serratia marcescens: _____

Escherichia coli: _____

Geobacillus stearothermophilus: _____

d. To get more precise results for the preceding graph, what would you do?

_____

_____

B. Short-Answer Questions

1. In what environments might the following organisms be found?

   a. hyperthermophile

   _____

   b. thermophile

   _____

    c. mesophile

_____

    d. psychrophile

_____

2. Differentiate between psychrophile and psychrotroph.

_____

_____

3. Why are psychrotrophic bacteria of concern to those in the food-service industry?

_____

_____

4. What is the optimum growth temperature for most human pathogens? Explain.

_____

_____

5. Name three cellular components involved in metabolism that are influenced by temperature changes.

_____

_____

pH and Microbial Growth

Learning Outcomes

After completing this exercise, you should be able to

1. Understand how pH affects the growth of bacteria.

2. Define the pH classes of bacteria, acidophile, neutrophile, and alkaliphile.

3. Demonstrate how different pH values affect the growth of microorganisms.

Another factor that exerts a strong influence on growth is the hydrogen ion concentration, designated by the term **pH** ($-\log_{10}[H^+]$). The hydrogen ion concentration affects proteins and other charged molecules in the cell. Each organism will have an optimal pH at which it grows best. If pH values exceed the optimum for an organism, the solubility of charged molecules can be adversely affected and molecules can precipitate out of solution. For example, the pH can directly affect the charge on amino acids in proteins and result in denaturation and loss of enzyme activity.

Microorganisms can be subdivided into groups based on their ability to grow at different pH values. Bacteria that grow at or near neutral pH are termed **neutrophiles.** Most bacteria are neutrophiles, although many can grow over a range of 2–3 pH units. Bacteria that grow at acidic pH values are **acidophiles.** For example, the chemolithotroph, *Thiobacillus thiooxidans,* is an acidophile that grows at pH 1; it derives its energy needs from the oxidation of sulfide, producing sulfuric acid that lowers the pH of its environment to a value of 1. *Thiobacillus* is often found in mining effluents that are enriched in metal sulfide ores. Most fungi and yeast are acidophiles that prefer to grow at pH values between 4 and 6. Media such as potato dextrose agar and Sabouraud's agar for cultivating fungi are adjusted to pH values around 5, which selects for and promotes the growth of these microorganisms. Bacteria that grow at alkaline pH values are termed **alkaliphiles.** True alkalinophilic bacteria are found growing in environments such as soda lakes and high-carbonate soils where the pH can reach 10 or above. Many of these organisms belong to the genus *Bacillus.* Some bacteria are alkaline tolerant, such as the opportunistic pathogen *Alcaligenes faecalis* that can cause urinary tract infections. This bacterium degrades urea to produce ammonia that increases the pH. It is interesting to note that even though the bacteria described above grow at extremes of pH, they maintain their cytoplasm at or near neutral pH to prevent damage and destruction to charged species and macromolecules.

Because pH can influence or inhibit the growth of microorganisms, it has been used in food preservation. Fermentation of foods can yield acids such as lactic acid and acetic acid, which lower the pH of the fermented food, thus preventing the growth of many microorganisms and the spoilage of the food. Examples are pickles, yogurt, and some cheeses. However, fungi can grow at acidic pH values, and they can often spoil fermented foods such as cheese.

In this exercise, we will test the degree of inhibition of microorganisms that results from media containing different pH concentrations. Note in the materials list that tubes of six different hydrogen concentrations are listed. Your instructor will indicate which ones, if not all, will be tested.

🕐 First Period

Materials

per student:
- 1 tube of nutrient broth of pH 3.0
- 1 tube of nutrient broth of pH 5.0
- 1 tube of nutrient broth of pH 7.0
- 1 tube of nutrient broth of pH 8.0
- 1 tube of nutrient broth of pH 9.0
- 1 tube of nutrient broth of pH 10.0

class materials:
- broth cultures of *Escherichia coli*
- broth cultures of *Staphylococcus aureus*
- broth cultures of *Alcaligenes faecalis**
- broth cultures of *Saccharomyces cerevisiae***

---

*Sporosarcina ureae can be used as a substitute for Alcaligenes faecalis.

**Candida glabrata is a good substitute for Saccharomyces cerevisiae.

1. Inoculate a tube of each of these broths with one organism. Use the organism following your assigned number from the table below:

STUDENT NUMBER	ORGANISM
1, 5, 9, 13, 17, 21, 25	*Escherichia coli*
2, 6, 10, 14, 18, 22, 26	*Staphylococcus aureus*
3, 7, 11, 15, 19, 23, 27	*Alcaligenes faecalis**
4, 8, 12, 16, 20, 24, 28	*Saccharomyces cerevisiae***

2. Incubate the tubes of *E. coli, S. aureus,* and *A. faecalis* at 37°C for 48 hours. Incubate the tubes of *S. ureae, C. glabrata,* and *S. cerevisiae* at 20°C for 48 to 72 hours.

① Second Period

Materials

- spectrophotometer
- 1 tube of sterile nutrient broth
- tubes of incubated cultures at various pHs

1. Use the tube of sterile broth to calibrate the spectrophotometer and measure the O.D. of each culture (page 145, Exercise 19). Record your results in the tables on Laboratory Report 26.
2. Plot the O.D. values in the graph on Laboratory Report 26 and answer all the questions.

26 pH and Microbial Growth

A. Results

1. If a spectrophotometer is available, dispense the cultures into labeled cuvettes and determine the O.D. values of each culture. To complete the tables, get the results of the other three organisms from other members of the class, and delete the substitution organisms in the tables that were not used.

   If no spectrophotometer is available, record only the visual reading as +, + +, + + +, and none.

pH	*Escherichia coli*		*Staphylococcus aureus*	
	Visual Reading	Spectrophotometer O.D.	Visual Reading	Spectrophotometer O.D.
3				
5				
7				
8				
9				
10				

pH	*Alcaligenes faecalis* or *Sporosarcina ureae*		*Saccharomyces cerevisiae* or *Candida glabrata*	
	Visual Reading	Spectrophotometer O.D.	Visual Reading	Spectrophotometer O.D.
3				
5				
7				
8				
9				
10				

2. Growth Curves

Once you have recorded all the O.D. values in the two tables, plot them on the following graph. Use different-colored lines for each species.

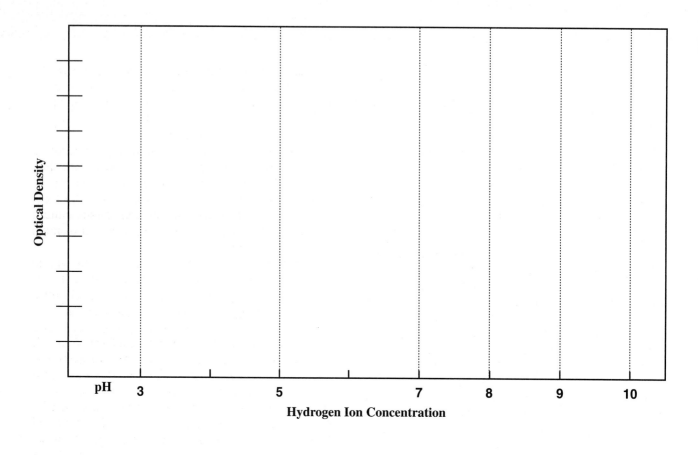

3. Which organism seems to grow best in acid media? _____

4. Which organism seems to grow best in alkaline media? _____

5. Which organism seems to tolerate the broadest pH range? _____

B. Short-Answer Questions

1. How does pH negatively affect the metabolism of microorganisms?

_____

_____

2. Define three groups of microorganisms in regard to their optimum pH for growth.

_____

_____

3. How would the pH of the culture medium be influenced by sugar fermentation? By urea hydrolysis?

_____

_____

Water Activity and Osmotic Pressure

Learning Outcomes

After completing this exercise, you should be able to

1. Understand the association of water activity to osmotic pressure and solute concentration.

2. Define the terms *plasmolysis, halophile, halotolerant,* and *osmophile.*

3. Demonstrate how different salt concentrations affect the growth of bacteria.

The growth of bacteria can be profoundly affected by the availability of water in an environment. The availability of water is defined by a physical parameter called the water activity, A_w. It is determined by measuring the ratio of the water vapor pressure of a solution to the water vapor pressure of pure water. The values for water activity vary between 0 and 1.0, and the closer the value is to 1.0, the more water is available to a cell for metabolic purposes. Water activity and hence its availability decrease with increases in the concentration of solutes such as salts. This results because water becomes involved in breaking ionic bonds and forming solvation shells around charged species to maintain them in solution.

In the process of **osmosis,** water diffuses from areas of low solute concentration where water is more plentiful to areas of high solute concentration where water is less available. Because there is normally a high concentration of nutrients in the cytoplasm relative to the outside of the cell, water will naturally diffuse into a cell. A medium where solute concentrations on the outside of the cell are lower than the cytoplasm is designated as **hypotonic** (figure 27.1). In general, bacteria are not harmed by hypotonic

solutions because the rigid cell wall protects the membrane from being damaged by the osmotic pressure exerted against it. It also prevents the membrane from disrupting when water diffuses across the cell membrane into the cytoplasm.

Environments where the solute concentration is the same inside and outside the cell are termed **isotonic.** Animal cells require isotonic environments or else cells will undergo lysis because only the fragile cell membrane surrounds the cell. Tissue culture media for growing animal cells provides an isotonic environment to prevent cell lysis.

Hypertonic environments exist when the solute concentration is greater on the outside of the cell relative to the cytoplasm, and this causes water to diffuse out of the cytoplasm. When this develops, the cell undergoes **plasmolysis,** resulting in a loss of water; dehydration of the cytoplasm; and shrinkage of the cell membrane away from the cell wall. In these situations, considerable and often irreversible damage can occur to the metabolic machinery of the cell. Low water activity and hypertonic environments have been used by humans for centuries to preserve food. Salted meat and fish, and jams and jellies with high sugar content resist contamination because very little water is available for cells to grow. Most bacteria that might contaminate these foods would undergo immediate plasmolysis.

Microorganisms can be grouped based on their ability to cope with low water activity and high osmotic pressure. Most bacteria grow best when the water activity is around 0.9 to 1.0. In contrast, **halophiles** require high concentrations of sodium chloride to grow. Examples are the halophilic bacteria that require 15–30% sodium chloride to grow and maintain the integrity of their cell walls. These bacteria, which belong to the Archaea, are found in salt lakes and brine solutions, and occasionally growing on salted fish. Some microorganisms are **halotolerant** and are capable of growth in moderate concentrations of salt. For example, *Staphylococcus aureus* can tolerate sodium chloride concentrations that approach 3 *M* or 11%.

Another group of organisms is the **osmophiles,** which are able to grow in environments where sugar concentrations are excessive. An example is *Xeromyces,* a yeast that can contaminate and spoil jams and jellies.

In this exercise, we will test the degree of inhibition of organisms that results with media containing

| Hypotonic | Isotonic | Hypertonic |

Figure 27.1 Osmotic variabilities.

different concentrations of sodium chloride. To accomplish this, you will streak three different organisms on four plates of media. The specific organisms used differ in their tolerance of salt concentrations. The salt concentrations will be 0.5, 5, 10, and 15%. After incubation for 48 hours and several more days, comparisons will be made of growth differences to determine their degrees of salt tolerance.

Materials

per student:
- 1 petri plate of nutrient agar (0.5% NaCl)
- 1 petri plate of nutrient agar (5% NaCl)
- 1 petri plate of nutrient agar (10% NaCl)
- 1 petri plate of milk salt agar (15% NaCl)

cultures:
- *Escherichia coli* (nutrient broth)
- *Staphylococcus aureus* (nutrient broth)
- *Halobacterium salinarium* (slant culture)

1. Mark the bottoms of the four petri plates, as indicated in figure 27.2.
2. Streak each organism in a straight line on the agar, using a wire loop.
3. Incubate all the plates for 48 hours at 37°C. Record your results in Laboratory Report 27.

4. Continue the incubation of the milk salt agar plate for several more days in the same manner, and record your results again in Laboratory Report 27.

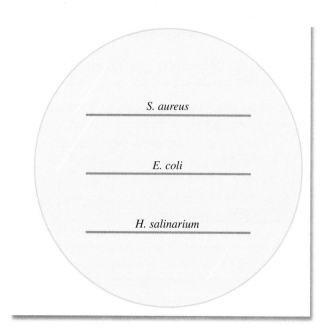

S. aureus

E. coli

H. salinarium

Figure 27.2 **Streak pattern.**

27   Water Activity and Osmotic Pressure

A. Results

1. Record the amount of growth of each organism at the different salt concentrations, using 0, +1, +2, and +3 to indicate degree of growth.

ORGANISM	SODIUM CHLORIDE CONCENTRATION							
	0.5%		5%		10%		15%	
	48 hr	96 hr	48 hr	96 hr	48 hr	96 hr	48 hr	96 hr
Escherichia coli								
Staphylococcus aureus								
Halobacterium salinarium								

2. Evaluate the salt tolerance of the above organisms.

   a. Tolerates very little salt: _____

   b. Tolerates a broad range of salt concentration: _____

   c. Grows only in the presence of high salt concentration: _____

3. How would you classify *Halobacterium salinarium* as to salt needs? Check one.

   _____ Obligate halophile   _____ Facultative halophile

B. Short-Answer Questions

1. Why are bacteria generally resistant to hypotonic environments, whereas animal cells are not?

   _____

   _____

2. How do hypertonic environments negatively affect most bacterial cells?

   _____

   _____

3. Why are staphylococci well suited for the colonization of skin?

   _____

   _____

4. Differentiate between halophiles and osmophiles. Which type would more likely cause spoilage of jams and jellies?

   _____

   _____

MICROBIOLOGY

Ultraviolet Light:
Lethal Effects

Learning Outcomes

After completing this exercise, you should be able to

1. Understand how UV light causes mutations and kills cells.

2. Demonstrate how endospore-forming bacteria are more resistant to UV light than are vegetative cells.

Ultraviolet (UV) light is nonionizing short wavelength radiation that falls between 4 nm and 400 nm in the visible spectrum (figure 28.1). In general, for electromagnetic radiation, the shorter the wavelength, the more damaging it is to cells; thus, UV light is much more germicidal than either visible light or infrared radiation. Most bacteria are killed by the effects of ultraviolet light, and UV light is routinely used to sterilize surfaces, such as the work areas of transfer hoods used for the inoculation of cultures. The primary lethal effects of ultraviolet light are due to its mutagenic properties. UV radiation at **260 nm** is the most germicidal because this wavelength is the specific wavelength at which DNA maximally absorbs UV light. When DNA absorbs UV light, it causes the formation of **pyrimidine dimers.** These form when a covalent bond is formed between two adjacent thymine or cytosine molecules in a DNA strand (figure 28.2). Dimers essentially cause the DNA molecule to become deformed so that the DNA polymerase cannot replicate DNA strands past the site of dimer formation, nor can genes past this point be transcribed.

Cells have evolved various repair mechanisms to deal with mutational changes in DNA in order to insure that fidelity of replication occurs. One system is the **SOS system,** which enzymatically removes the dimers and inserts in their place new pyrimidine molecules. Unlike DNA polymerase, enzymes of the SOS system can move beyond the point where dimers occur in the molecule. However, if the exposure of UV light is sufficient to cause massive numbers of dimers to form in the DNA of a cell, the SOS system is unable to effectively cope with this situation, and it begins to make errors by inserting incorrect bases for the damaged bases, eventually resulting in cell death.

The killing properties of UV light depend on several factors. Time of exposure is important as well as the presence of materials that will block the radiation from reaching cells. For example, glass and plastic can block UV radiation, and plastic lenses are an effective means to protect the eyes from UV damage. Endospores are more resistant to UV light than are vegetative cells for several reasons. First, the DNA of endospores is protected by small acid-soluble proteins that bind to DNA and alter its conformation, thereby protecting it from photochemical damage. Second, a unique spore photo-product is generated by UV light in endospores that functions in the enzymatic repair of damaged DNA during endospore germination.

In Exercise 28, you will examine the germicidal effects for UV light on *Bacillus megaterium,* an endospore former, and *Staphylococcus aureus,* a non-endospore former. One-half of each plate will be shielded from the radiation to provide a control comparison. *Bacillus megaterium* and *Staphylococcus aureus* will be used to provide a comparison of the relative resistance of vegetative cells and endospores.

Exposure to ultraviolet light may be accomplished with a lamp, as shown in figure 28.3, or with a UV box

Figure 28.1 **Lethal effects of light at various wavelengths.**

Figure 28.2   Formation of thymine dimers by UV light.

Figure 28.3   Plates are exposed to UV light with 50% coverage.

that has built-in ultraviolet lamps. The UV exposure effectiveness varies with the type of setup used. The exposure times given in table 28.1 work well for a specific type of mercury arc lamp. Note in the results that you may have to use different times. Your instructor will inform you as to whether you should write in new times that will be more suited to the equipment in your lab. Proceed as follows to do this experiment.

Materials

- petri plates of nutrient agar (one or more per student)
- ultraviolet lamp or UV exposure box
- timers
- cards (3″ × 5″)
- nutrient broth cultures of *Staphylococcus aureus* with swabs
- 4-day-old nutrient broth cultures of *Bacillus megaterium* with swabs

1. Refer to table 28.1 to determine which organism you will work with. You may be assigned more than one plate to inoculate. If different times are to be used, your instructor will inform you what times to write in. Since there are only 16 assignment numbers in the table, more student assignment numbers can be written in as designated by your instructor.
2. Label the bottoms of the plates with your assignment number and your initials.
3. Using a cotton-tipped swab that is in the culture tube, swab the entire surface of the agar in each plate. Before swabbing, express the excess culture from the swab against the inner wall of the tube.
4. Put on the protective goggles provided. Place the plates under the ultraviolet lamp *with the lids removed.* Cover one-half of each plate with a 3″ × 5″ card, as shown in figure 28.3. Label which side of the plate was covered. Note that if your number is 8 or 16, you will not remove the lid from your plate. The purpose of this exposure is to determine to what extent plastic protects cells from the effects of UV light.

Caution

Before exposing the plates to UV light, put on protective goggles. Avoid looking directly into the UV light source. These rays can cause cataracts and eye injury.

5. After exposing the plates for the correct time durations, re-cover them with their lids, and incubate them inverted at 37°C for 48 hours.

Laboratory Report

Record your observations in Laboratory Report 28 and answer all the questions.

Table 28.1   Student Inoculation Assignments

	EXPOSURE TIMES (STUDENT ASSIGNMENTS)							
S. aureus	1	2	3	4	5	6	7	8
	10 sec	20 sec	40 sec	80 sec	2.5 min	5 min	10 min	10 min*
B. megaterium	9	10	11	12	13	14	15	16
	1 min	2 min	4 min	8 min	15 min	30 min	60 min	60 min*

*These petri plates will be covered with dish covers during exposure.

28   Ultraviolet Light: Lethal Effects

A. Results

1. Your instructor will construct a table similar to the one below on the chalkboard for you to record your results. If substantial growth is present in the exposed area, record your results as + + +. If three or fewer colonies survived, record +. Moderate survival should be indicated as + +. No growth should be recorded as −. Record all information in the table.

ORGANISMS	EXPOSURE TIMES							
S. aureus	10 sec	20 sec	40 sec	80 sec	2.5 min	5 min	10 min	10*min
Survival								
B. megaterium	1 min	2 min	4 min	8 min	15 min	30 min	60 min	60*min
Survival								

* plates covered during exposure

2. How many times more resistant were *B. megaterium* spores than *S. aureus* vegetative cells?

   _____

   _____

3. Why was half of each plate covered with an index card?

   _____

   _____

4. What was the purpose of leaving the cover on one set of petri dishes?

   _____

   _____

B. Short-Answer Questions

1. Describe the damaging effects of UV radiation on living cells.

   _____

   _____

2. Why does exposure to UV radiation cause death in vegetative cells but not endospores?

_____

_____

3. At which wavelength is UV radiation most germicidal? Explain.

_____

_____

4. What limited protection do cells have against the damaging effects of UV radiation?

_____

_____

5. What types of damage to human tissues can result from prolonged exposures to UV radiation? What protective measures can be taken to limit these types of damage, both during the experiment and in everyday life?

_____

_____

6. Which *B. megaterium* culture, logarithmic or stationary phase, would show the best survival following exposure to UV radiation? Explain.

_____

_____

The Effects of Lysozyme on Bacterial Cells

Learning Outcomes

After completing this exercise, you should be able to

1. Understand the molecular structure of peptidoglycan in the bacterial cell.

2. Understand how lysozyme and autolysins degrade peptidoglycan to cause lysis of the bacterial cell.

3. Demonstrate how some common bacteria differ in their susceptibility to lysozyme.

Most bacterial cells are surrounded by a cell wall that contains a common and important component called **peptidoglycan,** which is only found in prokaryotes. Peptidoglycan is a polymer of alternating N-acetyl-glucosamine (NAG) and N-acetyl-muramic acid (NAM), which are linked covalently by a β (1–4) bond (figure 29.1a).

Each N-acetyl-muramic acid contains a short chain of four amino acids that serves to cross-link the adjacent polymers of the amino sugars, thus forming a lattice network or sac around the exterior of the bacterial cell that provides rigidity and also defines the shape of the cell (i.e., coccus, rod, or spiral) (figure 29.1b). There is some diversity in the structure of the peptide cross-link. For example, *Staphylococcus aureus* has an interbridge composed of five glycine residues that connects the tetrapeptides of the N-acetyl-muramic acid molecules. Other species of *Staphylococcus* may have both serine and glycine making up the interbridges. *Escherichia coli,* a gram-negative bacterium, lacks an interbridge, and in this case the tetrapeptides are directly linked through a peptide bond between an amino acid in one tetrapeptide and another amino acid in the adjacent chain.

The cell wall of gram-positive bacteria is composed of 90% peptidoglycan. In addition, most

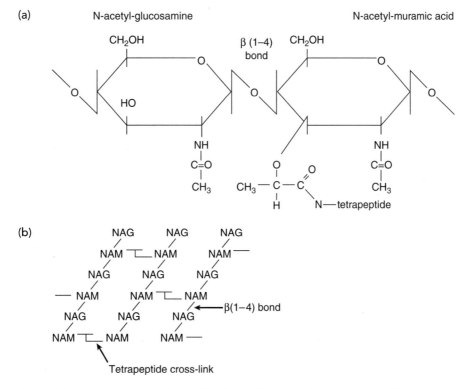

Figure 29.1 **(a) Structure of peptidoglycan showing the β (1–4) bond between N-acetyl-glucosamine (NAG) and N-acetyl-muramic acid (NAM). (b) Structure of peptidoglycan showing the alternating units of NAG and NAM bonded by the β (1–4) bond and cross-linked via the tetrapeptide component.**

gram-positive bacteria have **teichoic acids** in their cell walls. Teichoic acids are polymers of ribitol-phosphate or glycerol-phosphate that can bind covalently to peptidoglycan or to the cytoplasmic membrane. They are responsible for the net negative charge on gram-positive cells. They may function in the expansion of peptidoglycan during cell growth. Bacterial cell enzymes called **autolysins** partially open up the peptidoglycan polymer by breaking β (1–4) bonds between the N-acetyl-muramic acid and N-acetyl-glucosamine molecules so that new monomeric subunits of peptidoglycan can be inserted into the existing cell wall and allow for expansion. This process must be limited because if large gaps in peptidoglycan are produced by the autolysins, the cell membrane can rupture through the gaps. Teichoic acids may play a role in limiting the amount of degradation of peptidoglycan by the autolysins. It is interesting to note that teichoic acids are found in *Staphylococcus* but not in *Micrococcus*.

The cell wall of gram-negative bacteria in contrast to that of gram-positive bacteria is composed of a thin layer of peptidoglycan and an outer membrane that encloses the peptidoglycan. Thus, the peptidoglycan of gram-negative bacteria is not accessible from the external environment. The outer membrane of gram-negative bacteria forms an additional permeability barrier in these organisms. Gram-negative bacteria also lack teichoic acids in their cell walls.

Humans are born with intrinsic non-immune factors that protect them from infection by bacteria. One of these factors is **lysozyme,** which is found in most body fluids such as tears and saliva. Lysozyme, like the autolysins, degrades the β (1–4) bond between the amino sugar molecules in peptidoglycan, thus causing breaks in the lattice and weakening the cell wall. As a result, the solute pressure of the cytoplasm can cause the cell membrane to rupture through these breaks, resulting in cell lysis and death.

The presence of lysozyme in tears, for example, can protect our eyes from infection because bacteria that try to extablish an infection undergo lysis as a result of the lysozyme in tears that washes the surface of the eye. The egg white of hen's eggs also contains lysozyme where it also may play a protective function for the developing embryo.

Gram-negative bacteria are usually more resistant to lysozyme than are gram-positive bacteria because the gram-negative outer membrane prevents lysozyme from reaching the peptidoglycan layer. However, some gram-positive bacteria, such as *Staphylococcus aureus,* are resistant to lysozyme, presumably because of the presence of teichoic acids in their cell walls, which limits the action of autolysins and

lysozyme. In the following exercise you will study the effects of lysozyme on gram-positive and gram-negative bacteria. The source of lysozyme will be human saliva and the egg whites of hen's eggs.

Materials

per class of 30–40
- 2–3 hen's eggs

per pair of students:
- 3 plates of nutrient agar
- small vessels for collecting saliva
- sterile cotton swabs
- small sterile test tubes
- sterile Pasteur pipettes
- pipette bulbs

broth cultures of:
- *Escherichia coli*
- *Micrococcus luteus*
- *Staphylococcus aureus*

1. Using a marking pen, divide each plate into three sectors and label them 1, 2, and 3.
2. Using a sterile cotton swab, uniformly inoculate one plate with *E. coli.* Do this by first swabbing the plate in one direction with a cotton swab dipped in the culture. Then turn the plate 90 degrees and swab at a right angle to the first direction.
3. Repeat this procedure for the second and third plates, using *M. luteus* and *S. aureus.*
4. The instructor will break an egg and separate the white portion from the yolk. Using a sterile Pasteur pipette, collect a small amount of egg white and transfer it to a small test tube.
5. Collect saliva by expelling some into the small vessel provided.
6. Using a sterile Pasteur pipette, transfer one drop of egg white to sector 1 on the plate inoculated with *E. coli.* To sector 2, deliver 1 drop of saliva on this same plate. Sector 3 will serve as a growth control for the organism (figure 29.2).
7. Repeat this same procedure for the remaining plates inoculated with *M. luteus* and *S. aureus.*
8. Allow the liquids to absorb into the agar medium and then incubate the plates at 37°C for 24 hours.
9. In the next laboratory period, observe the plates. If the organism is affected by lysozyme, the cells exposed to the enzyme will undergo lysis, forming a clear area in the sector. Which culture was affected by the enzyme? Were all bacteria affected in the same way? Explain.

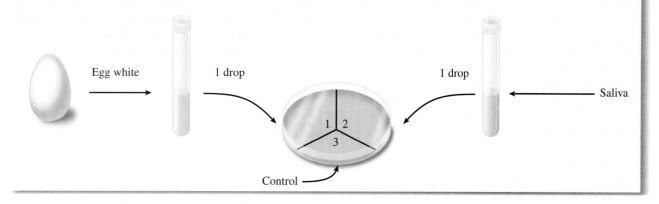

Figure 29.2   **Procedure for setting up the lysozyme plates.**

29

29 The Effects of Lysozyme on Bacterial Cells

A. Results

1. Record whether lysis occurred to the cells on the plate using $+++$ for lysis, $+$ for partial lysis, and $-$ for no lysis.

	SECTOR		
	Egg White	Saliva	Control
	1	2	3
Staphylococcus aureus			
Escherichia coli			
Micrococcus luteus			

2. Why did lysozyme affect the test organisms differently?

_____

_____

3. What was the purpose of including a quadrant to which no lysozyme was added?

_____

_____

B. Short-Answer Questions

1. What is the function of peptidoglycan in bacterial cells?

_____

_____

2. How does lysozyme specifically affect peptidoglycan?

_____

_____

3. Where in nature can lysozyme be found? Why is it produced in these environments?

_____

_____

4. Based on the results, what might one conclude about the types of bacteria that are involved in eye infections?

_____

_____

Evaluation of Alcohol:

Its Effectiveness as an Antiseptic

As an antiseptic, alcohol is widely used to swab the skin before inoculations or the drawing of blood from veins. This is to prevent the introduction of skin bacteria, especially pathogens, into tissue where they could cause disease. The number of bacteria on the skin will vary depending on location and available moisture. For example, approximately 1000 organisms per square centimeter may be present on the back, whereas greater than 10 million per square centimeter can be found in the moist armpits and groin.

Ethyl alcohol and isopropyl alcohol are effective antiseptics for preventing inoculation of skin bacteria but only when used in aqueous solutions of from 60% to 80%. These concentrations specifically kill bacterial cells by denaturing proteins and damaging cell membranes. Proteins are more soluble in aqueous solutions of alcohols, and hence they denature more readily when water is present. Alcohols are not, however, effective against endospores and some naked viruses (i.e., viruses that lack a lipid envelope). An advantage of alcohols is that they are relatively nontoxic to the human skin, but a disadvantage is that they are volatile and can lose their effectiveness due to evaporation.

An important question concerning alcohols is: How effective is alcohol in routine use? When the skin is swabbed prior to penetration, are all or mostly all of the surface skin bacteria killed? To determine alcohol effectiveness, as it might be used in routine skin disinfection, we are going to perform a very simple experiment that utilizes four thumbprints and a plate of enriched agar. Class results will be pooled to arrive at a statistical analysis.

Figure 30.1 illustrates the various steps in this test. Note that the petri plate is divided into four quadrants. On the left side of the plate, an unwashed left thumb is first pressed down on the agar in the lower-left quadrant of the plate. Next, the left thumb is pressed down on the upper-left quadrant. With the left thumb, we are trying to establish the percentage of bacteria that are removed by simple contact with the agar.

On the right side of the plate, an unwashed right thumb is pressed down on the lower-right quadrant of the plate. The next step is to either dip the right thumb into alcohol or to scrub it with an alcohol swab and dry it. Half of the class will use the dipping method and the other half will use alcohol swabs. Your instructor will indicate what your assignment

❷
Without touching any other surface the left thumb is pressed against the agar in quadrant B.

❻
The pad of the treated right thumb is pressed against the agar in the D quadrant.

❺
The alcohol-treated right thumb is allowed to completely air-dry.

❹
The pad of the right thumb is immersed in 70% alcohol or scrubbed with an alcohol swab for 10 seconds.

❶
The pad of the unwashed left thumb is momentarily pressed against the agar in quadrant A.

❸
The pad of the unwashed right thumb is momentarily pressed against the agar in quadrant C.

Figure 30.1 **Procedure for testing the effectiveness of alcohol on the skin.**

will be. The last step is to press the dried right thumb on the upper-right quadrant of the plate.

After the plate is inoculated, it is incubated at 37°C for 24 to 48 hours. Colony counts will establish the effectiveness of the alcohol.

Materials

- 1 petri plate of veal infusion agar
- small beaker
- 70% ethanol
- alcohol swab

1. Perform this experiment with unwashed hands.
2. With a marking pencil, mark the bottom of the petri plate with two perpendicular lines that divide it into four quadrants. Label the left quadrants **A** and **B** and the right quadrants **C** and **D**, as shown in figure 30.1. *(Keep in mind that when you turn the plates over to label them, the A and B quadrants will be on the right and C and D will be on the left.)*

3. Press the pad of your left thumb against the agar surface in the A quadrant.
4. Without touching any other surface, press the left thumb into the B quadrant.
5. Press the pad of your right thumb against the agar surface of the C quadrant.
6. Disinfect the right thumb by one of the two following methods:
   - dip the thumb into a beaker of 70% ethanol for 5 seconds, or
   - scrub the entire pad surface of the right thumb with an alcohol swab.
7. Allow the alcohol to completely evaporate from the skin.
8. Press the right thumb against the agar in the D quadrant.
9. Incubate the plate at 37°C for 24 to 48 hours.
10. Follow the instructions in Laboratory Report 30 for evaluating the plate and answer all of the questions.

30 Evaluation of Alcohol: Its Effectiveness as an Antiseptic

A. Results

1. Count the number of colonies that appear on each of the thumbprints and record them in the following table. If the number of colonies has increased in the second press, record a 0 in percent reduction. Calculate the percentages of reduction and record these data in the appropriate column. Use this formula:

$$\text{Percent reduction} = \frac{(\text{Colony count 1st press}) - (\text{Colony count 2nd press})}{(\text{Colony count 1st press})} \times 100$$

LEFT THUMB (Control)			RIGHT THUMB (Dipped)			RIGHT THUMB (Swabbed)		
Colony Count 1st Press	Colony Count 2nd Press	Percent Reduction	Colony Count 1st Press	Colony Count 2nd Press	Percent Reduction	Colony Count 1st Press	Colony Count 2nd Press	Percent Reduction
Av. % Reduction, Left (C)			Av. % Reduction, Right (D)			Av. % Reduction, Right (S)		

2. In general, what effect does alcohol have on the level of skin contaminants? _____

_____

3. Is there any difference between the effects of dipping versus swabbing? _____

Which method appears to be more effective? _____

4. There is definitely survival of some microorganisms even after alcohol treatment. Without staining or microscopic scrutiny, predict what types of microbes are growing on the medium where you made the right thumb impression after treatment. _____

_____

_____

_____

_____

B. Short-Answer Questions (see Exercise 32 for a definition of *antiseptic*)

1. For what purposes is alcohol a useful antiseptic?

_____

_____

2. What advantages does alcohol have over hand soap for antisepsis of the skin?

_____

_____

3. Why does treatment of human skin with alcohol not create a completely sterile environment?

_____

_____

Antimicrobic Sensitivity Testing:
The Kirby-Bauer Method

Learning Outcomes

After completing this exercise, you should be able to

1. Define the terms *antimicrobial, antibiotic, semi-synthetic,* and *synthetic* as they relate to agents used to treat bacterial infections.

2. Demonstrate the effectiveness of antimicrobials using the Kirby-Bauer method.

Antimicrobials and antibiotics have been our first line of defense in the battle to conquer infectious diseases. Diseases caused by *Staphylococcus, Streptococcus, Pseudomonas,* and *Mycobacterium* that once were fatal to humans became manageable and curable with the discovery of antibiotics. Alexander Fleming's observation in 1929 that a species of *Penicillium* growing on an agar plate with *Staphylococcus aureus* inhibited the bacterium led to the subsequent purification of penicillin by Howard Florey, Norman Heatley, and Ernst Chain and the development of penicillin as the first antibiotic to be used in the clinical treatment of bacterial infections. Other antibiotics were subsequently found and were added to the list that could be used to treat bacterial diseases. Prior to the discovery of antibiotics, very few options were available to treat infectious agents except for some antimicrobial chemical compounds such as salvarsan, an arsenic compound, and the sulfa drugs, which were used with some success. Today we have a vast number of compounds that are effective against bacterial, fungal, and some viral agents. By definition, **antimicrobials** are compounds that kill or inhibit microorganisms. **Antibiotics** are antimicrobials, usually of low molecular weight, produced by microorganisms that inhibit or kill other microorganisms. Two common examples are penicillin and streptomycin. Many times antibiotics are chemically altered to make them more effective in their mode of action. These are referred to as **semi-synthetics.** Some antimicrobials are chemically synthesized in the laboratory and are not produced by microbial biosynthesis. The sulfa drugs are **synthetics** that were used to treat some bacterial diseases before the discovery of penicillin.

Our perceived defeat of the infectious microbe, however, has been short-lived. An increasing problem in medicine at the present time is the development of antibiotic-resistant strains of bacteria such as *Staphylococcus aureus, Pseudomonas aeruginosa, Mycobacterium tuberculosis, Neisseria gonorrhoeae,* and *Enterococcus* that do not respond to the antibiotics that have been normally used to treat these bacteria. *Staphylococcus aureus* has become especially problematic because both hospital and community strains of MRSA (MRSA: methicillin-resistant *S. aureus*) have been isolated that do not respond to penicillin, and hence other and more expensive drugs such as vancomycin must be used to manage infections caused by these organisms. As might be expected, strains of *Staphylococcus* and *Enterococcus* have also developed resistance to vancomycin. In hospitals MRSA is responsible for many **health-care acquired infections (HAIs),** formerly known as nosocomial infections because they were acquired in hospitals. In 2002 the Centers for Disease Control (CDC) estimated that there were 1.7 million HAIs in the United States. Penicillin has long been the drug of choice to treat gonorrhea. However, penicillin-resistant *N. gonorrhoeae* strains have been isolated that require more expensive drugs to treat this sexually transmitted disease. Strains of *Mycobacterium tuberculosis* have been isolated from street people and drug addicts that do not respond to any antimicrobials available. Needless to say, antibiotic resistance is a major public health problem affecting the treatment and spread of disease.

When a patient seeks medical assistance for a bacterial infection, the physician will normally take a conservative approach and prescribe an antimicrobial without isolating the pathogen. The physician bases the choice of antimicrobial on symptoms presented by the patient, probable causative agents associated with the symptoms, and drugs known to be effective against these pathogens. Usually this approach is sufficient to treat the infection, but occasionally a pathogen will not respond to the prescribed antimicrobial. It is then crucial that the causative agent be isolated in pure culture, specifically identified, and tested for its sensitivity and resistance to various antimicrobial agents. Antimicrobial agents can vary in their effectiveness against various pathogenic bacteria. Some antimicrobial agents are narrow in their spectrum and may be more effective against gram-positive bacteria, while others are more effective against gram-negative bacteria. Broad-spectrum antimicrobials are effective against both kinds of organisms. Based on test results, an antimicrobial that the pathogen is sensitive to would then be prescribed to treat the infection.

Whether an antimicrobial has a broad or narrow spectrum depends appreciably on its mode of action and its ability to be transported into the cell. Antimicrobials have different modes of action, and they affect different aspects of bacterial cell metabolism. They can target cell wall synthesis (penicillin), DNA and RNA synthesis (cipro, rifampin), protein synthesis (tetracyclines, streptomycin), and vitamin synthesis (sulfa drugs). For example, antibiotics that target 70S ribosomes tend to be broad spectrum because all bacteria have 70S ribosomes, which are necessary for protein synthesis. Sulfa drugs are more narrow in spectrum because not all bacterial cells can synthesize the vitamin folic acid, whose synthesis is inhibited by this drug. Another factor that affects the ability of an antibiotic to function is permeability. The outer membrane of gram-negative bacteria acts as a permeability barrier and can restrict the entry of antimicrobials into the cell.

Table 31.1 summarizes the modes of action of selected antimicrobials.

The **Kirby-Bauer method** (figure 31.1) is used to determine the sensitivity or resistance of a bacterium to an antimicrobial. This is a standardized test procedure that is reliable, relatively simple, and yields results in as short a time as possible. It is performed by uniformly streaking a standardized inoculum of the test organism on Mueller-Hinton medium, and then paper disks containing specific concentrations of an antimicrobial or antibiotic are deposited on the agar surface. The antimicrobial diffuses out from the disk into the agar, forming a concentration gradient. If the agent inhibits or kills the test organism, there will be a zone around the disk where no growth occurs called the **zone of inhibition** (figure 31.2). This zone can vary, however, with the diffusibility of the agent, the size of the inoculum, the type of medium, and other factors. All of these factors were taken into consideration in developing this test. The Kirby-Bauer method is sanctioned by the U.S. FDA and the Subcommittee on Antimicrobial Susceptibility Testing of the National Committee for Clinical Laboratory Standards. Although time is insufficient here to consider all facets of this test, its basic procedure will be followed to test both antibiotics and antimicrobials.

The recommended medium in this test is Mueller-Hinton II agar. Its pH should be between 7.2 and 7.4, and it should be poured to a uniform thickness of 4 mm in the petri plate. This requires 60 ml in a 150 mm plate and 25 ml in a 100 mm plate. For certain fastidious microorganisms, 5% defibrinated sheep's blood is added to the medium.

Inoculation of the surface of the medium is made with a cotton swab from a broth culture. In clinical applications, the broth turbidity has to match a defined standard. Care must also be taken to express excess broth from the swab prior to inoculation.

High-potency disks are used that may be placed on the agar with a mechanical dispenser or sterile forceps. To secure the disks to the medium, it is necessary to press them down onto the agar.

Table 31.1   Modes of Action of Antibiotics

ANTIBIOTIC	CELLS AFFECTED	CELL TARGET/SPECIFIC SITE
Penicillin	G+; G-	Cell wall/β-lactamase, peptidoglycan synthesis—amino acid side chain
Vancomycin	G+	Cell wall/Peptidoglycan synthesis
Bacitracin	G+	Cell wall/Transport of peptidoglycan monomer
Isoniazid	*Mycobacterium tuberculosis*	Cell wall/Mycolic acid synthesis in *Mycobacterium*
Fluroquinolones	G+; G-	DNA/Topoisomerase unwinding of DNA in DNA synthesis
Rifamycins	G+; some G-	RNA/RNA polymerase in RNA synthesis
Tetracyclines	G+; G-	Protein synthesis/30S subunit of 70S ribosome
Streptomycin	G+; G-	Protein synthesis/30S subunit of 70S ribosome
Chloramphenicol	G+; G-	Protein synthesis/50S subunit of 70S ribosome
Sulfa drugs	G+; G-	Structural analogue of para-amino benzoic acid (PABA)—inhibit enzyme linking pteridine to PABA in folic acid synthesis

(1) The entire surface of a plate of nutrient medium is swabbed with organism to be tested.

(2) Handle of dispenser is pushed down to place 12 disks on the medium. In addition to dispensing disks, this dispenser also tamps disks onto medium.

(3) Individual cartridges (Difco) can be used to dispense single disks. Only 4 or 5 disks should be placed on small (100 mm) plates.

(4) After 18 hours of incubation, the zones of inhibition (diameters) are measured in millimeters. Significance of zones is determined from Kirby-Bauer chart.

Figure 31.1 **Antimicrobic sensitivity testing.**

After 16 to 18 hours of incubation, the plates are examined and the diameters of the zones of inhibition are measured to the nearest millimeter. Diameters for tested bacteria are compared to those in a table that are based on values obtained for ATCC (American Type Culture Collection) reference cultures that are sensitive to the specific antibiotic. Cultures are designated as resistant, sensitive, or intermediate in their response to the antibiotic. These designations

are derived by comparison to the response of the reference culture. (To determine the significance of the zone diameters, consult table 31.2.)

In this exercise, we will work with four microorganisms: *Staphylococcus aureus, Escherichia coli, Proteus vulgaris,* and *Pseudomonas aeruginosa.* Each student will inoculate one plate with one of the four organisms and place the disks on the medium by whichever method is available. Since each student

Figure 31.2 Kirby-Bauer plates for *Pseudomonas aeruginosa*, top left; *Staphylococcus aureus*, top right; *Escherichia coli*, bottom left; and *Proteus vugaris*, bottom right tested with: S - streptomycin; GM - gentamicin; C - chloramphenicol; Va - vancomycin; T - tetracycline.

© McGraw-Hill Education. Lisa Burgess, photographer

will be doing only a portion of the total experiment, student assignments will be made. Proceed as follows:

🕐 First Period

(Plate Preparation)

Materials

- 1 petri plate of Mueller-Hinton II agar
- nutrient broth cultures (with swabs) of *S. aureus*, *E. coli*, *P. vulgaris*, and *P. aeruginosa*
- disk dispenser (BBL or Difco)
- cartridges of disks (BBL or Difco)
- forceps and Bunsen burner
- zone interpretation charts (Difco or BBL)
- metric ruler

1. Select the organisms you are going to work with from the following table.

ORGANISM	STUDENT NUMBER
S. aureus	1, 5, 9, 13, 17, 21, 25
E. coli	2, 6, 10, 14, 18, 22, 26
P. vulgaris	3, 7, 11, 15, 19, 23, 27
P. aeruginosa	4, 8, 12, 16, 20, 24, 28

Table 31.2 Zones of Inhibition in the Kirby-Bauer Method of Antimicrobic Sensitivity Testing

ANTIBIOTIC	CODE	POTENCY	Zone of Inhibition (mm)		
			RESISTANT	INTERMEDIATE	SENSITIVE
Amikacin	AN-30	30 μg			
Enterobacteriaceae					
P. aeruginosa, Acinetobacter					
staphylococci			≤14	15–16	≥17
Amoxicillin/Clavulinic acid	AmC-30	20/10 μg			
Enterobacteriaceae			≤13	14–17	≥18
Staphylococcus spp.			≤19	—	≥20
Haemophilus spp.			≤19	—	≥20
Ampicillin	AM-10	10 μg			
Enterobacteriaceae			≤13	14–16	≥17
Staphylococcus spp.			≤28	—	≥29
Enterococcus spp.			≤16	—	≥17
Listeria monocytogenes			≤19	—	≥20
Haemophilus spp.			≤18	19–21	≥22
β-hemolytic streptococci			—	—	≥24
Aziocillin	AZ-75	75 μg			
P. aeruginosa			≤17	—	≥18
Bacitracin	B-10	10 units	≤8	9–12	≥13

Table 31.2   Zones of Inhibition in the Kirby-Bauer Method of Antimicrobic Sensitivity Testing (continued)

ANTIBIOTIC	CODE	POTENCY	Zone of Inhibition (mm)		
			RESISTANT	INTERMEDIATE	SENSITIVE
Carbenicillin	CB-100	100 μg			
Enterobacteriaceae and					
Acinetobacter			≤19	20–22	≥22
P. aeruginosa			≤13	14–16	≥17
Cefaclor	CEC-30	30 μg			
Enterobacteriaceae and					
staphylococci			≤14	15–17	≥18
Haemophilus spp.			≤16	17–19	≥20
Cefazolin	CZ-30	30 μg			
Enterobacteriaceae and					
staphylococci			≤14	15–17	≥18
Cephalothin	CF-30	30 μg			
Enterobacteriaceae and					
staphylococci			≤14	15–17	≥18
Chloramphenicol	C-30	30 μg			
Enterobacteriaceae,					
P. aeruginosa, Acinetobactor,					
staphylococci, enterococci,					
and *V. cholerae*			≤12	13–17	≥18
Haemophilus spp.			≤25	26–28	≥29
S. pneumoniae			≤20	—	≥21
Streptococci			≤17	18–20	≥21
Ciprofloxacin	CIP-5	5 μg			
Enterobacteriaceae,					
P. aeruginosa, Acinetobacter,					
staphylococci, and enterococci			≤15	16–20	≥21
Haemophilus spp.			—	—	≥21
N. gonorrhoeae			≤27	28–40	≥41
Clarithromycin	CLR-15	15 μg			
Staphylococcus spp.			≤13	14–17	≥18
Haemophilus spp.			≤10	11–12	≥13
S. pneumoniae and					
other streptococci			≤16	17–20	≥21
Clindamycin	CC-2	2 μg			
Staphylococcus spp.			≤14	15–20	≥21
S. pneumoniae and					
other streptococci			≤15	16–18	≥19

(continued)

Table 31.2   Zones of Inhibition in the Kirby-Bauer Method of Antimicrobic Sensitivity Testing (continued)

ANTIBIOTIC	CODE	POTENCY	Zone of Inhibition (mm)		
			RESISTANT	INTERMEDIATE	SENSITIVE
Doxycycline	D-30	30 µg			
Enterobacteriaceae,					
P. aeruginosa, Acinetobactor,					
staphylococci, and enterococci			≤12	13–15	≥16
Erythromycin	E-15	15 µg			
Staphylococcus spp. and					
enterococci			≤13	14–22	≥23
S. pneumoniae and					
other streptococci			≤15	16–20	≥21
Gentamicin	GM-120	120 µg			
Enterobacteriaceae,					
P. aeruginosa, Acinetobacter,					
and staphylococci			≤12	13–14	≥15
Imipenem	IPM-10	10 µg			
Enterobacteriaceae,					
P. aeruginosa, Acinetobacter,					
and staphylococci			≤13	14–15	≥16
Haemophilus spp.			—	—	≥16
Kanamycin	K-30	30 µm			
Enterobacteriaceae and					
staphylococci			≤13	13–17	≥18
Lomefloxacin	LOM-10	10 µg			
Enterobacteriaceae,					
P. aeruginosa, Acinetobacter,					
and staphylococci			≤18	19–21	≥22
Haemophilus spp.			—	—	≥22
N. gonorrhoeae			≤26	27–37	≥38
Loracarbef	LOR-30	30 µg			
Enterobacteriaceae and					
staphylococci			≤14	15–17	≥18
Haemophilus spp.			≤15	16–18	≥19
Meziocillin	MZ-75	75 µg			
Enterobacteriaceae and					
Acinetobacter			≤17	18–20	≥21
P. aeruginosa			≤15	—	≥16

Table 31.2 Zones of Inhibition in the Kirby-Bauer Method of Antimicrobic Sensitivity Testing (continued)

ANTIBIOTIC	CODE	POTENCY	Zone of Inhibition (mm)		
			RESISTANT	INTERMEDIATE	SENSITIVE
Minocycline	MI-30	30 μg			
Enterobacteriaceae,					
P. aeruginosa, Acinetobacter,					
staphylococci, and enterococci			≤14	15–18	≥19
Moxalactam	MOX-30	30 μg			
Enterobacteriaceae,					
P. aeruginosa, Acinetobacter,					
and staphylococci			≤14	15–22	≥23
Nafcillin	NF-1	1 μg			
Staphylococcus aureus			≤10	11–12	≥13
Nalidixic Acid	NA-30	30 μg			
Enterobacteriaceae			≤13	14–18	≥19
Neomycin	N-30	30 μg	≤12	13–16	≥17
Netilmicin	NET-30	30 μg			
Enterobacteriaceae,					
P. aeruginosa, Acinetobacter,					
and staphylococci			≤12	13–14	≥15
Norfloxacin	NCR-10	10 μg			
Enterobacteriaceae,					
P. aeruginosa, Acinetobacter,					
staphylococci, and entercocci			≤12	13–16	≥17
Novobiocin	NB-30	30 μg	≤17	18–21	≥22
Oxacillin	OX-1	1 μg			
Staphylococcus aureus			≤10	11–12	≥13
staphylococcus (coagulase negative)			≤17	—	≥18
Penicillin	P-10	10 units			
Staphylococcus spp.			≤28	—	≥29
Enterococcus spp.			≤14	—	≥15
L. monocytogenes			≤19	20–27	≥28
N. gonorrhoeae			≤26	27–46	≥47
β-hemolytic streptococci			—	—	≥24
Piperacillin	PIP-100	100 μg			
Enterobacteriaceae, and					
Acinetobacter			≤17	18–20	≥21
P. aeruginosa			≤17	—	≥18

(continued)

Table 31.2 Zones of Inhibition in the Kirby-Bauer Method of Antimicrobic Sensitivity Testing (continued)

ANTIBIOTIC	CODE	POTENCY	Zone of Inhibition (mm)		
			RESISTANT	INTERMEDIATE	SENSITIVE
Polymyxin B	PB-300	300 U	≤8	9–11	≥12
Rifampin	RA-5	5 μg			
Staphylococcus spp.					
Enterococcus spp. and *Haemophilis* spp.			≤16	17–19	≥20
S. pneumoniae			≤16	17–18	≥19
Spectinomycin	SPT-100	100 μg			
N. gonorrhoeae			≤14	15–17	≥18
Streptomycin	S-300	300 μg			
Enterobacteriaceae			≤11	12–14	≥15
Sulfisoxazole	G-25	250 μg			
Enterobacteriaceae,					
P. aeruginosa, Acinetobacter,					
V. cholerae, and staphylococci			≤12	13–16	≥17
Tetracycline	Te-30	30 μm			
Enterobacteriaceae,					
P. aeruginosa, Acinetobacter,					
V. cholerae, staphylococci, and					
enterococci			≤14	15–18	≥19
Haemophilus spp.			≤25	26–28	≥29
N. gonorrhoeae			≤30	31–37	≥38
S. pneumoniae and other streptococci			≤18	19–22	≥23
Tobramycin	NN-10	10 μg			
Enterobacteriaceae,					
P. aeruginosa, Acinetobacter,					
and staphylococci			≤12	13–14	≥15
Trimethoprim	TMP-5	5 μg			
Enterobacteriaceae and					
staphylococci			≤10	11–15	≥16
Vancomycin	Va-30	30 μg			
Staphylococcus spp.			—	—	≥15
Enterococcus spp.			≤14	15–16	≥17
S. pneumoniae and other streptococci			—	—	≥17

Courtesy and © Becton, Dickinson and Company

2. Label your plate with the name of your organism.

3. Inoculate the surface of the medium with the swab after expressing excess fluid from the swab by pressing and rotating the swab against the inside walls of the tube above the fluid level. Cover the surface of the agar evenly by swabbing in three directions. A final sweep should be made of the agar rim with the swab.

4. Allow **3 to 5 minutes** for the agar surface to dry before applying disks.

5. Dispense disks as follows:

   a. If an automatic dispenser is used, remove the lid from the plate, place the dispenser over the plate, and push down firmly on the plunger. With the sterile tip of forceps, tap each disk lightly to secure it to medium.

   b. If forceps are used, sterilize them first by flaming before picking up the disks. Keep each disk at least 15 mm from the edge of the plate. Place no more than 13 on a 150 mm plate, no more than 5 on a 100 mm plate. Apply light pressure to each disk on the agar with the tip of a sterile forceps or inoculating loop to secure it to the medium (see figure 31.1).

6. Invert and incubate the plate for 16 to 18 hours at 37°C.

⏱ Second Period

(Interpretation)

After incubation, measure the zone diameters with a metric ruler to the nearest whole millimeter (figure 31.3). The disk is included in the measurement. The zone of complete inhibition is determined without magnification. Ignore faint growth or tiny colonies that can be detected by very close scrutiny. Large colonies growing within the clear zone might represent resistant variants or a mixed inoculum and may require reidentification and retesting in clinical situations. Ignore the "swarming" characteristics of *Proteus,* measuring only to the margin of heavy growth.

Record the zone measurements on the table of Laboratory Report 31 and on the chart on the demonstration table, which has been provided by the instructor.

Use table 31.2 for identifying the various disks.

To determine which antibiotics your organism is sensitive to (S), or resistant to (R), or intermediate (I), consult table 31.2. It is important to note that the significance of a zone of inhibition varies with the type of organism. If you cannot find your antibiotic on the chart, consult a chart that is supplied on the demonstration table or bulletin board. Table 31.2 is incomplete.

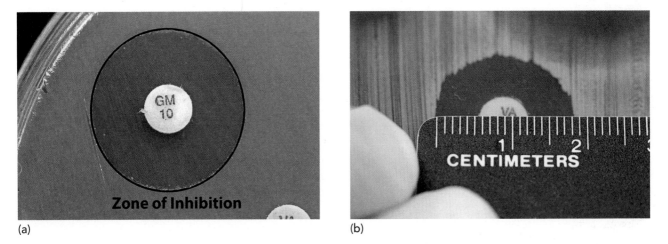

(a)   (b)

Figure 31.3   (a) **Zone of inhibition for gentamicin on *Pseudomonas aeruginosa*. (b) Measuring the zone of inhibition.**
© McGraw-Hill Education. Lisa Burgess, photographer

31 Antimicrobic Sensitivity Testing: The Kirby-Bauer Method

A. Results

1. List the antimicrobics that were used for each organism. Consult table 31.2 to identify the various disks. After measuring and recording the zone diameters, consult table 31.2 for interpretation. Record the degrees of sensitivity (R, I, or S) in the sensitivity column. Exchange data with other class members to complete the entire chart.

	ANTIMICROBIC	ZONE DIA.	RATING (R, I, S)	ANTIMICROBIC	ZONE DIA.	RATING (R, I, S)
S. aureus						
P. aeruginosa						
Proteus vulgaris						
E. coli						

2. Which antimicrobics would be suitable for the control of the following organisms?

   S. aureus: _____

   E. coli: _____

   P. vulgaris: _____

   P. aeruginosa: _____

B. Short-Answer Questions

1. Differentiate between the following and provide one example of each:

   a. antibiotics and antimicrobial drugs

   _____

   b. broad- and narrow-spectrum antibiotics

   _____

2. What factors influence the size of the zone of inhibition for an antibiotic?

   _____

   _____

3. Why are certain gram-negative bacteria more resistant than gram-positive bacteria to antibiotics that attack cytoplasmic targets?

   _____

   _____

4. Why are gram-positive bacteria typically more resistant than gram-negative bacteria to antibiotics that disrupt plasma membranes, such as polymyxin B?

   _____

   _____

5. If a bacterial isolate shows intermediate to moderate resistance to an antibiotic, how might this antibiotic still be successfully used in the treatment of this microbe?

   _____

   _____

6. What specific medium must be used in testing the effectiveness of antibiotics?

   _____

   _____

Evaluation of Antiseptics:
The Filter Paper Disk Method

Learning Outcomes

After completing this exercise, you should be able to

1. Define the terms *antiseptic, disinfectant, sterilant, sporocide, sanitizer, bacteriostatic,* and *bacteriocidal.*

2. Demonstrate the relative effectiveness of common antiseptics and disinfectants such as bleach, hydrogen peroxide, and mouthwash using the paper disk method.

Every day, we use a number of different chemical agents to control or kill microorganisms. Hospitals disinfect contaminated areas with strong oxidizing agents such as sodium hypochlorite, which is identical to common household bleach. Alcohol is applied by a medical professional to the skin before giving an injection, and betadine—an organic form of iodine—is swabbed onto the skin before surgery to ensure that skin bacteria such as *Staphylococcus aureus* will not cause postsurgical infections. Cities and municipalities add chlorine to water supplies to prevent the spread of potential pathogens such as *Salmonella typhi* in the drinking water. Chemical agents are added to food to retard spoilage by microbes, to increase the shelf life of food products in the supermarket, and to prevent the growth of pathogenic bacteria such as *Clostridium botulinum,* which causes a deadly form of food poisoning.

Chemical agents that are used to control microorganisms can be defined as antiseptics or disinfectants. **Antiseptics** are substances such as alcohol or betadine that inhibit microbial growth or kill microorganisms and are gentle enough to be applied to living tissue. However, these compounds do not destroy endospores. They are used for hand washing, treating surface wounds on the skin, and preparing the skin for invasive procedures such as incisions and inoculations to prevent infections by opportunistic skin bacteria. **Disinfectants** are chemical agents that are applied to inanimate objects such as floors, walls, and tabletops to kill microorganisms. They are usually more harsh than antiseptics and are therefore damaging to living tissue. Some disinfectants are classified as **sterilants** or **sporocides,** which means that they destroy all microbial life, including endospores. An example is ethylene oxide, a gas that is used to sterilize heat-sensitive objects such as plastic petri plates, plastic pipettes, plastic syringes, and some surgical instruments that cannot be exposed to high temperatures. Agents that reduce microbial numbers to a safe level but do not completely eliminate all microbes are defined as **sanitizers.** These agents are used in the food industry to treat cooking equipment such as dishes and utensils. If a particular agent only inhibits the growth of bacterial cells but does not kill them, we say that the agent is **bacteriostatic.** Were the agent to be removed, growth of the bacterial cells would resume. In contrast, agents that kill bacterial cells are termed **bacteriocidal.** Agents can also be specific for certain groups of microorganisms; thus, there are bactericides, fungicides, and viricides, and their use would depend on which group needed to be controlled. Some antiseptics are classified as drugs and are therefore regulated by the Food and Drug Administration.

It is difficult to compare the relative effectiveness of disinfectants and antiseptics because they can have vastly different modes of action and chemical properties such as solubility, which can influence the effectiveness of the agent under test conditions. However, by modifying the method used in the Kirby-Bauer technique for evaluating antibiotics, we can use this procedure to determine if an antiseptic or disinfectant inhibits the growth of a test organism and compare the relative efficacy of one agent against another. In this procedure, a plate of suitable medium is streaked with the test organism. Filter paper disks, the same size as antibiotic disks, are impregnated with the agent by dipping the disk in a germicide or disinfectant and placing the disk on an inoculated plate. The plate is incubated for 48 hours. The agent will diffuse from the disk into the agar, forming a concentration gradient. If the substance is inhibitory, a clear zone of inhibition will surround the disk where no growth has occurred. The size of the zone can be used to quantitatively compare one agent's effectiveness against other chemical substances.

In this exercise, you will measure the relative effectiveness of various disinfectants and antiseptics against *Staphylococcus aureus,* a gram-positive bacterium, and *Pseudomonas aeruginosa,* a gram-negative bacterium. Suggested chemical agents to be tested are: bleach (1:10 dilution), 5% phenol, 5% aqueous iodine, betadine, commercial mouthwashes, amphyl, and 5% formaldehyde.

Table 32.1   **Student Assignments**

CHEMICAL AGENT		STUDENT NUMBER	
		S. aureus	*P. aeruginosa*
3% Hydrogen peroxide		1, 15, 29	2, 16, 30
5% Lysol		3, 17, 31	4, 18, 32
5% Iodine		5, 19, 33	6, 20, 34
1:10 Bleach		7, 21	8, 22
Mouthwash		9, 23	10, 24
Betadine		11, 25	12, 26
Amphyl		13, 27	14, 28

🕐 First Period

(Disk Application)

Materials

per student:

- 1 nutrient agar pour and 1 petri plate per organism
- broth culture of *S. aureus* and *P. aeruginosa on* demonstration table
- petri dish containing sterile disks of filter paper ($\frac{1}{2}$″ diameter)
- forceps and Bunsen burner
- chemical agents in small beakers (3% hydrogen peroxide, 5% Lysol, 5% aqueous iodine, 1:10 bleach, mouthwash, betadine, amphyl)

1. Consult table 32.1 to determine your assignment.
2. Liquefy a nutrient agar pour in a water bath and cool to 50°C.
3. Label the bottom of a petri plate with the names of the organism and chemical agent.
4. Inoculate the agar pour with a loopful of *S. aureus* and aseptically pour it into a sterile petri plate, allowing the medium to solidify. (Alternatively,

a plate of solidified medium can be uniformly streaked with the test organism.)

5. Sterilize the forceps by dipping it in alcohol and flaming (*careful: flames and alcohol*). Using the flamed forceps, pick up a sterile filter paper disk and dip it halfway into a disinfectant. Place the disk on the inoculated plate and gently press the disk onto the agar surface. Place no more than four disks on a 150 mm plate.
6. Repeat the same procedure for *Pseudomonas aeruginosa*.
7. Incubate the plates for 48 hours at 37°C.

🕐 Second Period

(Evaluation)

1. Measure the zone of inhibition from the edge of the disk to the edge of the growth for the test organism (see illustration 5, figure 32.1 and figure 31.3b).
2. Evaluate the antiseptics and disinfectants based on the zones of inhibition for each agent. Complete the table in Laboratory Report 32.

(1) Liquefied nutrient agar is inoculated with one loopful of organisms.

(2) Seeded nutrient agar is poured into plate and allowed to solidify.

(3) Sterile disk is dipped halfway into agent. If completely submerged it will be too wet.

(4) Impregnated disk is placed in center of nutrient agar and pressed down lightly to secure it.

(5) After 24 to 48 hours of incubation the zone of inhibition is measured on bottom of plate. Note that measurement is between disk edge and growth.

Figure 32.1   **Filter paper disk method of evaluating an antiseptic.**

32 Evaluation of Antiseptics: The Filter Paper Disk Method

A. Results

1. With a millimeter scale, measure the zones of inhibition between the edge of the filter paper disk and the organisms. Record this information. Exchange your plates with other students' plates to complete the measurements for all chemical agents.

DISINFECTANT/ANTISEPTIC	MILLIMETERS OF INHIBITION	
	Staphylococcus aureus	*Pseudomonas aeruginosa*
3% Hydrogen peroxide		
5% Lysol		
5% Iodine		
1:10 Bleach		
Mouthwash		
Betadine		
Amphyl		

2. Which chemical was the most effective for inhibiting the growth of *S. aureus?* Of *P. aeruginosa?*

   _____

   _____

3. Which chemical was the least effective for inhibiting the growth of *S. aureus?* Of *P. aeruginosa?*

   _____

   _____

4. What do your results indicate about the relative chemical resistances of these two species?

   _____

   _____

B. Short-Answer Questions

1. Differentiate between antiseptic and disinfectant. Include examples of each in your answer. Indicate whether any chemicals can be used as both.

   _____

   _____

2. What factors influence the size of the zone of inhibition produced by a chemical?

   _____

   _____

3. How might the physical differences between gram-positive and gram-negative bacteria contribute to differences in chemical resistances?

   _____

   _____

Effectiveness of Hand Scrubbing

Learning Outcomes

After completing this exercise, you should be able to

1. Define the microorganisms that normally occur on the human skin.

2. Understand how these organisms may lead to health-care-associated infections.

3. Demonstrate how scrubbing with soap will diminish some of the bacteria on the human skin.

The importance of hand washing in preventing the spread of disease is accredited to the observations of Ignaz Semmelweis at the Lying-In Hospital in Vienna in 1846 and 1847. Semmelweis was the head of obstetrics, and he noted that the number of cases of childbirth fever (puerperal sepsis) was primarily the result of the lack of sanitary practices. He observed that medical students and physicians would go directly from dissection and autopsy rooms to a patient's bedside and assist in deliveries without washing their hands. In these wards, the death rate from childbirth fever was very high, approaching 20% in some cases. In contrast, other women were attended to by midwives and nurses who were not allowed in autopsy rooms and who were more sanitary in handling patients. When women were assisted in deliveries by midwives and nurses, the death rate was considerably less. As a result of his observations, Semmelweis instituted a policy whereby physicians and medical students had to disinfect their hands with a solution of chloride of lime (bleach) prior to examining obstetric patients or assisting in deliveries. His efforts resulted in a significant decrease of puerperal sepsis, down to about 1% for women treated by physicians and medical students. But this success was short lived, as complaints by doctors and medical students to the director of the hospital forced Semmelweis to abandon this practice, and death rates for childbirth fever once again increased. Despondent, Semmelweis resigned his post and returned to his native Hungary.

Today, it is routine practice to wash one's hands prior to the examination of a patient and to do a complete surgical scrub prior to surgery. Scrubbing is the most important way to prevent infections in hospitals and physician's offices, and is an effective way to remove some opportunistic pathogens that occur on the skin, such as *Staphylococcus aureus.* The failure of medical professionals to wash their hands before examining patients has resulted in a serious increase in health-care-associated infections in patients, especially in hospitals. Hand washing is also important in day care centers and for food preparers who work in private and public kitchens. There individuals can transmit enteric pathogens such as bacteria and viruses to susceptible individuals.

The human skin is inhabited by a diverse group of microorganisms. They protect us from invasion by pathogens and hence contribute to our overall health. The normal flora that dwells on the skin can be placed into three main groups:

Diphtheroids: These organisms are represented by *Corynebacterium* and *Propionibacterium.* They prefer to grow in oily regions of the skin such as the head, chest, and back. They degrade fatty secretions such as sebum produced by the hair follicles. The corynebacteria that subsist on skin are nonpathogenic but are related to the human pathogen *Corynebacterium diphtheriae* in their variable morphology. The most common skin diphtheroid is *Propionibacterium acnes,* an obligate anaerobe that lives within hair follicles where oxygen is limited. It degrades sebum, the fatty secretion responsible for moisturizing the skin.

Staphylococci: These bacteria are the second major group of bacteria found associated with the skin. They are coagulase-negative, salt-tolerant, and tend to grow well in dry areas where they also utilize skin secretions for growth. The presence of staphylococci on the skin is thought to be beneficial because they produce antimicrobial compounds that inhibit the growth of other gram-positive bacteria, and they use available nutrients that prevent pathogen colonization of the skin. *Staphylococcus epidermidis* is an example of this group. It is estimated, however, that about 25% of the U.S. population also carries coagulase-positive *Staphylococcus aureus* as part of the normal flora of skin, especially in the nostrils. Less than 2% of the staphylococcus in these cases are methicillin-resistant *S. aureus* (MRSA), a major problem in hospital-acquired (nosocomial) infections because of its antibiotic resistance. Recently it has been shown that two different genetic strains of MRSA exist, a hospital-acquired strain and a community-acquired strain. The latter accounts for about 79% of MRSA infections and has been especially prevalent among infections that occur in young athletes using school gyms and training facilities.

Fungi: The fungi are represented by *Malassezia furfur,* a small, nonpathogenic yeast that utilizes fatty substances for growth. It often grows on the face, especially around the nose, where it can cause a harmless flaking of the skin. The spores of transient saprophytic fungi can collect on the skin and be grown if the skin is sampled for microorganisms. Some fungi and yeasts can cause opportunistic infections. The dermatophytes are fungi that cause infections of the hair, skin, and nails such as athlete's foot.

In addition to the normal flora, there are transient bacteria that can occur on the skin. Organisms such as endospore formers may be cultured by swabbing the skin because their endospores are present. However, they are not part of the persistent population that inhabits the skin. Other bacteria may be present because the skin has become temporarily contaminated with them. Most are easily removed by washing because they are contaminants on the skin, and antiseptic soaps are effective in killing them. The normal flora is much more difficult to remove by washing because these organisms reside in hair follicles and are entrenched in the skin, making them very difficult to remove or kill.

In this exercise, the class will evaluate the effectiveness of length of time in the removal of organisms from the hands using a surgical scrub technique. One member of the class will be selected to perform the scrub. Another student will assist by supplying the soap, brushes, and basins, as needed. During the scrub, at 2-minute intervals, the hands will be scrubbed into a basin of sterile water. Bacterial counts will be made of these basins to determine the effectiveness of the previous 2-minute scrub in reducing the bacterial flora of the hands. Members of the class not involved in the scrub procedure will make the inoculations from the basins for the plate counts.

Scrub Procedure

The two members of the class who are chosen to perform the surgical scrub will set up their materials near a sink for convenience. As one student performs the scrub, the other will assist in reading the instructions and providing materials as needed. The basic steps,

which are illustrated in figure 33.1, are also described in detail below. Before beginning the scrub, both students should read all the steps carefully.

Materials

- 5 sterile surgical scrub brushes, individually wrapped
- 5 basins (or 2000 ml beakers), containing 1000 ml each of sterile water. These basins should be covered to prevent contamination.
- 1 dispenser of green soap
- 1 tube of hand lotion

Step 1 To get some idea of the number of transient organisms on the hands, the scrubber will scrub all surfaces of each hand with a sterile surgical scrub brush for 30 seconds in basin A. No green soap will be used for this step. The successful performance of this step will depend on:

- spending the same amount of time on each hand (30 seconds),
- maintaining the same amount of activity on each hand, and
- scrubbing under the fingernails, as well as working over their surfaces.

After completion of this 60-second scrub, notify Group A that their basin is ready for the inoculations.

Step 2 Using the *same* brush as above, begin scrubbing with green soap for 2 minutes, using cool tap water to moisten and rinse the hands. One minute is devoted to each hand.

The assistant will make one application of green soap to each hand as it is being scrubbed.

Rinse both hands for 5 seconds under tap water at the completion of the scrub.

Discard the brush.

Note: This same procedure will be followed exactly in steps 4, 6, and 8 of figure 33.1.

Step 3 With a *fresh* sterile brush, scrub the hands in basin B in a manner that is identical to step 1. Don't use soap. Notify Group B when this basin is ready.

Note: Exactly the same procedure is used in steps 5, 7, and 9 of figure 33.1, using basins C, D, and E.

Remember: It is important to use a fresh sterile brush for the preparation of each of these basins.

After Scrubbing After all scrubbing has been completed, the scrubber should dry his or her hands and apply hand lotion.

Making the Pour Plates

While the scrub is being performed, the rest of the class will be divided into five groups (A, B, C, D, and

(1) Sixty-second hand scrub into basin A. No soap.

(2) Two-minute soap scrub with running water.

(3) Sixty-second hand scrub into basin B. No soap.

(4) Same as 2.

(5) Sixty-second hand scrub into basin C. No soap.

(6) Same as 2.

(7) Sixty-second hand scrub into basin D. No soap.

(8) Same as 2.

(9) Sixty-second hand scrub into basin E. No soap.

Figure 33.1 **Hand scrubbing routine.**

E) by the instructor. Each group will make six plate inoculations from one of the five basins (A, B, C, D, or E). It is the function of these groups to determine the bacterial count per milliliter in each basin. In this way, we hope to determine, in a relative way, the effectiveness of scrubbing in bringing down the total bacterial count of the skin.

Materials

- 30 veal infusion agar pours—6 per group
- 1 ml pipettes
- 30 sterile petri plates—6 per group
- 70% alcohol
- L-shaped glass stirring rod (optional)

1. Liquefy six pours of veal infusion agar and cool to 50°C. While the medium is being liquefied, label two sets of plates each: 0.1 ml, 0.2 ml, and 0.4 ml. Also, indicate your group designation on the plate.
2. As soon as the scrubber has prepared your basin, take it to your table and make your inoculations as follows:
   a. Stir the water in the basin with a pipette or an L-shaped stirring rod for 15 seconds. If the stirring rod is used (figure 33.2), sterilize it before using by immersing it in 70% alcohol and flaming. *For consistency of results all groups should use the same method of stirring.*
   b. Deliver the proper amounts of water from the basin to the six petri plates with a sterile

Figure 33.3 **Scrub water for count is distributed to six petri plates in amounts as shown.**

Figure 33.2 **An alternative method of stirring utilizes an L-shaped glass stirring rod.**

serological pipette. Refer to figure 33.3. If a pipette was used for stirring, it may be used for the deliveries.

  c. Pour a tube of veal infusion agar, cooled to 50°C, into each plate, rotate to get good distribution of organisms, and allow to cool.

  d. Incubate the plates at 37°C for 24 hours.

3. After the plates have been incubated, select the pair that has the best colony distribution with no fewer than 30 or more than 300 colonies. Count the colonies on the two plates and record your counts on the chart on the chalkboard.

4. After all data are on the chalkboard, complete the table and graph on Laboratory Report 33.

33   Effectiveness of Hand Scrubbing

A. Results

1. The instructor will draw a table on the chalkboard similar to the one below. Examine the six plates that your group inoculated from the basin of water. Select the two plates of a specific dilution that have approximately 30 to 300 colonies and count all of the colonies of each plate with a colony counter. Record the counts for each plate and their averages on the chalkboard. Once all the groups have recorded their counts, record the dilution factors for each group in the proper column. To calculate the organisms per milliliter, multiply the average count by the dilution factor.

GROUP	0.1 ml COUNT		0.2 ml COUNT		0.4 ml COUNT		DILUTION FACTOR*	ORGANISMS PER MILLILITER
	Per Plate	Average	Per Plate	Average	Per Plate	Average		
A								
B								
C								
D								
E								

*Dilution factors: 0.1 ml = 10; 0.2 ml = 5; 0.4 ml = 2.5

2. Graph
   After you have completed this tabulation, plot the number of organisms per milliliter that was present in each basin.

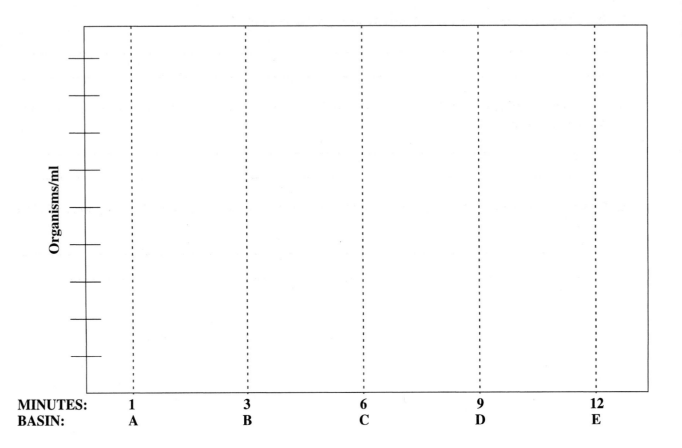

MINUTES:	1	3	6	9	12
BASIN:	A	B	C	D	E

3. What conclusions can be derived from this exercise?

   _____

   _____

   _____

4. What might be an explanation of a higher count in basin D than in B, ruling out contamination or faulty techniques?

   _____

   _____

B. Short-Answer Questions

   1. Name the three types of microbes most commonly associated with skin.

      _____

      _____

2. How are normal skin flora beneficial to the host?

   _____

   _____

3. What important opportunistic pathogen is associated with skin?

   _____

   _____

4. Differentiate between normal flora and transient bacteria found on skin. Which type is more difficult to remove? Explain.

   _____

   _____

5. Why is it so important that surgeons scrub their hands prior to surgery even though they wear surgical gloves?

   _____

   _____

Identification of Unknown Bacteria

One of the most interesting experiences in introductory microbiology is to attempt to identify an unknown microorganism that has been assigned to you as a laboratory problem. The next six exercises pertain to this phase of microbiological work. You will be given one or more cultures of bacteria to identify. The only information that might be given to you about your unknowns will pertain to their sources and habitats. All the information needed for identification will have to be acquired by you through independent study.

Although you will be engrossed in trying to identify an unknown organism, there is a more fundamental underlying objective of this series of exercises that goes far beyond simply identifying an unknown. That objective is to gain an understanding of the cultural and physiological characteristics of bacteria. Physiological characteristics will be determined with a series of biochemical tests that you will perform on the organisms. Although correctly identifying the unknowns that are given to you is very important, it is just as important that you understand the chemistry of the tests that you perform on the organisms.

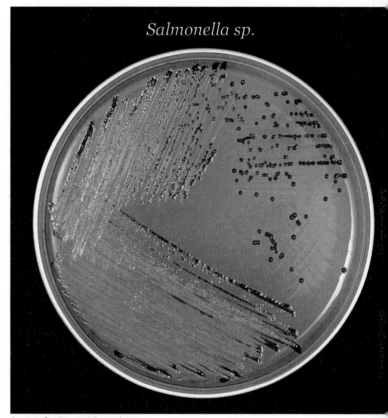

Salmonella sp.

Centers for Disease Control

The first step in the identification procedure is to accumulate information that pertains to the organisms' morphological, cultural, and physiological (biochemical) characteristics. This involves making different kinds of slides for cellular studies and the inoculation of various types of media to note the growth characteristics and types of enzymes produced. As this information is accumulated, it is recorded in an orderly manner on descriptive charts, which are located in the back of the manual.

After sufficient information has been recorded, the next step is to consult a taxonomic key, which enables one to identify the organism. For this final step, *Bergey's Manual of Systematic Bacteriology* will be used. Copies of volumes 1 and 2 of this book will be available in the laboratory, library, or both.

Success in this endeavor will require meticulous techniques, intelligent interpretation, and careful record keeping. Your mastery of aseptic methods in the handling of cultures and the performance of inoculations will show up clearly in your results. Contamination of your cultures with unwanted organisms will yield false results, making identification hazardous speculation. If you have reason to doubt the validity of the results of a specific test, repeat it; ***don't rely on chance!*** As soon as you have made an observation or completed a test, record the information on the descriptive chart. Do not trust your memory—record data immediately.

Morphological Study of an Unknown Bacterium

Learning Outcomes

After completing this exercise, you should be able to

1. Establish working and reserve stock cultures of your unknown organism(s).
2. Determine the optimum growth temperature for your unknown.
3. Determine the Gram reaction, microscopic morphology, motility, and some cultural characteristics of your unknown.

The first step in the study of your unknown bacterium is to set up stock cultures that will be used in the subsequent exercises. Your reserve stock culture will not be used for making slides or inoculating tests. It will be stored in the refrigerator in case your working stock becomes contaminated and you need to make a fresh working stock. The working stock will be used to inoculate the various tests that you will perform to identify your unknown bacterium. It is crucial that you practice good aseptic technique when inoculating from your working stock in order to avoid contaminating the culture. If it becomes contaminated or loses viability, you can prepare a fresh culture from the reserve stock culture that you have maintained in the refrigerator.

Identifying your unknown will be a kind of "microbiological adventure" that will test the skills and knowledge that you have acquired thus far. You will gather a great deal of information regarding your unknown by performing staining reactions and numerous metabolic tests. The Gram stain will play a very critical role in the process because it will eliminate thousands of possible organisms. The results of these tests will be compared to flowcharts provided in this manual and to information in *Bergey's Manual*. From your "detective" work, you will be able to ascertain the identity of the unknown that you were given. To set up the stock cultures, proceed in the following way (see figure 34.1).

Stock Cultures

You will receive a broth culture or an agar slant of your unknown bacterium. From this culture you will prepare your working stock and your reserve stock cultures. From the working stock, you will be able to determine such things as cell morphology, the Gram reaction of the unknown, and, in some cases, whether the culture forms any pigment. You can also determine other morphological characteristics such as the presence of a glycocalyx, endospores, or cytoplasmic granules.

Materials

- nutrient agar or tryptone agar slants

⏱ First Period

1. Label the agar slants with the code number of the unknown, your name, lab section, and date.
2. Inoculate both slants with your unknown organism. Begin your streak at the bottom of the slant and move the inoculating loop toward the top of the slant in a straight motion. Remember to practice good aseptic technique.
3. Place the respective tubes in the appropriate test tube racks labeled with the two incubation temperatures, 20°C and 37°C (figure 34.1). Also label the test tubes with the incubation temperature. Incubate the slants for 18 to 24 hours.

⏱ Second Period

1. Examine the slants. Look for growth. Some organisms produce sparse growth, and you must examine the cultures closely to determine if growth is present. Is either culture producing a pigment and, if so, is the pigment associated with the cells or has it diffused into the agar? Remember, however, that pigment production could require longer incubation times.
2. Determine which incubation temperature produced the best growth. If no growth occurred on either slant, your original culture could be nonviable or

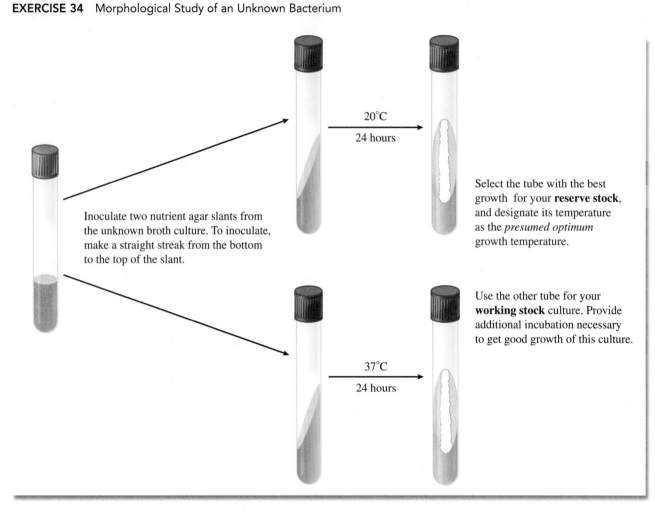

Inoculate two nutrient agar slants from the unknown broth culture. To inoculate, make a straight streak from the bottom to the top of the slant.

20°C
24 hours

Select the tube with the best growth for your **reserve stock**, and designate its temperature as the *presumed optimum* growth temperature.

Use the other tube for your **working stock** culture. Provide additional incubation necessary to get good growth of this culture.

37°C
24 hours

Figure 34.1 **Stock culture procedure.**

more time is needed for growth of the culture to occur. A third possibility is that neither temperature supported growth. Think through the possibilities and decide what course of action you need to take.

3. If growth occurred on the slant, pick the tube with the best growth and designate it as your **reserve stock culture.** Store the reserve stock in the refrigerator. Cultures stored in this manner are viable for several weeks. Do not use the reserve stock to make inoculations of the various media you will employ or to make stains. **Do not store the culture in your desk.**

4. Designate the second culture as your **working stock culture.** This culture will be the source of the inoculum for the various tests and stains that you will perform (figure 34.2). If the culture is 18 to 24 hours old, it can be used to perform the Gram stain. If not, you will have to prepare a fresh slant from the working stock to do the Gram stain.

As soon as morphological information is acquired, be sure to record your observations on the descriptive chart on page 245. Proceed as follows:

Materials

- Gram-staining kit
- spore-staining kit
- acid-fast staining kit
- Loeffler's methylene blue stain
- nigrosin or india ink
- tubes of nutrient broth and nutrient agar
- gummed labels for test tubes

New Inoculations

For all of these staining techniques, you will need 24- to 48-hour cultures of your unknown. If your working stock slant is a fresh culture, use it. If you don't have a fresh broth culture of your unknown, inoculate a tube of nutrient broth and incubate it at its estimated optimum temperature for 24 hours.

Gram's Stain

Once you have a good Gram stain of your organism, you can determine several characteristics of your unknown. First, you should be able to determine the

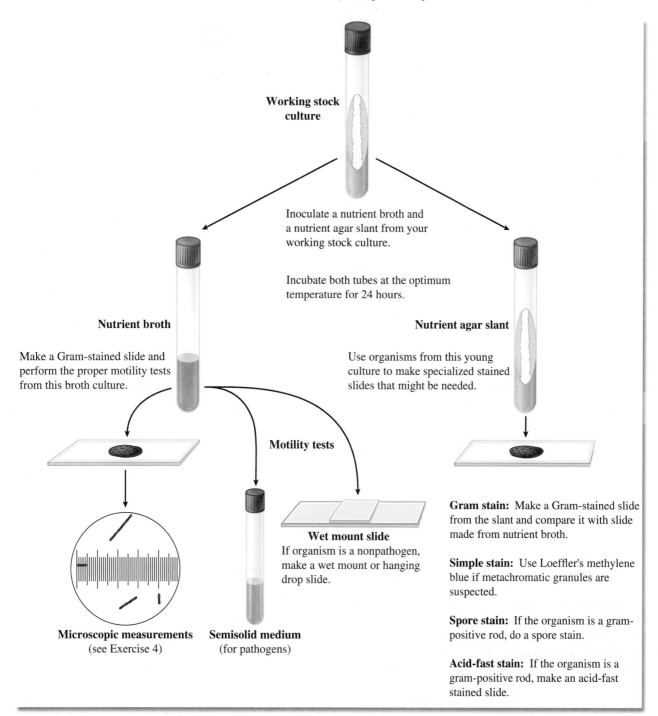

Working stock culture

Inoculate a nutrient broth and a nutrient agar slant from your working stock culture.

Incubate both tubes at the optimum temperature for 24 hours.

Nutrient broth

Make a Gram-stained slide and perform the proper motility tests from this broth culture.

Nutrient agar slant

Use organisms from this young culture to make specialized stained slides that might be needed.

Motility tests

Wet mount slide
If organism is a nonpathogen, make a wet mount or hanging drop slide.

Microscopic measurements
(see Exercise 4)

Semisolid medium
(for pathogens)

Gram stain: Make a Gram-stained slide from the slant and compare it with slide made from nutrient broth.

Simple stain: Use Loeffler's methylene blue if metachromatic granules are suspected.

Spore stain: If the organism is a gram-positive rod, do a spore stain.

Acid-fast stain: If the organism is a gram-positive rod, make an acid-fast stained slide.

Figure 34.2 **Procedure for morphological study.**

morphology of your organism: rod, coccus, spiral, etc. Furthermore, if you have prepared a thin smear, you can ascertain something about the arrangement of the cells. Note whether the cells occur singly, in pairs, in masses, or in chains. For example, the streptococci occur in chains, whereas the staphylococci occur in masses much like a bunch of grapes. The Gram stain can also be used to determine the size of your organism. Refer to Exercise 4.

Note: The results from the Gram stain can be verified by performing the following test: Place one drop of 3% KOH on a microscope slide and transfer a loopful of your unknown cells to the KOH solution. While observing the slide edge-on at eye level, mix the cells and KOH solution and slowly raise the loop from the cells. Gram-negative cells will lyse in the KOH solution, releasing their DNA and causing the liquid to become very viscous. Often, "strings" of DNA can be seen adhering to the loop as it is raised from the slide. Gram-positive cells do not undergo lysis in the KOH solution, and hence an increase in viscosity or DNA strings will not be seen in these cells.

Keep in mind that short rods with round ends (coccobacilli) look like cocci. If you have what seems to be a coccobacillus, examine many cells before you make a final decision. Also, keep in mind that *while rod-shaped organisms frequently appear as cocci under certain growth conditions, cocci rarely appear as rods.* (*Streptococcus mutans* is unique in forming rods under certain conditions.) Thus, it is generally safe to assume that if you have a slide on which you see both coccus-like cells and short rods, the organism is probably rod-shaped. This assumption is valid, however, only if you are not working with a contaminated culture!

Record the shape of the organism and its reaction to the stain on the descriptive chart on page 245.

Motility

For nonpathogens, the wet mount or hanging drop prepared from a broth culture is the preferred way to determine motility. Refer to Exercise 17. For pathogens, SIM medium can be used to ascertain motility (Exercise 17). Inoculate the culture by stabbing the "deep" with an inoculating needle. If growth occurs only along the line of inoculation, the organism is nonmotile. In contrast, turbidity throughout the tube would indicate that your organism is motile. In general, cocci are nonmotile, whereas rods can be either motile or nonmotile.

Endospores

If your unknown is a gram-positive rod, it may be an endospore-former. Endospores, however, do not usually occur in cocci or in gram-negative rods. Examination of your Gram-stained slide made from the agar slant should provide a clue since endospores show up as transparent oval structures in Gram-stained preparations. Endospores can also be seen on unstained organisms if studied with phase-contrast optics.

If there seems to be evidence that the organism is a spore-former, make a slide using one of the spore-staining techniques you used in Exercise 15. *Since some spore-formers require at least a week's time of incubation before forming spores, it is prudent to double-check for spores in older cultures.*

Record on the descriptive chart whether the spore is terminal, subterminal, or in the middle of the rod.

Acid-Fast Staining

The mycobacteria and some species of *Nocardia* are acid-fast. For these bacteria, the presence of acid-fastness can interfere with the Gram stain, causing these bacteria to stain gram-negative. Performing the acid-fast stain will sort out part of this problem. Do not depend solely on the Gram stain as the results can be misleading, especially for the acid-fast bacteria.

If your unknown is a gram-positive, non–spore-forming rod, it could be an acid-fast bacterium. Acid-fastness can vary with culture age, but most cultures display this property after 2 days of incubation. For best results, do not do this stain on old cultures. Refer to Exercise 16 for the staining procedure.

Other Structures

If the cytoplasm in the Gram-stained cells appears uneven, you may want to do a simple stain with Loeffler's methylene blue (Exercise 11) to determine the presence of metachromatic granules (volutin), which are storage granules of polyphosphate.

Although a capsule stain (Exercise 13) may be performed at this time, it might be better to wait until a later date when you have the organism growing on blood agar. Capsules usually are more apparent when the organisms are grown on this medium.

Laboratory Report

There is no Laboratory Report to fill out for this exercise. All information is recorded on the descriptive chart.

Descriptive Chart

STUDENT: _____

LAB SECTION: _____

Habitat: _____ Culture No.: _____

Source: _____

Organism: _____

MORPHOLOGICAL CHARACTERISTICS	PHYSIOLOGICAL CHARACTERISTICS	

MORPHOLOGICAL CHARACTERISTICS

Cell shape:

Arrangement:

Size:

Spores:

Gram's Stain:

Motility:

Capsules:

Special Stains:

CULTURAL CHARACTERISTICS

Colonies:

Nutrient Agar:

Blood Agar:

Agar Slant:

Nutrient Broth:

Gelatin Stab:

Oxygen Requirements:

Optimum Temp.:

PHYSIOLOGICAL CHARACTERISTICS

	TESTS	RESULTS
Fermentation	Glucose	
	Lactose	
	Sucrose	
	Mannitol	
Hydrolysis	Gelatin Liquefaction	
	Starch	
	Casein	
	Fat	
IMViC	Indole	
	Methyl Red	
	Voges-Proskauer (acetylmethylcarbinol)	
	Citrate Utilization	
	Nitrate Reduction	
	H_2S Production	
	Urease	
	Catalase	
	Oxidase	
	DNase	
	Phenylalanase	

	REACTION	TIME
Litmus Milk	Acid	_____
	Alkaline	_____
	Coagulation	_____
	Reduction	_____
	Peptonization	_____
	No Change	_____

Cultural Characteristics

Learning Outcomes

After completing this exercise, you should be able to

1. Determine the cultural characteristics of your unknown grown on nutrient agar, gelatin, and in thioglycollate broth.
2. Define the colony characteristics of your unknown such as elevation, edge, and any pigment production.

The cultural characteristics of an organism pertain to its macroscopic appearance on different kinds of media. In *Bergey's Manual,* you will find descriptive terms used by bacteriologists for recording cultural characteristics. For the general description of colonies, nutrient agar or any complex, rich medium is useful for this purpose. The nature of the growth in a nutrient broth can vary, and this too can be a source of certain information about an organism. Thioglycollate medium (Exercise 24) can be used to determine the oxygen requirements of an organism: Where do the cells grow in a tube of this medium? Some media, such as blood agar, are "differential," demonstrating the hemolytic capability of an organism. In the following exercise, you will inoculate your unknown into different media to determine its cultural characteristics in the various media.

🕐 First Period

(Inoculations)

During this period, one nutrient agar plate, one nutrient gelatin deep, two nutrient broths, and one tube of fluid thioglycollate medium will be inoculated. Inoculations will be made with the original broth culture of your unknown. The reason for inoculating two tubes of nutrient broth here is to recheck the optimum growth temperature of your unknown. In Exercise 34, you incubated your nutrient agar slants at 20°C and 37°C. It may well be that the optimum growth temperature is closer to 30°C. It is to check out this intermediate temperature that an extra nutrient broth is being inoculated. Proceed as follows:

Materials

for each unknown:
- 1 nutrient agar pour
- 1 nutrient gelatin deep
- 2 nutrient broths
- 1 fluid thioglycollate medium (FTM)
- 1 petri plate

1. Pour a petri plate of nutrient agar for each unknown and streak it with a method that will give good isolation of colonies. Use the original broth culture for streaking.
2. Inoculate the tubes of nutrient broth with a loop.
3. Make a stab inoculation into the gelatin deep by stabbing the inoculating needle (straight wire) directly down into the medium to the bottom of the tube and pulling it straight out. The medium must not be disturbed laterally.
4. Inoculate the tube of FTM with a loopful of your unknown. Mix the organisms throughout the tube by rolling the tube between your palms.
5. Place all tubes except one nutrient broth into a test tube rack and incubate for 24 hours at the temperature that seemed optimal in Exercise 34. Incubate the remaining tube of nutrient broth separately at 30°C. Incubate the agar plate, inverted, at the presumed best temperature.

🕐 Second Period

(Evaluation)

After the cultures have been properly incubated, *carry them to your desk in a careful manner* to avoid disturbing the growth pattern in the nutrient broths and FTM. Before studying any of the tubes or plates, place the tube of nutrient gelatin in an ice water bath. It will be studied later. Proceed as follows to study each type of medium and record the proper descriptive terminology on the descriptive chart on page 245.

Materials

- reserve stock agar slant of unknown
- spectrophotometer and cuvettes
- hand lens
- ice water bath near sink

Nutrient Agar Slant (Reserve Stock)

Examine your reserve stock agar slant of your unknown that has been stored in the refrigerator since

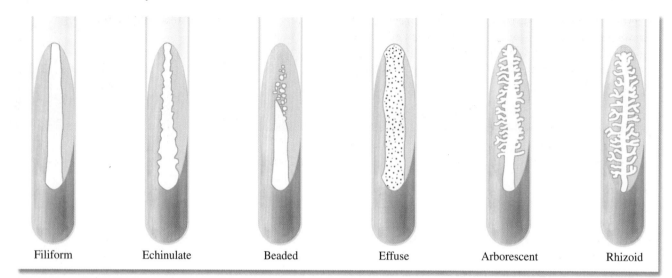

| Filiform | Echinulate | Beaded | Effuse | Arborescent | Rhizoid |

Figure 35.1 Types of bacterial growth on nutrient agar slants.

the last laboratory period. Evaluate it in terms of the following criteria:

Amount of Growth The abundance of growth may be described as *none, slight, moderate,* and *abundant.*

Color Pigments can be associated with a colony, for example, prodigiosin, the red pigment made by *Serratia marcescens* when grown at 27°C. However, pigments can be produced by an organism that diffuses into the medium, causing the medium to be colored, such as the case for the green fluorescent pigment produced by *Pseudomonas fluorescens*. To check for diffusable pigments, hold your plate up to the light and observe the color of the medium in the plate. Most bacteria, however, do not produce pigments, and their colonies are white or buff colored.

Opacity Organisms that grow prolifically on the surface of a medium will appear more opaque than those that exhibit a small amount of growth. Degrees of opacity may be expressed in terms of *opaque, transparent,* and *translucent* (partially transparent).

Form The gross appearance of different types of growth are illustrated in figure 35.1. The following descriptions of each type will help in differentiation:

> *Filiform:* characterized by uniform growth along the line of inoculation
> *Echinulate:* margins of growth exhibit toothed appearance
> *Beaded:* separate or semiconfluent colonies along the line of inoculation
> *Effuse:* growth is thin, veil-like, unusually spreading
> *Arborescent:* branched, treelike growth
> *Rhizoid:* rootlike appearance

Nutrient Broth

The nature of growth on the surface, subsurface, and bottom of the tube is significant in nutrient broth cultures. Describe your cultures as thoroughly as possible on the descriptive chart with respect to these characteristics:

Surface Figure 35.2 illustrates different types of surface growth. A *pellicle* type of surface differs from the *membranous* type in that the latter is much thinner. A *flocculent* surface is made up of floating adherent masses of bacteria.

Subsurface Below the surface, the broth may be described as *turbid* if it is cloudy, *granular* if specific small particles can be seen, *flocculent* if small masses are floating around, and *flaky* if large particles are in suspension.

Sediment The amount of sediment in the bottom of the tube may vary from none to a great deal. To

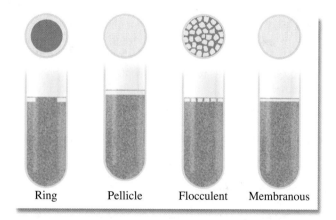

| Ring | Pellicle | Flocculent | Membranous |

Figure 35.2 Types of surface growth in nutrient broth.

describe the type of sediment, agitate the tube, putting the material in suspension. The type of sediment can be described as *granular, flocculent, flaky,* and *viscid.* Test for viscosity by probing the bottom of the tube with a sterile inoculating loop.

Amount of Growth   To determine the amount of growth, it is necessary to shake the tube to disperse the organisms. Terms such as *slight, moderate,* and *abundant* adequately describe the amount.

Temperature Requirements   To determine which temperature produces better growth, transfer the contents of the nutrient broth tubes to separate cuvettes and measure the optical density (absorbance) with a spectrophotometer. Because the cultures may be too turbid to measure, you may have to dilute the cultures with water before taking the readings. Record in the descriptive chart which temperature produces better growth for your organism. This temperature will be closer to the one needed for optimum growth of your organism.

Fluid Thioglycollate Medium

The growth pattern of your bacterium in fluid thioglycollate medium will give some indication of the oxygen requirement of your organism. Examine your FTM tube and compare the growth pattern of your organism with that of figure 24.5 on page 181. More than likely, your bacterium will be either aerobic, microaerophilic, or a facultative anaerobe. Strict anaerobes such as *Clostridium* require special culture conditions for growth.

Gelatin Stab

Some bacteria produce **proteases,** enzymes that degrade proteins. Determine if your unknown produces proteases by examining the nutrient gelatin tube that you inoculated with your unknown. After incubation, place the culture in an ice bath and allow it to stand for several minutes. Remove the tube and tilt it several times from side to side to ascertain if liquefaction has occurred. Any degraded gelatin will remain liquid after being placed in the ice bath. If liquefaction has not occurred, the contents of the tube will be a solid. Also be sure to note if your organism can grow in gelatin since some bacteria are unable to do so. Check the configuration with figure 35.3 to see if any of the illustrations match your tube. A description of each type follows:

> *Crateriform:* saucer-shaped liquefaction
> *Napiform:* turnip-like
> *Infundibular:* funnel-like or inverted cone
> *Saccate:* elongate sac, tubular, cylindrical
> *Stratiform:* liquefied to the walls of the tube in the upper region

Note: The configuration of liquefaction is not as significant as the mere fact that liquefaction takes place.

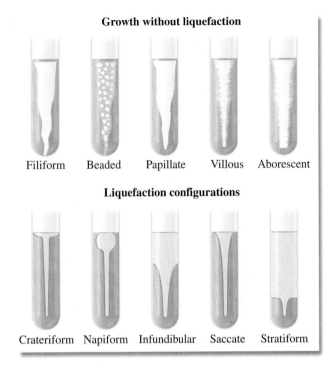

Growth without liquefaction

Filiform   Beaded   Papillate   Villous   Aborescent

Liquefaction configurations

Crateriform   Napiform   Infundibular   Saccate   Stratiform

Figure 35.3   Growth in gelatin stabs.

If your organism liquefies gelatin, but you are unable to determine the exact configuration, don't worry about it. However, be sure to record on the descriptive chart the *presence* or *absence* of protease production.

Another important point: Some organisms produce protease at a very slow rate. Tubes that are negative should be incubated for another 4 or 5 days to see if protease is produced slowly.

Type of Growth   (No Liquefaction) If no liquefaction has occurred, check the tube to see if the organism grows in nutrient gelatin (some do, some don't). If growth has occurred, compare the growth with the top of the illustration in figure 35.3. It should be pointed out, however, that, from an identification standpoint, the nature of growth in gelatin is not very important.

Nutrient Agar Plate

Colonies grown on plates of nutrient agar should be studied with respect to size, color, opacity, form, elevation, and margin. With a dissecting microscope or hand lens, study individual colonies carefully. Refer to figure 35.4 for descriptive terminology. Record your observations on the descriptive chart.

Laboratory Report

There is no Laboratory Report for this exercise. Record all information on the descriptive chart on page 245.

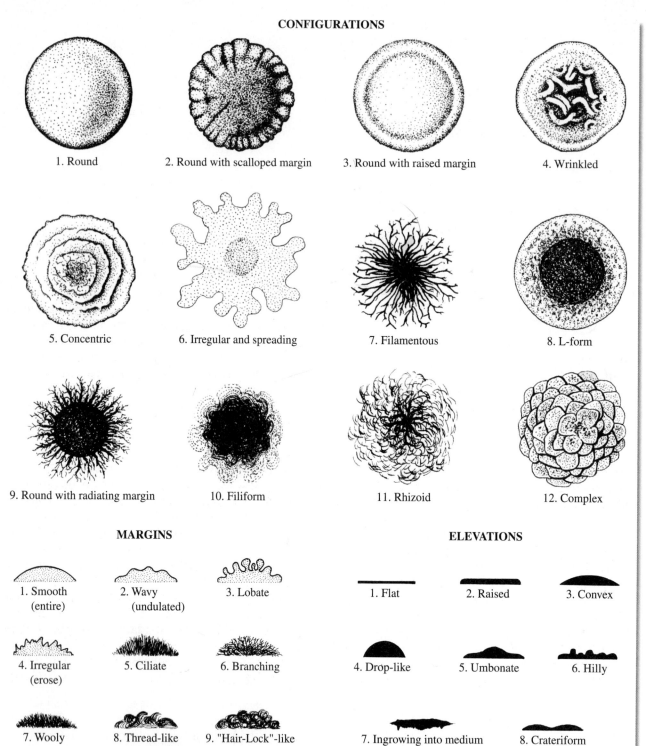

CONFIGURATIONS

1. Round

2. Round with scalloped margin

3. Round with raised margin

4. Wrinkled

5. Concentric

6. Irregular and spreading

7. Filamentous

8. L-form

9. Round with radiating margin

10. Filiform

11. Rhizoid

12. Complex

MARGINS

1. Smooth (entire)

2. Wavy (undulated)

3. Lobate

4. Irregular (erose)

5. Ciliate

6. Branching

7. Wooly

8. Thread-like

9. "Hair-Lock"-like

ELEVATIONS

1. Flat

2. Raised

3. Convex

4. Drop-like

5. Umbonate

6. Hilly

7. Ingrowing into medium

8. Crateriform

Figure 35.4 **Colony characteristics.**

Physiological Characteristics:
Oxidation and Fermentation Tests

Learning Outcomes

After completing this exercise, you should be able to

1. Understand the difference between oxidative reactions and fermentation reactions used to characterize bacteria.

2. Determine if your unknown is metabolically oxidative or fermentative by performing the appropriate diagnostic tests.

3. Further characterize your unknown using additional oxidative or carbohydrate fermentation tests.

4. Acquire a knowledge of enzymatic reactions, substrates, and end products for the biochemical basis for tests that you perform.

The sum total of the chemical reactions that occur in a cell are referred to as metabolism, and the individual chemical reactions that make up the metabolic pathways in a cell are catalyzed by protein molecules called **enzymes.** Most enzymes function inside the cell where metabolic pathways carry out the breakdown **(catabolism)** of food materials and the biosynthesis of cell constituents **(anabolism).** Because bacteria cannot carry out phagocytosis owing to their rigid cell walls, they excrete **exoenzymes** that function outside the cell to degrade large macromolecules. For example, exoenzymes break down proteins and polysaccharides into amino acids and monosaccharides, respectively, which are then transported into the cell

for metabolic needs. Protease, DNase, and amylase are examples of exoenzymes (figure 36.1).

Some enzymes are assisted in catalytic reactions by **coenzymes.** The latter transfer small molecules from one molecule to another. For example, the coenzymes NAD^+ and FAD transfer protons and coenzyme A transfers acetate groups. Most coenzymes are derivatives of vitamins. As examples, NAD^+ is synthesized from niacin, and FAD comes from folic acid. Coenzymes are only required by a cell in catalytic amounts, however, and when an enzymatic reaction catalyzes an oxidation step that reduces NAD^+ to NADH, the coenzyme must be converted back into its oxidized form if the metabolic pathway is to continue to function. Many of the reactions that define respiration and fermentation are concerned with regenerating coenzymes such as NAD^+ and FAD.

The primary goal of catabolism is the production of energy, which is needed for biosynthesis and growth. Bacteria can obtain their energy needs by two different metabolic means, respiration or fermentation. In respiration, organic molecules are completely degraded to carbon dioxide and water. ATP is generated by the energy created from a proton gradient that is established across the cell membrane when protons are transported from the cytoplasm to the outside of the cell. The shuttling of electrons down an electron transport chain involving cytochromes facilitates the movement of the protons to the outside of the cell. This process is called **oxidative phosphorylation** and, in the process, reduced coenzyme NADH generated in metabolic reactions is converted back to NAD^+ because oxygen acts as the terminal electron

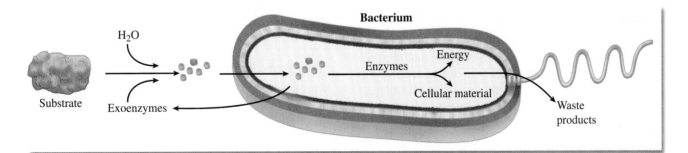

Figure 36.1 Note that the hydrolytic exoenzymes split larger molecules into smaller ones, utilizing water in the process. The smaller molecules are then assimilated by the cell to be acted upon by endoenzymes to produce energy and cellular material.

acceptor and is converted to water. In contrast, fermentation is the partial breakdown of organic molecules to alcohols, aldehydes, acids, and gases such as carbon dioxide and hydrogen. In this process, organic molecules in metabolic pathways serve as terminal electron acceptors and become the end products in a fermentation pathway. Reactions that carry out oxidation steps and utilize NAD^+ in metabolic pathways are coupled to reactions that use NADH to reduce the metabolic intermediates. An example is the oxidation of glyceraldehyde-3-phosphate in glycolysis being coupled with the formation of lactate from pyruvate when *Streptococcus lactis* ferments glucose.

In fermentation, ATP is synthesized by **substrate level phosphorylation,** in which metabolic intermediates in pathways directly transfer high-energy phosphates to ADP to synthesize ATP. In the glycolytic fermentation of glucose, ATP is formed when phosphoenolpyruvate transfers a high-energy phosphate to ADP and pyruvate is formed. In general, fermentation is much less efficient in producing energy relative to respiration because the use of metabolic intermediates as electron acceptors leaves most of the available energy in molecules that form the end products. Some bacteria are capable of growing both by respiration and fermentation. *Escherichia coli* is a facultative aerobe that will grow by respiratory means if oxygen is present but will switch metabolic gears in anaerobic conditions and grow by fermentation.

Sugars, particularly glucose, are compounds most widely used by fermenting organisms. However, other compounds such as organic acids, amino acids, and fats are fermented by bacteria. Butter becomes rancid because bacteria ferment butterfat, producing volatile and odoriferous organic acids. The end products of a particular fermentation are like a "fingerprint" for an organism and can be used in its identification. For example, *Escherichia coli* can be differentiated from *Enterobacter aerogenes* because the primary fermentation end products for *E. coli* are mixed organic acids, whereas *E. aerogenes* produces acetylmethylcarbinol, a neutral end product.

Tests to Be Performed

Two different kinds of tests will be performed in this exercise: (1) **fermentation tests** to determine if your unknown is capable of carrying out various fermentation reactions and (2) **oxidative tests** to determine if your unknown carries out respiratory metabolism. One test, the O/F glucose test, is designed to differentiate between these two modes of metabolism and ascertain if the organism is oxidative, fermentative, or capable

of both kinds of metabolism. The fermentation tests to be done are the O/F glucose, specific sugar fermentations, mixed-acid fermentation (methyl red [MR] test), butanediol fermentation (Voges-Proskauer [VP] test), and citrate test (figure 36.2). The oxidative tests to be performed are: the oxidase, catalase, and nitrate tests (figure 36.3). If the O/F glucose test determines that your organism is oxidative and not capable of fermenting sugars, then your bacterium cannot be identified by fermentation tests, and you will have to rely on other tests to identify your unknown.

The performance of these tests on your unknown may involve a considerable number of inoculations because a set of positive test controls are needed to which you will compare your unknown bacterium (figure 36.4). Although photographs of the various tests are provided in this manual, seeing the actual test results will be much more meaningful. Also keep in mind that some bacteria may not give precisely the same results as listed in *Bergey's Manual* as an isolate can often differ from its description in the manual.

As you perform the various tests, try to determine which tests may define a specific group of organisms. Some tests may be specific in an identification of an unknown, while others may not be specific and therefore not useful in determining the identify of your unknown. Keep in mind that *it is not routine practice to perform all the tests in identifying an unknown.* Although your goal is to identify your unknown, it is also important for you to learn how to perform the various tests and how to interpret them. The use of an unknown bacterium to identify simply makes it more of a challenge. In actual practice in hospitals and clinical laboratories, biochemical tests are used very selectively. The "shotgun" method of using all the tests is to be avoided because it is wasteful and can lead to confusing results.

Fermentation Tests
🕒 First Period

Inoculation should be set up for positive test controls and for your unknown. The media for each set of inoculations are listed separately under each heading.

Unknown Inoculations

The first biochemical test in determining the identity of your organism will be to ascertain whether your organism is oxidative or fermentative. For this, you will inoculate O/F glucose tubes with your unknown, a fermentative and oxidative organism, *Escherichia coli,* and an oxidative organism, *Pseudomonas aeruginosa.*

Unknown

Inoculations:
A minimum of eight tubes
inoculated with the unknown.

→ **Oxidative tests**

Fermentation tests

| O/F Glucose | O/F Glucose | Glucose | Lactose | Mannitol | MR-VP 1 | MR-VP 2 | Citrate |

Evaluation and tests

O/F glucose: See table 36.1 for interpretation.

Glucose Lactose Mannitol: If yellow, acid has been produced. If a bubble is present in the Durham tube, gas was produced.

MR-VP 1: Do methyl red test.

MR-VP 2: Do Voges-Proskauer test.

Citrate: If tube is blue, citrate was utilized.

Positive test results

MR: **Red color** is positive for mixed acid fermentation (MR test).

VP: **Pink** or **red color** is positive for butanediol production (VP test).

Figure 36.2   **Procedure for performing fermentation test.**

Oxidative Test

Figure 36.3 Procedure for performing oxidative tests.

Table 36.1 Interpretation of the O/F Glucose Test

RESULT		INTERPRETATION
ANAEROBIC	AEROBIC	
Yellow	Yellow	Oxidative and fermenta-tive metabolism
Green	Yellow	Oxidative metabolism
Green	Green	Sugar not metabolized (nonsaccharolytic)

Specific Fermentation Reactions

If your organism was found to be fermentative in its metabolism, it will be important to determine which specific sugars are fermented or which fermentation pathways are used for growth. The fermentation of specific sugars or the end products of fermentation pathways are important phenotypic characteristics used to identify bacteria in *Bergey's Manual.* The following fermentation tests will be studied in this exercise: (1) Durham tube sugar fermentations, (2) mixed-acid fermentation (methyl red test), (3) 2,3-butanediol fermentation (Voges-Proskauer test), and (4) citrate test.

Unknown Inoculations

First Period Figure 36.2 illustrates the procedure for inoculating the fermentation tests with your unknown. Your instructor may suggest other carbohydrates to be tested, and therefore blanks have been provided in the materials section for this purpose. *Different colored tube caps may be used to distinguish the different carbohydrates, and so be sure to record the cap color with the sugars given below.*

Materials

- 6 O/F glucose tubes
- sterile mineral oil

1. Each unknown organism and each test organism will be inoculated into two tubes of O/F glucose by stabbing with an inoculating needle.
2. To one of the tubes for each organism, aseptically deliver about 1 ml of sterile mineral oil after you have inoculated the tube. The mineral oil will establish anaerobic conditions in the tube. The tube without the mineral oil will be aerobic; therefore, be sure to loosen the cap about a quarter of a turn to allow access to the air.
3. Incubate the tubes at 37°C for 24 hours.
4. Record the results and compare them to the data in table 36.1 and the results in figure 36.5. **Note:** If your tubes do not show any color change from the uninoculated control at 24 hours, incubate them for an additional 48 hours and read them again.

Materials: (*for each unknown*)

- carbohydrate broths with Durham tubes and phenol red indicator with the following sugars:
  1 glucose tube
  1 lactose tube
  1 mannitol tube
- 2 MR-VP broth tubes
- 1 Simmon's citrate tube
- 1 trypticase soy agar (TSA) plate
- 1 nitrate broth tube

1. Label each tube with the number of your unknown and an identifying letter, as shown in figure 36.2.
2. Label one-half of a TSA plate with your unknown and the other half with *Pseudomonas aeruginosa.* This plate will be used in the next section for oxidative tests to determine if your organism

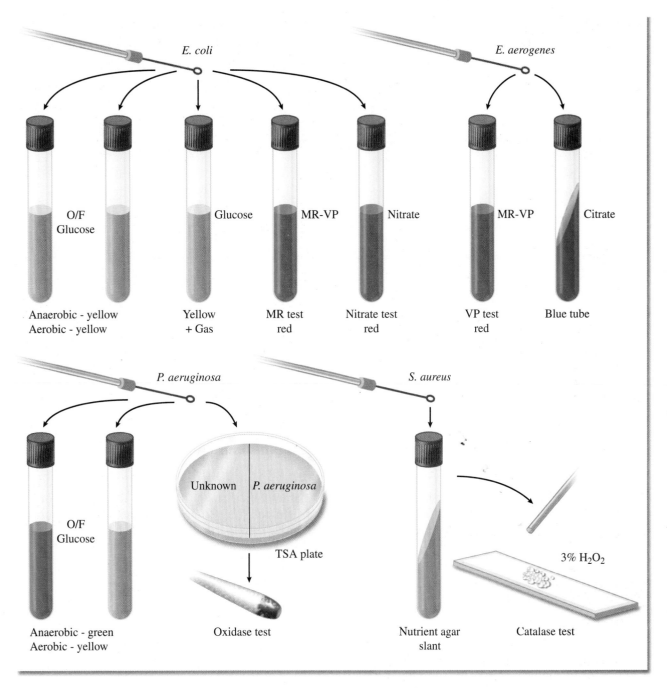

Figure 36.4   Procedure for doing positive test controls for fermentation and oxidative tests.

produces the respiratory enzyme cytochrome oxidase (figure 36.3).

3. Inoculate the Durham sugar tubes and the MR-VP broth tubes with your unknown.

4. Using an inoculating needle, first inoculate the Simmon's citrate slant by streaking the slant, and then stab the center of the slant about three-quarters of the way down into the butt of the tube.

5. Incubate the carbohydrate tubes, the Simmon's citrate tube, and the TSA plate for 24 hours.

6. Incubate the MR-VP broth tubes for 3–5 days.

Test Control Inoculations

Figure 36.4 illustrates the procedure for inoculating the test control tubes.

Materials

- 4 O/F glucose deeps
- 1 glucose broth with Durham tube and phenol red indicator
- 2 MR-VP broth tubes
- 1 Simmon's citrate tube

Figure 36.5   **O/F glucose test. (a) Fermentative and oxidative; (b) oxidative; (c) glucose not metabolized or inert.**
© McGraw-Hill Education. Auburn University Photographics Services

1. Label each tube with the organism inoculated:

   O/F glucose deeps
   glucose broth
   MR-VP broth
   MR-VP broth
   Simmon's citrate

2. Inoculate each of the test tubes with the appropriate organism as listed.

3. Incubate the glucose tube and the Simmon's citrate tube at 37°C for 24 hours.

4. Incubate the MR-VP broth tubes at 37°C for 3–5 days.

Note: Positive gas production should only be recorded when at least 10% of the medium has been displaced from the Durham tube.

Each sugar broth is supplemented with a specific carbohydrate at a concentration of 0.5% as well as beef extract or peptone to satisfy the nitrogen requirements of most bacteria. It is reasonable to assume that your unknown may ferment other sugars, but glucose, lactose, and mannitol are reasonable choices to start with as they are important in differentiating some of the medically important bacteria. Your instructor may suggest additional carbohydrates to be tested now or later that will assist in the identification of your organism.

ⓘ Second Period

(Test Evaluations)

After 24 to 49 hours' incubation, arrange all your tubes with the unknown tubes in one row and the test controls in another. As you interpret the results, record the information in the descriptive chart on page 245. Do not trust your memory. Any result that is not properly recorded will have to be repeated.

Carbohydrates in Durham Tubes

If an organism ferments a sugar, acid is usually produced, and gas may also be an end product of the fermentation. The presence of acid is indicated by a color change in the pH indicator, phenol red, from red at alkaline pH values to yellow at acidic pH values. The production of gas such as hydrogen and carbon dioxide is revealed by the displacement of medium from the Durham tube (figures 36.6 and 36.7d).

Inverted vial
(with gas)

Figure 36.6   **Durham fermentation tube.**

Figure 36.7 **Durham tubes:**
(a) **uninoculated control;**
(b) **alkaline; (c) acid, no gas;**
(d) **acid, gas.**
© McGraw-Hill Education. Lisa Burgess, photographer

(a)   (b)   (c)   (d)

Interpretation of the Results

Examine the glucose tube inoculated with *E. coli.* Note that the phenol red has turned from red to yellow, indicating the presence of acids from the fermentation of the glucose. Also note if medium has been displaced from the Durham tube. If at least 10% of the liquid has been displaced, it means that gas has been formed from the fermentation of the sugar. Figure 36.7 illustrates the difference between a positive and a negative test for both acid and gas production.

Now examine the test tubes with the test sugars, glucose, lactose, and mannitol that you inoculated with your unknown organism. Record the results for acid and gas production, comparing them to the positive control tubes. If there was no color change, record "none" in the descriptive chart. No color change is usually consistent with an oxidative organism. Keep in mind that a negative test result for your unknown is just as important as a positive result.

Mixed-Acid Fermentation

(Methyl Red Test)

An important test in differentiating some of the gram-negative intestinal bacteria is that of mixed-acid fermentation. Genera such as *Escherichia, Salmonella, Proteus,* and *Aeromonas* ferment glucose to produce a number of organic acids such as lactic, acetic, succinic, and formic acids. In addition CO_2, H_2, and ethanol are also produced in this fermentation.

The amount of acid produced is sufficient to lower the pH of the MR-VP broth to 5.0 or less.

To test for the presence of these acids, the pH indicator, methyl red, is added to the medium, which turns red if acid is present. A positive methyl red test indicates that the organism has carried out *mixed-acid fermentation.* The bacteria that are mixed-acid fermenters also generally produce gas because they produce the enzyme **formic hydrogenlyase,** which splits formic acid to produce CO_2 and H_2.

Medium MR-VP medium is a glucose broth that is buffered with peptone and dipotassium phosphate.

Test Procedure Perform the methyl red test first on the control, *E. coli,* and then on your unknown.

Materials

- dropping bottle of methyl red indicator

1. Add 3–4 drops of methyl red indicator to one of the MR-VP control tubes inoculated with *E. coli.* The tube should become red immediately. A positive tube is shown in figure 36.8b.
2. Repeat this procedure for one of the MR-VP broth tubes inoculated with your unknown organism. If your tube does not become red but remains unchanged (figure 36.8a), your unknown is methyl red negative.
3. Record your results in the descriptive chart on page 245.

(a)   (b)   (c)   (d)

Figure 36.8   MR-VP: (a) methyl red negative; (b) methyl red positive; (c) Voges-Proskauer negative; (d) Voges-Proskauer positive.
© McGraw-Hill Education. Lisa Burgess, photographer

2,3-Butanediol Fermentation

(Voges-Proskauer Test)

Some of the gram-negative intestinal bacteria do not carry out mixed-acid fermentation, but rather they ferment glucose to produce limited amounts of some organic acids and primarily a more neutral end product, 2,3-butanediol. All species of *Enterobacter* and *Serratia*, as well as some species of *Erwinia* and *Aeromonas*, carry out the butanediol fermentation. There are also some species of *Bacillus* that produce butanediol when grown on glucose. If an organism produces butanediol and is positive for the Voges-Proskauer (VP) test, it is usually negative for the methyl red test. The methyl red test and the Voges-Proskauer test are important tests for differentiating the gram-negative bacteria.

The neutral end product, 2,3-butanediol, is not detected directly but must be converted to acetoin by oxidation of the 2,3-butanediol. The acetoin reacts with Barritt's reagent, which consists of alpha-naphthol and KOH. The reagent is added to a **3- to 5-day old culture** grown in MR-VP medium and vigorously shaken to oxidize the 2,3-butanediol to acetoin. The tube is allowed to stand at room temperature for 30 minutes, during which time the tube will turn pink to red if acetoin is present (figure 36.8d).

Test Procedure   Perform the VP test on the control MR-VP broth tube inoculated with *Enterobacter aerogenes* and on the second MR-VP broth tube inoculated with your unknown organism. Follow this procedure:

Materials

- Barritt's reagent
- 2 pipettes (1 ml size)
- 2 empty test tubes

1. Label one empty test tube for your unknown and the other for *E. aerogenes* (positive control).
2. Pipette 1 ml of culture from your unknown to its respective tube and 1 ml of *E. aerogenes* to its respective tube. Use separate pipettes for each transfer.
3. Add 18 drops (about 0.5 ml) of Barritt's reagent A (alpha-naphthol) to each of the tubes containing 1 ml of culture.
4. Add 18 drops (0.5 ml) of Barritt's reagent B (KOH) to each of the test tubes.
5. Cap or cover the mouth of each test tube and shake the tubes vigorously. Allow the tubes to stand for 30 minutes. In this time, the tube with *E. aerogenes* should turn pink to red. Compare this to your unknown. *Vigorous shaking is necessary to oxidize the 2,3-butanediol to acetoin, which reacts with Barritt's reagents to give the red color.*

   Figure 36.8d shows a positive VP result, which is pink to red.
6. Record your results on the descriptive chart on page 245.

Citrate Test

Some bacteria are capable of using citrate as a sole carbon source. Normally citrate is oxidatively

Figure 36.9 **Citrate utilization (a) positive test, Prussian blue; (b) uninoculated control.**
© McGraw-Hill Education. Lisa Burgess, photographer

Figure 36.10 **Left-hand swab shows a purple reaction due to oxidase production. Right-hand swab shows a culture that is oxidase negative.**
© McGraw-Hill Education. Auburn University Photographics Services

metabolized by the Krebs cycle. However, some bacteria such as *Enterobacter aerogenes* and *Salmonella typhimurium* can cleave citrate to produce oxaloacetate and pyruvate. These intermediates are then fermented to produce several end products such as formate, acetate, lactate, acetoin, and CO_2. The medium also contains ammonium salts that serve as a sole nitrogen source for growth. Organisms degrading citrate must also use the ammonium salts, and in the process, they produce ammonia that causes the medium to become alkaline. Under alkaline conditions, the pH indicator in the medium turns from dark green to a deep Prussian blue, indicating the utilization of citrate.

Materials

- Simmon's citrate tubes

1. Examine the citrate slants. The slant inoculated with *E. aerogenes* will be a deep Prussian blue because the organism has utilized citrate. Compare this tube to your unknown. If the tube for your unknown has remained green, citrate was not utilized (figure 36.9).
2. Record your results in the descriptive chart.

Oxidative Tests (Refer to Figure 36.10)

Oxidase Test

The oxidase test assays for the presence of cytochrome oxidase, an enzyme in the electron transport chain. This enzyme catalyzes the transfer of electrons from reduced cytochrome *c* to molecular oxygen, producing oxidized cytochrome *c* and water.

Cytochrome oxidase occurs in bacteria that carry out respiration where oxygen is the terminal electron acceptor; hence, the test differentiates between those bacteria that have cytochrome oxidase and use oxygen as a terminal electron acceptor from those that can use oxygen as a terminal electron acceptor but have other types of terminal oxidases. The enzyme is detected by the use of an artificial electron acceptor, N,N,N′,N′-tetramethyl-*p*-phenylenediamine, which changes from yellow to purple when electrons are transferred from reduced cytochrome *c* to the artificial acceptor.

The oxidase test will differentiate most species of oxidase-positive *Pseudomonas* from the Enterobacteriaceae, which are oxidase negative. The artificial acceptor is somewhat unstable and can oxidize if left exposed to air for prolonged periods of time. In this exercise, you will use a commercially prepared reagent stored in glass ampules that are broken just prior to use.

Materials

- TSA plate streaked with your unknown on one-half and *Pseudomonas aeruginosa* streaked on the other half (figure 36.3).
- ampule of 1% oxidase reagent, N,N,N′,N′-tetramethyl-*p*-phenylenediamine dihydrochloride (Difco)
- sterile swabs
- Whatman no. 2 filter paper

1. Grasp an ampule of oxidase reagent between your thumb and forefinger. Hold the ampule so that it

is pointed away from you and squeeze the ampule until the glass breaks. Tap the ampule gently on the tabletop several times.

2. Touch a sterile swab to the growth of *Pseudomonas aeruginosa* on the TSA plate. Deliver several drops of oxidase reagent to the cells on the swab. (**Note:** You do not have to remove the cap of the oxidase reagent as it has a small hole for delivery of the reagent.)

   Alternatively: Transfer growth from the TSA plate to a piece of filter paper and add several drops of reagent to the cells on the paper.

3. A positive culture will cause the reagent to turn from yellow to purple in 10 to 30 seconds. A change after 30 seconds is considered a negative reaction (figure 36.10).

4. Repeat the test procedure for your unknown organism and record the results in the descriptive chart.

Catalase Test

When aerobic bacteria grow by respiration, they use oxygen as a terminal electron acceptor, converting it to water. However, they also produce hydrogen peroxide as a by-product of this reaction. Hydrogen peroxide is a highly reactive oxidizing agent that can damage enzymes, nucleic acids, and other essential molecules in the bacterial cell. To avoid this damage, aerobes produce the enzyme **catalase,** which degrades hydrogen peroxide into harmless oxygen and water.

$$2H_2O_2 \xrightarrow{\text{catalase}} 2H_2O + O_2$$

Strict anaerobes and aerotolerant bacteria such as *Streptococcus* lack this enzyme, and hence they are unable to deal with the hydrogen peroxide produced in aerobic environments. The presence of catalase is one way to differentiate these bacteria from aerobes or facultative aerobes, both of which produce catalase. For example, catalase production can be used to differentiate aerobic staphylococci from streptococci and enterococci, which lack this enzyme.

Test Procedure To determine if catalase is produced, a small amount of growth is transferred from a plate or slant, using a wooden stick, to a clean microscope slide. A couple of drops of 3% hydrogen peroxide are added to the cells on the slide. If catalase is produced,

Figure 36.11 Catalase test: vigorous evolution of oxygen occurs when catalase is present.
© McGraw-Hill Education. Lisa Burgess, photographer

there will be vigorous bubbling due to the breakdown of hydrogen peroxide and the production of oxygen gas (figure 36.11).

Note: Do not use a wire loop to transfer and mix the cells as iron can cause the hydrogen peroxide to break down, releasing oxygen. Also, do not perform the catalase test on cells growing on blood agar since blood contains catalase.

Materials

- 3% hydrogen peroxide
- nutrient agar slant tube with *Staphylococcus aureus* and your unknown on the TSA plate

1. Using the end of a wooden swab, transfer some cells from the *S. aureus* culture to the surface of a clean microscope slide.

2. Add 2 to 3 drops of 3% hydrogen peroxide to the cells, mix with the wooden stick, and observe for vigorous bubbling (figure 36.11).

3. Repeat the same procedure for your test organism and record your results in the descriptive chart on page 245.

Nitrate Reduction

Some facultative anaerobes can use nitrate as a terminal electron acceptor in a type of anaerobic respiration called **nitrate respiration.** Bacteria such as *Paracoccus* and some *Pseudomonas* and *Bacillus* reduce nitrate to a gaseous end product such as N_2O or N_2. Other bacteria such as *Escherichia coli* partially reduce nitrate to nitrite. Several enzymes are involved in the reduction of nitrate, one of which is nitrate

Figure 36.12 Nitrate test: Left-hand tube shows red color due to nitrate reduction. Middle tube shows reduction of nitrate to nitrogen gas that is trapped in the Durham tube. Right tube is an uninoculated control.
© McGraw-Hill Education. Auburn University Photographics Services

reductase, which catalyzes the transfer of electrons from cytochrome *b* to nitrate, reducing it to nitrite. The enzymes involved in nitrate reduction are inducible and are only produced if nitrate is present and anaerobic conditions exist for growth. The chemical reaction for the reduction of nitrate to nitrite is as follows:

$$NO_3^- + 2e^- + 2H^+ \xrightarrow{\text{nitrate reductase}} NO_2^- + H_2O$$

Test Procedure The ability of bacteria to reduce nitrate can be determined by assaying for the end products of nitrate reduction: gas or nitrite. Cultures are grown in beef extract medium containing potassium nitrate. Gases produced from nitrate reduction are captured in Durham tubes placed in the nitrate medium. Partial reduction of nitrate to nitrite is assayed for by adding sulfanilic acid (reagent A) followed by dimethyl-alpha-naphthylamine (reagent B). If nitrite is produced by reduction, it will form a chemical complex with the sulfanilic acid and the dimethyl-alpha-naphthylamine to give a dark red color (figure

36.12). A negative test could mean that nitrate was not reduced or that some other reduced form of nitrogen was produced, such as ammonia or hydroxlyamine. As a check, zinc powder is added to the test medium. Zinc metal will chemically reduce nitrate to nitrite, causing the medium to turn dark red as result of the formation of the chemical complex. If a nongaseous product such as ammonia was produced, no color will develop after the addition of the zinc metal.

Materials

- nitrate broth cultures with Durham tubes of the unknown organism and the test control *E. coli*
- nitrate test reagents: reagent A—sulfanilic acid; reagent B—dimethyl-alpha-naphthylamine
- zinc powder

1. Examine the nitrate broth of your unknown. If gas has been displaced in the Durham tube, it means that your organism has reduced nitrate to a gaseous end product, such as nitrogen gas. If no gas is present, reduction may have resulted in the formation of nitrite or the formation of a nongaseous end product.

2. To test for the presence of nitrite, first assay the test control *E. coli* culture by adding 2 to 3 drops of reagent A and 2 to 3 drops of reagent B to the nitrate broth culture of the organism. A deep red color will develop immediately (figure 36.12).

Caution

Avoid skin contact with solution B. Dimethyl-alpha-naphthylamine is carcinogenic.

3. Repeat this same test procedure for your unknown bacterium. If a red color fails to appear, your organism did not reduce nitrate or it may have produced a nongaseous end product of nitrate reduction.

   Zinc Test: To the negative culture, add a pinch of zinc powder and shake the tube vigorously. If a red color develops in the tube, nitrate was reduced by the zinc metal, indicating a negative test for nitrate reduction. If no color develops, a nongaseous end product may have been formed, which means your unknown reduced nitrate.

4. Record your results on the descriptive chart on page 245.

MICROBIOLOGY

Physiological Characteristics:
Hydrolytic and Degradative Reactions

Learning Outcomes

After completing this exercise, you should be able to

1. Perform tests to determine if your unknown has the ability to degrade or hydrolyze starch, proteins, lipids, tryptophan, urea, and phenylalanine.
2. Understand the biochemical basis of these tests.

Because bacteria have a rigid cell wall, they are unable to surround and engulf their food by the process of phagocytosis, which is characteristic of higher cells. To acquire nutrients, bacteria excrete a variety of hydrolytic and degradative exoenzymes that degrade large macromolecules into smaller units that can be transported into the cell for metabolic purposes. For example, amylases and cellulases degrade starch and cellulose, respectively, into simple sugars that are then transported into the cell where they are metabolized by fermentation or oxidation. A variety of proteases degrade proteins, such as casein and gelatin, and poly-peptides into amino acids. Triglycerides are degraded into fatty acids and glycerol by various lipases. Some-times bacteria also hydrolyze small molecules because they can thereby acquire carbon compounds for meta-bolic purposes. For example, tryptophan is split into pyruvate and indole by the enzyme tryptophanase. The pyruvate is metabolized, but the indole ring accu-mulates in the growth medium because it cannot be broken down. The accumulation of indole is the basis for a biochemical test that differentiates bacteria that produce trytophanase from those that do not produce the enzyme. Some bacteria hydrolyze urea to produce carbon dioxide and ammonia, thereby causing the pH to become alkaline. The change in pH is detected by a color change in a pH indicator. *Proteus* and other bac-teria can oxidatively deaminate phenylalanine to pro-duce phenylpyruvic acid. The latter can be detected with ferric chloride.

The presence of various hydrolytic and degrada-tive enzymes can be used as a basis for identifying bacteria. In this exercise, you will perform biochemi-cal tests for detecting hydrolytic and degradative reac-tions carried out by bacteria. In each case, you will compare your unknown to reactions carried out by test control organisms.

First Period

(Inoculations)

If each student is working with only one unknown organism, students can work in pairs to share petri plates. Note in figure 37.1 how each plate can serve for two unknowns and a test control organism streaked down the middle of the plate. If each student is work-ing with two unknowns, the plate should not be shared. Whether or not materials will be shared will depend on the availability of materials.

Materials

per pair of students with one unknown each or for one student with two unknowns:

- 1 starch agar plate
- 1 skim milk agar plate
- 1 spirit blue agar plate
- 3 urea slants or broths
- 3 tryptone broths
- phenylalanine agar
- nutrient broth cultures: *Bacillus subtilis, Escherichia coli, Staphylococcus aureus,* and *Proteus vulgaris*

1. Label and streak the three different agar plates, as shown in figure 37.1. Note that straight-line streaks are made on each plate. Indicate, also, the type of medium in each plate.
2. Label a urea slant or urea broth tube with *P. vul-garis* and a tryptone broth tube with *E. coli*. These are the test control tubes for urea and tryptophan hydrolysis, respectively. Inoculate each with the respective organism.
3. For each unknown, label one urea slant or broth and one tryptone tube with the code number of your unknown. Inoculate each tube with your unknown.
4. Label one phenylalanine agar slant with your unknown number and a second slant with *P. vulgaris.* Inoculate each slant with the respective organism.
5. Incubate the test control cultures at 37°C. Incu-bate the unknowns at the optimum temperatures that you determined for them.

Unknown inoculations:
Starch, skim milk and spirit blue agar plates, tryptone broth urea, and phenylalanine agar slants are inoculated with the unknown. Note that straight–line streaks are used on the plates and that test control organisms and unknown organisms are inoculated on the same plate for comparison.

Incubation:
The plates and the tube media should be incubated at the optimum temperature determined for your unknown for 48 hours.

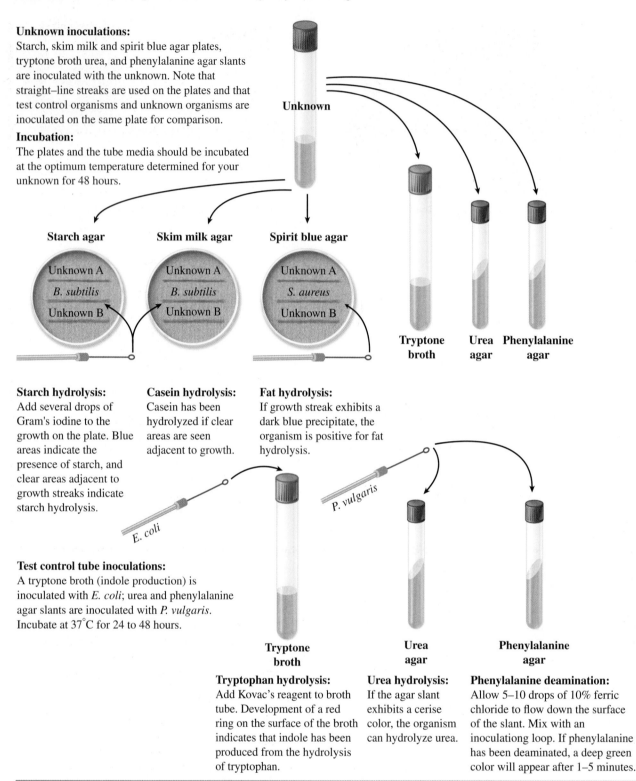

Starch hydrolysis:
Add several drops of Gram's iodine to the growth on the plate. Blue areas indicate the presence of starch, and clear areas adjacent to growth streaks indicate starch hydrolysis.

Casein hydrolysis:
Casein has been hydrolyzed if clear areas are seen adjacent to growth.

Fat hydrolysis:
If growth streak exhibits a dark blue precipitate, the organism is positive for fat hydrolysis.

Test control tube inoculations:
A tryptone broth (indole production) is inoculated with *E. coli*; urea and phenylalanine agar slants are inoculated with *P. vulgaris*. Incubate at 37°C for 24 to 48 hours.

Tryptophan hydrolysis:
Add Kovac's reagent to broth tube. Development of a red ring on the surface of the broth indicates that indole has been produced from the hydrolysis of tryptophan.

Urea hydrolysis:
If the agar slant exhibits a cerise color, the organism can hydrolyze urea.

Phenylalanine deamination:
Allow 5–10 drops of 10% ferric chloride to flow down the surface of the slant. Mix with an inoculationg loop. If phenylalanine has been deaminated, a deep green color will appear after 1–5 minutes.

Figure 37.1   Procedure for doing hydrolysis tests on unknowns.

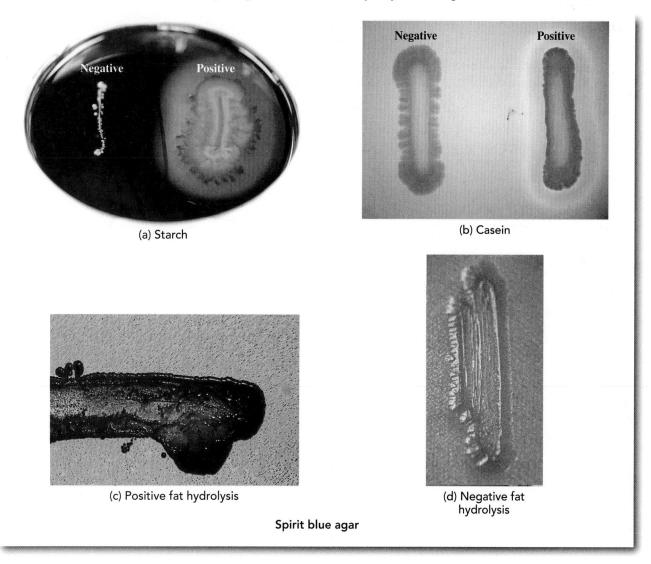

Figure 37.2 **Hydrolysis test plates: Starch, casein, fat.**
© McGraw-Hill Education. Lisa Burgess, photographer.

⏱ Second Period

(Evaluation of Tests)

After 24 to 48 hours incubation, compare your unknown and test controls, recording all data on the descriptive charts on page 245.

Starch Hydrolysis

The starch macromolecule consists of two constituents: (1) amylose, a straight chain polymer of 200 to 300 glucose molecules and (2) amylopectin, a larger branched polymer of glucose. Bacteria that hydrolyze starch produce *amylases* that degrade the starch molecule into molecules of maltose, glucose, and dextrins.

Starch hydrolysis is detected by adding Gram's iodine to starch medium. Iodine complexes with the starch macromolecule and causes the medium to turn blue. However, if the starch has been degraded, the medium adjacent to the bacterial growth will be clear after the addition of the iodine (figure 37.2a). Record the results on the descriptive chart on page 245.

Casein Hydrolysis

Casein is the predominant protein in milk, and its presence causes milk to have its characteristic white color. Many bacteria produce *proteases,* which are enzymes that degrade protein molecules such as casein into peptides and amino acids. This process is referred to as *proteolysis.* For example, the protease that degrades casein is a lysyl endopeptidase that cleaves at lysine residues.

Examine the growth on the skim milk agar. Note the **clear zone** surrounding the bacterium in the right streak in figure 37.2b, which illustrates casein hydrolysis. Compare your unknown to figure 37.2b. Record the results on the descriptive chart on page 245.

Fat Hydrolysis

Fats or triglycerides are composed of a glycerol molecule to which fatty acid molecules are covalently bonded through ester bonds. Triglycerides are primarily fat storage products in higher organisms such as animals. Some bacteria produce enzymes called *lipases* that cleave the fatty acids from glycerol. The fatty acids and glycerol can then be used for metabolic purposes such as synthesizing phospholipids for membrane construction or for catabolism to produce energy. The decomposition of triglycerides and the breakdown of the fatty acids into short-chain volatile organic acids is the reason why butter or margarine becomes rancid.

Spirit blue agar contains peptone as a source of carbon, nitrogen, and vitamins. It also contains tributyrin, a simple, natural animal triglyceride that serves as a substrate for lipases. Release of the fatty acids from tributyrin via lipase activity results in the lowering of the pH of the agar to produce a **dark blue precipitate.** However, some bacteria do not completely hydrolyze all the fatty acids from the tributyrin, and as a result, the pH is not sufficiently lowered to give the dark blue precipitate. In this case, all you notice may be simply the depletion of fat or oil droplets in the agar to indicate lipase activity.

Examine the growth of *S. aureus* on the plate. You should be able to see the dark blue reaction, as shown in figure 37.2c. Compare this to your unknown. *If your unknown appears negative, hold the plate up toward the light and look for a region near the growth where oil droplets are depleted. If you see the depletion of oil droplets, record this as a positive test in the descriptive chart.*

Tryptophan Degradation

Some bacteria have the ability to degrade the amino acid tryptophan to produce indole, ammonia, and pyruvic acid. The pyruvic acid can then be used by an organism for various metabolic purposes. The enzyme responsible for the cleavage of tryptophan is *tryptophanase*. The degradation of tryptophan by the enzyme can be detected with Kovac's reagent, which

forms a deep red color if indole is present. Tryptone broth (1%) is used for the test because it contains high amounts of tryptophan. Tryptone is derived from casein by a pancreatic digestion of the protein.

Materials

- Kovac's reagent
- tryptone broth cultures of *E. coli* and your unknown

To test for indole and therefore the activity of tryptophanase, add 10 to 12 drops of Kovac's reagent to the tryptone broth culture of *E. coli*. A red organic layer should form on top of the culture, as shown in figure 37.3. Repeat the test for your unknown culture and record the results on the descriptive chart.

Figure 37.3 **Tryptophan hydrolysis. The left tube shows the presence of indole (red band) at the top of the tube. The right-hand tube is an uninoculated control.**
© McGraw-Hill Education. Auburn University Photographics Services

PPA test

Figure 37.5   **Left-hand tube exhibits a positive reaction (green). Other tube is negative.**

Figure 37.4   **Urease test. Tube on the left is positive (*Proteus*); tube on the right is negative.**
© McGraw-Hill Education. Auburn University Photographics Services

Urea Hydrolysis

Urea is a waste product of animal metabolism that is broken down by a number of bacteria. The enzyme responsible for urea hydrolysis is *urease,* which splits the molecule into carbon dioxide and ammonia. Urease is produced by some of the gram-negative enteric bacteria such as *Proteus, Providencia,* and *Morganella,* which can be differentiated from other gram-negative enteric bacteria by this test. Refer to the separation outline in figure 39.2.

Urea medium contains yeast extract, urea, a buffer, and the pH indicator phenol red. Urea is unstable and is broken down by heating under steam pressure at 15 psi. Therefore, the medium is prepared by adding filter-sterilized urea to the base medium after autoclaving it.

When urease is produced by an organism, the resulting ammonia causes the pH to become alkaline. As the pH increases, the phenol red changes from yellow (pH 6.8) to a bright pink or cerise color (pH 8.1 or greater). See figure 37.4.

Examine the urea slant inoculated with *Proteus vulgaris* and compare it to your unknown. *If your urea slant is negative, continue the incubation for an additional*

7 days to check for slow urease production. Record your results in the descriptive chart on page 245.

Phenylalanine Deamination

Gram-negative bacteria such as *Proteus, Morganella,* and *Providencia* can oxidatively deaminate the amino acid phenylalanine to produce phenylpyruvic acid and ammonia. The reaction is catalyzed by the enzyme *phenylalanine deaminase,* a flavoprotein oxidase.

The enzyme can be detected by the addition of 10% ferric chloride, which forms a green-colored complex with α-keto acids such as phenylpyruvic acid. The test is useful in differentiating the above bacteria from other Enterobacteriaceae.

Materials

- dropping bottle of 10% ferric chloride

Allow 5 to 10 drops of 10% ferric chloride to flow down the slant of the test control organism, *P. vulgaris*. To facilitate the reaction, use an inoculating loop to emulsify the culture on the slant with the test reagent. A deep **green color** should appear in 1 to 5 minutes. Refer to figure 37.5. Repeat the test procedure for your unknown. Record your results in the descriptive chart on page 245.

Physiological Characteristics:
Multiple Test Media

Learning Outcomes

After completing this exercise, you should be able to

1. Inoculate the multiple test media Kligler's iron agar, SIM agar, and litmus milk and interpret the results.

2. Understand the biochemical basis for the test results.

Some media are designed to give multiple test results. These include: Kligler's iron agar, which determines fermentation reactions for glucose and lactose and the production of hydrogen sulfide; SIM, which determines hydrogen sulfide and indole production and motility; and litmus milk, which detects fermentation, proteolysis, and other reactions in milk. In addition, the IMViC tests will be discussed; these are an important group of tests used in differentiating some gram-negative enteric bacteria.

First Period

(Inoculations)

As before, test control cultures are included in this exercise. For economy of materials, one set of test control cultures will be made by students working in pairs.

Materials

- nutrient broth cultures of *Proteus vulgaris*, *Staphylococcus aureus*, and *Escherichia coli* for each unknown per student
- 1 Kligler's iron agar slant per unknown and 1 per test control culture
- 1 SIM deep per unknown and 3 for the test control cultures
- 1 litmus milk tube per unknown

1. Label one tube of Kligler's iron agar with *P. vulgaris* and additional tubes with your unknown numbers. Inoculate each tube by swabbing and then stabbing with an inoculating loop.

2. Label the SIM deeps with *P. vulgaris*, *S. aureus*, *E. coli*, and your unknown number.

3. Label one tube of litmus milk with your unknown number. (**Note:** A test control culture for litmus

milk will not be made. Interpretation of results will be made based on figure 38.3.)

4. Incubate the test control cultures at 37°C and the unknown cultures at their optimum temperatures for 24 to 48 hours.

Second Period

(Evaluation of Tests)

After the 24 to 48 hours of incubation, examine the tubes and evaluate the results based on the following discussion. Record the test results in the descriptive chart.

Kligler's Iron Agar

Kligler's iron agar is a multiple test medium that will detect the fermentation of glucose and lactose and the production of hydrogen sulfide resulting from the breakdown of the amino acid cysteine. It contains 0.1% glucose, 1% lactose, peptone, ferrous salts, and phenol red as a pH indicator. It is prepared as a slant and is inoculated by streaking the slant and stabbing the butt of the tube. The medium is useful in the differentiation of the gram-negative enteric bacteria.

Fermentation Reactions

The following are the possible results for the fermentation of the carbohydrates in the medium (see figure 38.1a–c).

1. Alkaline (red) slant/acid (yellow) butt (figure 38.1a): This means that only glucose was utilized. The organism utilized the low concentration of glucose initially and then degraded the peptone in the medium. The slant is alkaline (red) because glucose was degraded aerobically, and the ammonia released from peptone utilization caused the aerobic slant to become alkaline. However, the butt is yellow (acid) because glucose was fermented anaerobically to produce enough acids to cause the acidic reaction in the butt. If gas is produced, it will be evident by the splitting of the medium and the formation of gas bubbles in the agar slant.

2. Acid (yellow) slant/acid (yellow) butt (figure 38.1b): The organism has fermented both glucose and lactose, producing acids that cause the pH

(a)　　　(b)　　　(c)　　　(d)　　　(e)

Figure 38.1　Fermentation reactions and hydrogen sulfide production on Kligler's iron agar. (a) Alkaline/acid; (b) acid/acid with gas; (c) alkaline/no change; (d) uninoculated; (e) hydrogen sulfide production.
© McGraw-Hill Education. Auburn University Photographic Services.

indicator to turn yellow. Lactose is present in 10 times (1%) the concentration of glucose (0.1%), and sufficient acid is produced to cause both the slant and butt to be acidic. However, the tubes must be read at 24 hours because they can revert to alkaline in 48 hours if the lactose becomes depleted and the peptones are utilized, producing ammonia.

3. Alkaline (red) slant/alkaline (red) butt; alkaline (red) slant/no change butt (figure 38.1c): No fermentation of either sugar has occurred. Some enteric bacteria can use the peptones both aerobically and anaerobically, causing both the slant and butt to become alkaline. Others can only use the peptone aerobically, producing an alkaline slant but no change in the butt.

Hydrogen Sulfide

Bacteria such as *Proteus vulgaris* can degrade the amino acid cysteine to produce pyruvic acid, ammonia, and hydrogen sulfide. The initial step in the reaction pathway is the removal of sulfide from cysteine, which is catalyzed by *cysteine desulfurase*. This enzyme also requires the coenzyme pyridoxal phosphate for activity.

$$\begin{array}{l} CH_2 \cdot SH \\ | \\ CH \cdot NH_2 \\ | \\ COOH \end{array} + H_2O \xrightarrow{\text{Cysteine desulfurase}} \begin{array}{l} CH_3 \\ | \\ C{=}O \\ | \\ COOH \end{array} + H_2S\uparrow + NH_3$$

Cysteine　　　　　　　　　Pyruvic　Hydrogen　Ammonia
　　　　　　　　　　　　　　acid　　sulfide　　gas

Kligler's iron agar contains ferrous salts that will react with the hydrogen sulfide liberated by cysteine desulfurase to produce an insoluble black precipitate,

ferrous sulfide. **Note:** Bacteria such as *Proteus* that produce sulfide can obscure the fermentation reaction in the butt of the tube. If sulfide is produced, an acid reaction has occurred in the butt even if it cannot be observed.

Examine the Kligler's iron agar and record the results for the slant and butt of the tubes for the test organisms. Compare the results of your unknown to figure 38.1, which includes an uninoculated control (figure 38.1d). *Proteus* will produce hydrogen sulfide and will cause the tube butt to turn black (figure 38.1e). Record your results in the descriptive chart.

SIM Medium

SIM medium is a multiple test medium that detects the production of hydrogen sulfide and indole and determines if an organism is motile or not. The medium contains hydrolyzed casein, ferrous salts, and agar (0.7%), which makes the medium semisolid. It is inoculated by stabbing. The breakdown of tryptophan in the medium will produce indole, which can be detected by adding Kovac's reagent. If cysteine is degraded, hydrogen sulfide will be released, which will combine with the ferrous salts to produce a black precipitate in the tube.

Because the medium contains a low concentration of agar (0.7%), bacteria that are motile can swim in the medium. Nonmotile bacteria, such as *Staphylococcus aureus,* will only grow along the line of inoculation (figure 38.2a). Compare to uninoculated control (figure 38.2b). In contrast, motility is determined by diffuse growth out from the line of inoculation or by turbidity throughout the tube (figure 38.2c).

Examine the tubes with control bacteria. *E. coli* is motile and will therefore cause the tube to be turbid

Figure 38.2   Reactions on SIM medium: (a) Motility negative; (b) uninoculated control; (c) motility positive; (d) indole production; (e) hydrogen sulfide production.
Parts (a)–(c), (e): © McGraw-Hill Education. Lisa Burgess, photographer; part (d): © McGraw-Hill Education. Auburn University Photographic Services.

(figure 38.2c). Adding Kovac's reagent to the top of the tube results in a red ring, indicating the presence of indole (figure 38.2d). The SIM tube inoculated with *P. vulgaris* has a black precipitate due to the production of hydrogen sulfide (figure 38.2e). Compare these results with those of your unknown and record them in the descriptive chart.

The IMViC Tests

Sometimes a grouping of tests can be used to differentiate organisms. For example, *E. coli* can be differentiated from *E. aerogenes* by comparing four tests collectively called the IMViC (*I*: indole test; *M*: methyl red test; *V*: Voges-Proskauer test; and *C*: citrate test) tests. The results for the two organisms appear below. The two bacteria give exactly opposite reactions for the tests.

	I	**M**	**V**	**C**
E. coli	+	+	–	–
E. aerogenes	–	–	+	+

These two bacteria are very similar in morphological and physiological characteristics. The IMViC tests can be valuable when testing water for sewage contamination because they can confirm *E. coli,* which is always associated with sewage.

If your organism is a gram-negative rod and a facultative anaerobe, group these tests and see how your organism fits the combination of tests.

Litmus Milk

Litmus milk contains 10% powdered skim milk and the pH indicator litmus. The medium is adjusted to pH 6.8 and has a purplish blue color before inoculation.

Milk contains the proteins casein, lactalbumin, and lactoglobulin as well as the disaccharide lactose, and they provide an excellent growth medium for many microorganisms. Bacteria that ferment the lactose, resulting in acid production, will cause the litmus to turn pink. Other bacteria digest the milk proteins using proteases. This results in the release of ammonia, causing the litmus to turn purple because of the alkaline condition. Some of the proteolytic bacteria can also cause the milk proteins to precipitate and coagulate, thus forming a clot. Clotting can occur because of the production of acid or because of the release of the enzyme *rennin* that converts casein to paracasein.

Certain facultative bacteria can cause the reduction of the litmus dye to a colorless or leuco form. The color change is due to a drop in oxygen levels in the tube that accompanies the production of acids. The reduction of litmus can also occur when bacteria use the dye as an alternative electron acceptor. In these cases, the litmus is acting as a oxidation-reduction indicator as well as a pH indicator.

Figures 38.3 and 38.4 indicate the various color changes and reactions in litmus milk. It should be noted that some of the reactions take 4–5 days to fully develop, and therefore cultures should be incubated for this period of time. However, it is important to check the cultures every 24 hours for changes.

Litmus Milk Reactions

Acid, no clot   Fermentation of lactose and/or dextrose produces acids that cause the litmus to turn pink (figure 38.3a).

Alkaline   Breakdown of milk proteins such as lactalbumin results in the release of ammonia and amines that cause an alkaline reaction and blue color (figure 38.3b).

Figure 38.3 **Litmus milk reactions:** (a) acid, no clot; (b) alkaline; (c) peptonization; (d) reduction; (e) uninoculated control.

© McGraw-Hill Education. Lisa Burgess, photographer

(a)   (b)   (c)   (d)   (e)

Peptonization Digestion of milk proteins is evidenced by a clearing of the medium (figure 38.3c).

Reduction Reductase enzymes cause the removal of oxygen and the decolorization of the litmus (figure 38.3d).

Stormy Fermentation and Coagulation Acid production can cause the precipitation of milk proteins resulting in the formation of a curd or clot (figure 38.4b and c).

Record the results of the litmus milk reactions for your unknown on the descriptive chart on page 245.

Laboratory Report

Complete the Laboratory Report 36–38, which summarizes all the physiological tests performed on your unknown in the last three exercises.

(a)   (b)   (c)

Figure 38.4 **Litmus milk reactions:** (a) Acid without clot; (b) acid with clot; (c) stormy fermentation.

© McGraw-Hill Education. Lisa Burgess, photographer.

36–38 Physiological Characteristics of Bacteria

A. Results

Place the results of the physiological tests in the Descriptive Chart (page 245).

B. Short-Answer Questions

1. Why is it important to complete morphological and cultural characterizations before pursuing physiological testing?

   _____

   _____

2. In regard to taxonomic classification of bacteria, what is the relationship between physiological and genetic differentiation of bacteria?

   _____

   _____

3. What is the function of bacterial exoenzymes?

   _____

   _____

4. Differentiate between the following:

   a. anabolism and catabolism

   _____

   b. fermentation and respiration

   _____

5. Why is the catalase test useful for the differentiation of staphylococci from streptococci?

   _____

   _____

6. End products of biochemical reactions are often acids or alkalies.

   a. How are these products typically detected in a culture medium?

   _____

   b. Name two tests for reactions that produce acid end products.

   _____

    c. Name two tests for reactions that produce alkaline end products.

_____

7. End products of biochemical reactions are sometimes gases.

    a. What types of gases can be produced as a result of sugar fermentation?

_____

    b. How are these gases detected in fermentation reactions?

_____

    c. Name a nonfermentative test in which gas production indicates a positive test result.

_____

8. The tests for the hydrolysis of starch, casein, and triglycerides are similar in terms of setup. How do the methods for the detection of hydrolysis differ between these three tests?

_____

_____

9. Which test is used to differentiate between fermentative and oxidative metabolism?

_____

_____

10. For which of the five hydrolysis tests (starch, casein, triglycerides, tryptophan, and urea) would a positive test result be expected for:

    a. the etiologic agent of acne, *Propionibacterium acnes*? Explain.

_____

    b. the etiologic agent of gastric ulcers, *Helicobacter pylori*? Why would this activity be advantageous in this environment?

_____

11. Kligler's iron agar and SIM are multiple test media.

    a. What test do these media have in common?

_____

    b. What components of the media are included for this test?

_____

    c. Both media are stabbed but for different reasons. Explain.

_____

12. What difficulties does one encounter when trying to differentiate bacteria on the basis of physiological tests?

_____

_____

13. In addition to the morphological, cultural, and physiological tests performed on an unknown, what tests can be conducted to further assist in its identification?

_____

_____

C. Matching Questions

1. *MEDIA.* Match the name of the medium with the physiological test. A medium may be used more than once. Tests may require more than one answer.

   a. Kligler's iron agar
   b. MR-VP broth
   c. phenol red lactose
   d. SIM medium
   e. Simmon's citrate agar
   f. skim milk agar
   g. spirit blue agar
   h. tryptone broth

   _____ 2,3-butanediol fermentation
   _____ carbohydrate fermentation
   _____ casein hydrolysis
   _____ citrate utilization
   _____ hydrogen sulfide production
   _____ mixed-acid fermentation
   _____ triglyceride hydrolysis
   _____ tryptophan degradation

2. *REAGENTS.* Match the name of the reagent with the physiological test. Tests may require more than one answer.

   a. alpha-naphthol
   b. dimethyl-alpha-naphthylamine
   c. ferric chloride
   d. Gram's iodine
   e. hydrogen peroxide
   f. Kovac's reagent
   g. methyl red
   h. N,N,N′,N′-tetramethyl-*p*-phenylenediamine dihydrochloride
   i. potassium hydroxide
   j. sulfanilic acid

_____ 2,3-butanediol fermentation

_____ catalase test

_____ mixed-acid fermentation

_____ nitrate reduction

_____ oxidase test

_____ phenylalanine deamination

_____ starch hydrolysis

_____ tryptophan degradation

3. *ENZYMES*. Match the name of the enzyme with the biochemical reaction. Enzymes may be used more than once.

   a. amylase

   b. cysteine desulfurase

   c. lipase

   d. protease

   e. tryptophanase

   f. urease

_____ casein hydrolysis

_____ gelatin liquefaction

_____ hydrogen sulfide production

_____ indole

_____ starch hydrolysis

_____ triglyceride hydrolysis

_____ urea hydrolysis

4. *PRODUCTS*. Match the name of the product with the biochemical reaction. Products may be used more than once. Tests may require more than one answer.

   a. 2,3-butanediol

   b. ammonia

   c. fatty acids

   d. indole

   e. molecular oxygen

   f. phenylpyruvic acid

_____ catalase

_____ phenylalanine deamination

_____ triglyceride hydrolysis

_____ tryptophan degradation

_____ urea hydrolysis

_____ Voges-Proskauer test

Use of *Bergey's Manual*

Learning Outcomes

After completing this exercise, you should be able to

1. Use the results of growth tests, staining reactions, and biochemical tests to determine the identity of an unknown bacterium by applying information from *Bergey's Manual*.

2. Construct a dichotomous key to differentiate bacteria using phenotypic information given in a data table.

The most widely used source of information for the classification and identification of bacteria is *Bergey's Manual*. One of the purposes of this exercise is to help you use the information in *Bergey's Manual* to identify the unknown bacterium you tested throughout Exercises 34–38. You will be comparing the recorded morphological, cultural, and physiological characteristics from your descriptive chart to information in this respected resource.

Bergey's Manual

In 1923, the first edition of *Bergey's Manual of Determinative Bacteriology* was published by the American Society for Microbiology, based on work by David H. Bergey and four other collaborators. This initial reference attempted to organize and describe bacteria that had been identified in the laboratory. Over the next 50 years, seven revisions were published, continuing to combine both systematic and determinative information.

Between 1984 and 1989, the first edition of *Bergey's Manual of Systematic Bacteriology* was published, expanding the scope of the manual and focusing on systematic information. This four-volume set separates bacteria into groups by phenotypic characteristics such as Gram stain reaction, cell shape, oxygen requirements, and metabolic properties, but it also includes detailed information on the relationships between organisms. Volumes 1 and 2 of this edition contain information about most of the organisms used in undergraduate microbiology laboratory exercises.

The ninth edition of *Bergey's Manual of Determinative Bacteriology* was published in 1994 and represented a major change from the eight previous editions. Because of the publication of the systematic volumes, this edition of the *Determinative* manual contains only phenotypic information, designed to serve solely as a reference for identifying unknown bacteria. It is a single volume and separates bacteria into four major categories and 35 groups within those categories. The manual contains numerous tables differentiating the genera in each group and subsequent tables differentiating species within each genus.

More recently, a second edition of *Bergey's Manual of Systematic Bacteriology* was published from 2001 to 2012. This five-volume set is very different from its first edition and the ninth edition of the *Determinative* manual. Many new bacterial species have been identified since the 1980s when the previous edition was published. Newer genetic analyses, such as 16S rRNA analysis, were used to classify bacteria in this edition, and it presents organisms based on current understanding of their evolutionary relatedness. Volume 1 presents members of Domain Archaea. Volume 2 includes the gram-negative *Proteobacteria*, such as *Neisseria*, *Pseudomonas*, and the *Enterobacteriaceae*, whereas volume 3 contains the gram-positive *Firmicutes*, including *Bacillus*, *Clostridium*, *Streptococcus*, and *Staphylococcus*. Volume 4 covers many different phyla of bacteria, including the spirochetes and chlamydias. Volume 5 contains other gram-positive bacteria with a high guanine and cytosine concentration in their DNA. This volume presents genera such as *Corynebacterium*, *Micrococcus*, and *Mycobacterium*. Most of the organisms used in undergraduate laboratories can be found in volumes 2, 3, and 5.

Presumptive Identification of an Unknown Bacterium

For this exercise, you can use any edition of *Bergey's Manual* available in the laboratory or the library. The table of contents or the index can be used to help you locate the section that provides information about your microbe. For example, if you are working in the ninth edition of the *Determinative* manual, you start at the table of contents to find the group that best fits the basic information you know about your organism.

If you know you have an endospore-forming gram-positive rod, you go to the section on Group 18 starting on page 559. Here you find detailed information on each genus in this group, and you can identify your unknown species using the other test information you have gathered.

If *Bergey's Manual* is not readily available, you can still determine the genus of your unknown organism using the separation outlines in figures 39.1 through 39.3 and the additional information about each genus provided below. Seven different groups of gram-positive bacteria are separated in figure 39.1. Four different groups of gram-negative bacteria are separated in figure 39.2, and further differentiation is shown in figure 39.3.

Once you think you have identified your unknown, record that information in part A of the Laboratory Report for this exercise. Check with your instructor to see if your identification is correct. If you did not correctly identify your unknown, consider the possibilities given in the Problem Analysis section of this exercise.

Endospore-Forming, Gram-Positive Rods   Three major genera are listed in this group. Most members are motile, and differentiation is based on their respiratory metabolism.

Bacillus: Although most of these organisms are aerobic, some are facultative aerobes. Catalase is usually produced.

Clostridium: While most of members of this genus are strict anaerobes, some may grow in the presence of oxygen. Catalase is not usually produced.

Sporolactobacillus: Microaerophilic and catalase-negative. Nitrates are not reduced and indole is not formed. Spore formation occurs very infrequently (1% of cells). Three species are now recognized for *Sporolactobacillus*.

Acid-Fast Rods   This group consists of only one genus.

Mycobacterium: There are 54 species in this genus. Differentiation of species within this group depends to some extent on whether the organism is classified as a slow or a fast grower.

Regular, Non-Endospore-Forming, Gram-Positive Rods
The characteristics of three major genera are described for this group.

Lactobacillus: Non-endospore-forming rods, varying from long and slender to coryneform (club-shaped) coccobacilli. Chain formation is common. Only rarely motile. Facultative anaerobic or microaerophilic. Catalase-negative. Nitrate usually not reduced. Gelatin not liquefied. Indole and H_2S not produced.

Listeria: Regular, short rods with rounded ends; occur singly and in short chains. Aerobic

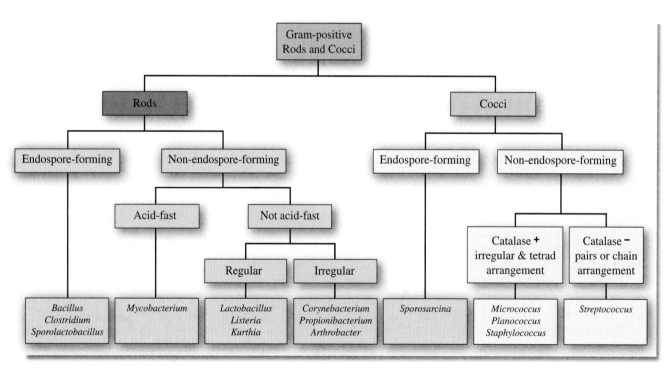

Figure 39.1   **Separation outline for gram-positive rods and cocci.**

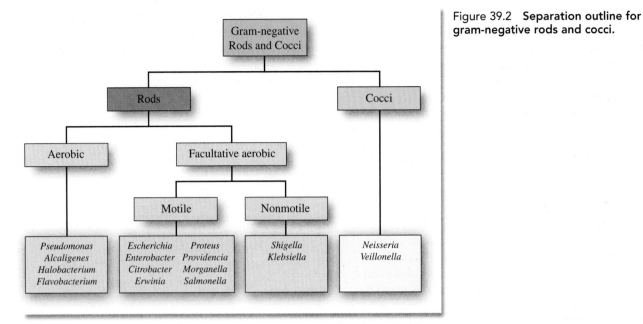

Figure 39.2 **Separation outline for gram-negative rods and cocci.**

Figure 39.3 **Separation outline for facultatively anaerobic gram-negative rods.**

and facultative anaerobic. Motile when grown at 20–25°C. Catalase-positive and oxidase-negative. Methyl red positive. Voges-Proskauer positive. Negative for citrate utilization, indole production, urea hydrolysis, gelatinase production, and casein hydrolysis.

Kurthia: Regular rods, 2–4 micrometers long with rounded ends; in chains in young cultures and coccoidal in older cultures. Strictly aerobic. Catalase-positive, oxidase-negative. Also

negative for gelatinase production and nitrate reduction. Only two species in this genus.

Irregular, Non-Endospore-Forming, Gram-Positive Rods
Although there are many genera that fit into this group, only three genera are described here.

Corynebacterium: Straight to slightly curved rods with tapered ends. Sometimes club-shaped. Palisade arrangements common due to snapping division of cells. Metachromatic granules formed. Facultative anaerobic. Catalase-positive. Most species produce acid from glucose and some other sugars. Often produce pellicle in broth.

Propionibacterium: Pleomorphic rods, often diphtheroid or club-shaped with one end rounded and the other tapered or pointed. Cells may be coccoid, bifid (forked, divided), or even branched. Nonmotile. Some produce clumps of cells with "Chinese character" arrangements. Anaerobic to aerotolerant. Generally catalase-positive. Produce large amounts of propionic and acetic acids. All produce acid from glucose.

Arthrobacter: Gram-positive rod and coccoid forms. Pleomorphic. Growth often starts out as rods, followed by shortening as growth continues, and finally becoming coccoidal. Some V-shaped and angular forms; branching by some. Rods usually nonmotile; some motile. Oxidative, never fermentative. Catalase-positive. Little or no gas produced from glucose or other sugars. Type species is *Arthrobacter globiformis.*

Endospore-Forming Cocci This group consists of unusual cocci.

Sporosarcina: Cells are spherical or oval when single. Cells may adhere to each other after division, producing tetrads or packets of eight or more. Endospores are formed. Strictly aerobic. Generally motile.

Catalase-Positive, Gram-Positive Cocci Our concern here is with only three genera in this group. Oxygen requirements and cellular arrangement are the principal factors in differentiating the genera. Most of these genera are not closely related.

Micrococcus: Spheres, occurring as singles, pairs, irregular clusters, tetrads, or cubical packets. Usually nonmotile. Strict aerobes (one species is facultative anaerobic). Catalase- and oxidase-positive. Most species produce carotenoid pigments. All species will grow in media containing 5% NaCl.

Planococcus: Spheres, occurring singly, in pairs, in groups of three cells, occasionally in tetrads. Although cells are generally gram-positive, they may be gram-variable. Motility is present. Catalase- and gelatinase-positive. Carbohydrates not utilized. Do not hydrolyze starch or reduce nitrate.

Staphylococcus: Spheres, occurring as singles, pairs, and irregular clusters. Nonmotile. Facultative anaerobes. Usually catalase-positive.

Most strains grow in media with 10% NaCl. Susceptible to lysis by lysostaphin. Glucose fermentation: acid, no gas. Coagulase production by some. Refer to Exercise 51 for species differentiation.

Catalase-Negative, Gram-Positive Cocci Note that the single genus of this group is included in the same section of *Bergey's Manual* as other gram-positive cocci.

Streptococcus: Spherical to ovoid cells that occur in pairs or chains when grown in liquid media. Some species, notably *S. mutans,* will develop short rods when grown under certain circumstances. Facultative anaerobes. Catalase-negative. Carbohydrates are fermented to produce lactic acid without gas production. Many species are commensals or parasites of humans or animals. Refer to Exercise 52 for species differentiation of pathogens.

Aerobic Gram-Negative Rods Although there are many genera of gram-negative aerobic rod-shaped bacteria, only four genera are likely to be encountered in this exercise.

Pseudomonas: Generally motile. Strict aerobes. Catalase-positive. Some species produce soluble fluorescent pigments that diffuse into the agar of a slant.

Alcaligenes: Rods, coccal rods, or cocci. Motile. Obligate aerobes with some strains capable of anaerobic respiration in presence of nitrate or nitrite.

Halobacterium: Cells may be rod- or disk-shaped. Cells divide by constriction. Most are strict aerobes; a few are facultative anaerobes. Catalase- and oxidase-positive. Colonies are pink, red, or red to orange. Gelatinase not produced. Most species require high NaCl concentrations in media. Cell lysis occurs in hypotonic solutions.

Flavobacterium: Gram-negative rods with parallel sides and rounded ends. Nonmotile. Oxidative. Catalase-, oxidase-, and phosphatase-positive. Growth on solid media is typically pigmented yellow or orange. Nonpigmented strains do exist.

Facultatively Anaerobic Gram-Negative Rods If your unknown appears to fall into this group, use the separation outline in figure 39.3 to determine the genus. Another useful separation outline is provided in figure 53.1. *Keep in mind, when using these separation outlines, that there are some minor exceptions in the applications of these tests.* The diversity of species within a particular genus often presents some problematical exceptions to the rule.

Gram-Negative Cocci   These genera are morphologically quite similar, yet physiologically quite different.

> ***Neisseria:*** Cocci, occurring singly, but more often in pairs (diplococci); adjacent sides are flattened. One species (*N. elongata*) consists of short rods. Nonmotile. Except for *N. elongata,* all species are oxidase- and catalase-positive. Aerobic.
>
> ***Veillonella:*** Cocci, appearing as diplococci, masses, and short chains. Diplococci have flattening at adjacent surfaces. Nonmotile. All are oxidase- and catalase-negative. Nitrate is reduced to nitrite. Anaerobic.

Problem Analysis

If you have identified your unknown by following the above procedures, congratulations! Not everyone succeeds at their first attempt. If you are having difficulty, consider the following possibilities:

- You may have been given the wrong unknown! Although this is a remote possibility, it does happen at times. Occasionally, clerical errors are made when unknowns are put together.
- Your organism may be giving you a false negative result on a test. This may be due to an incorrectly prepared medium, faulty test reagents, or improper testing technique.
- Your unknown organism may not match the description exactly as stated in *Bergey's Manual*. The information given in *Bergey's Manual* is based on the fact that 90% of organisms within a species conform, whereas 10% of a species can deviate from the given descriptions. *In other words, test results, as stated in the manual, may not always apply to your unknown.*
- Your culture may be contaminated. If you are not working with a pure culture, all tests are unreliable.
- You may not have performed enough tests. Check the various tables in *Bergey's Manual* to see if there is some other test that will be helpful. In addition, double-check the tables to make sure that you have read them correctly.

Confirmation of Results

There are several ways to confirm your presumptive identification. One method is to apply serological techniques, if your organism is one for which typing serum is available. Another alternative is to use one of the miniature multitest systems that are described in the next section of this manual. Your instructor will indicate which of these alternatives, if any, will be available.

Challenge Exercise

To further practice the techniques and tests you have already learned to use in identifying an unknown bacterium, here is another challenging activity. In this part of the exercise, your instructor will give you a broth culture of one of the organisms listed in Table 39.1. Your task is to correctly identify this unknown bacterium in the least number of steps, using the least amount of materials. Use the following general procedure to complete this assignment.

1. Streak your unknown for isolation onto two agar plates. Incubate one at 25°C and one at 37°C.
2. After incubation, record the preferred temperature for your organism in part B of the Laboratory Report, and use this temperature when incubating all further tests.
3. From one colony on your plate, make two pure culture slants to use throughout this exercise. One can serve as your working culture, and the other can serve as a reserve culture in case of contamination during the process.
4. From that same colony, prepare a Gram stain of the bacterium and observe microscopically. Record the Gram reaction and the cellular morphology of your organism.
5. Using the relevant information provided in Table 39.1, construct a dichotomous key to use as you attempt to identify your unknown. First, find a test that divides your type of organism (gram-positive or gram-negative) into two groups. Then, continue by dividing each group into two groups until all remaining candidates for the unknown organism are separated from each other on your outline. Complete this task on a separate sheet of paper, and ask your instructor to review your separation outline before you conduct any further tests.
6. Run the tests indicated on your outline one-by-one until you have eliminated all but one organism. This should be your unknown. Consider using multiple test media as appropriate to limit the amount of materials and time for this process. As you work, record all of the tests that you conduct and the results in the Laboratory Report table.
7. Conduct a confirmatory test on your organism. This should be a test that you have not already used, and it is always best to choose one where you expect a positive result for your unknown.
8. Record the identity of your unknown species and check your answer with your instructor.

Laboratory Report

Record the identity of your unknown in part B of Laboratory Report 39 and answer all of the questions.

Table 39.1   Characteristics of Common Gram-Positive and Gram-Negative Bacterial Species

SPECIES	SHAPE	GRAM REACTION	OXIDASE	CATALASE	METHYL RED	VOGES-PROSKAUER	LACTOSE FERMENTATION	GLUCOSE FERMENTATION	CITRATE	H₂S	NITRATE REDUCTION	INDOLE	UREA	MOTILITY
Alcaligenes faecalis	Rod	Neg	Pos	Pos	Neg	Neg	Neg	Neg	Pos	Neg	Neg	Neg	Neg	Pos
Enterobacter aerogenes	Rod	Neg	Neg	Pos	Neg	Pos	AG	AG	Pos	Neg	Pos	Neg	Neg	Pos
Escherichia coli	Rod	Neg	Neg	Pos	Pos	Neg	AG	AG	Neg	Neg	Pos	Pos	Neg	Pos
Klebsiella pneumoniae	Rod	Neg	Neg	Pos	Neg	Pos	AG	AG	Pos	Neg	Pos	Neg	Pos	Neg
Proteus vulgaris	Rod	Neg	Neg	Pos	Pos	Neg	Neg	A or AG	Neg	Pos	Pos	Pos	Pos	Pos
Pseudomonas aeruginosa	Rod	Neg	Pos	Pos	Neg	Neg	Neg	Neg	Pos	Neg	Pos	Neg	Neg	Pos
Salmonella typhimurium	Rod	Neg	Neg	Pos	Pos	Neg	Neg	A or AG	Pos	Pos	Pos	Neg	Neg	Pos
Serratia marcescens	Rod	Neg	Neg	Pos	Neg	Pos	Neg	A or AG	Pos	Neg	Pos	Neg	Neg	Pos
SPECIES	**SHAPE**	**GRAM REACTION**	**OXIDASE**	**CATALASE**	**METHYL RED**	**VOGES-PROSKAUER**	**LACTOSE FERMENTATION**	**GLUCOSE FERMENTATION**	**CITRATE**	**H₂S**	**NITRATE REDUCTION**	**INDOLE**	**UREA**	**MOTILITY**
Enterococcus faecalis	Coccus	Pos	Neg	Neg	Pos	Neg	A or AG	A or AG	Neg	Neg	Neg	Neg	Neg	Neg
Micrococcus luteus	Coccus	Pos	Pos	Pos	Neg	Neg	Neg	Neg	Neg	Neg	Pos	Neg	Neg	Neg
Staphylococcus aureus	Coccus	Pos	Neg	Pos	Pos	Pos	A or AG	A or AG	Neg	Neg	Pos	Neg	Pos	Neg
Staphylococcus epidermidis	Coccus	Pos	Neg	Pos	Neg	Neg	Neg	A or AG	Neg	Neg	Neg	Neg	Pos	Neg

39   Use of *Bergey's Manual*

A. Identification of an Unknown Bacterium (from Exercises 34–38)

1. Unknown bacterial species: _____

2. What tests/characteristics were the most helpful in your identification of this organism?

   _____

   _____

3. What is the significance of your unknown organism? Use your textbook or other reliable resource to learn more about its relevance.

   _____

   _____

4. Recent advances in DNA technology have provided other techniques for microbial identification. Are the methods you used in these exercises still valuable, and if so, when might they be used?

   _____

   _____

B. Challenge Exercise

1. Record initial information you obtained for your unknown bacterium.

   a. Preferred temperature: _____

   b. Cell shape/arrangement: _____

   c. Gram reaction: _____

2. Record the conducted tests and their results in this table.

TEST	RESULT

3. Unknown bacterial species: _____

4. What confirmation test did you choose to run on your organism? What was the expected result, and did you obtain that result on your unknown?

   _____

   _____

5. If the confirmation test did not match your expected result, suggest reasons why your initial identification may have been wrong.

   _____

   _____

Miniaturized Multitest Systems

Having run a multitude of tests in Exercises 34 through 39 in an attempt to identify an unknown, you undoubtedly have become aware of the tremendous amount of media, glassware, and preparation time that is involved just to set up the tests. And then, after performing all of the tests and meticulously following all the instructions, you discover that finding the specific organism in "Encyclopedia Bergey" is not exactly the simplest task you have accomplished in this course. The thought must arise occasionally: "There's got to be an easier way!" Fortunately, there are *miniaturized multitest systems*.

Miniaturized systems have the following advantages over the macromethods you have used to study the physiological characteristics of your unknown: (1) minimum media preparation, (2) simplicity of performance, (3) reliability, (4) rapid results, and (5) uniform results. These advantages have resulted in widespread acceptance of these systems by microbiologists.

Since it is not possible to describe all of the systems that are available, only four have been selected here: two by bioMérieux Vitek Products and two by Becton-Dickinson. All four of these products are designed specifically to provide rapid identification of medically important organisms, often within 5 hours. Each method consists of a plastic tube or strip that contains many different media to be inoculated and incubated. To facilitate rapid identification, these systems utilize numerical coding systems that can be applied to charts or computer programs.

© Corbis RF

The four multitest systems described in this unit have been selected to provide several options. Exercises 40 and 41 pertain to the identification of gram-negative, *oxidase-negative* bacteria (Enterobacteriaceae). Exercise 42 (Oxi/Ferm Tube) is used for identifying gram-negative, *oxidase-positive* bacteria. Exercise 43 (Staph-Ident) is a rapid system for the differentiation of the staphylococci.

As convenient as these systems are, one must not assume that the conventional macromethods of Part 8 are becoming obsolete. Macromethods must still be used for culture studies and confirmatory tests; confirmatory tests by macromethods are often necessary when a particular test on a miniaturized system is in question. Another point to keep in mind is that all of the miniaturized multitest systems have been developed for the identification of **medically important** microorganisms. If one is trying to identify a saprophytic organism of the soil, water, or some other habitat, there is no substitute for the conventional methods.

If these systems are available to you in this laboratory, they may be used to confirm your conclusions that were drawn in Part 8 or they may be used in conjunction with some of the exercises in Part 11. Your instructor will indicate what applications will be made.

Enterobacteriaceae Identification:

The API 20E System

The **API 20E System** is a miniaturized version of conventional tests that is used for the identification of members of the family Enterobacteriaceae and other gram-negative bacteria. This system utilizes a plastic strip (figure 40.1) with 20 separate compartments. Each compartment consists of a depression, or *cupule*, and a small *tube* that contains a specific dehydrated medium (see illustration 4, figure 40.2). The system has a capacity of 23 biochemical tests.

To inoculate each compartment, it is necessary to first make up a saline suspension of the unknown organism; then, with the aid of a Pasteur pipette, fill each compartment with the bacterial suspension. The cupule receives the suspension and allows it to flow into the tube of medium. The dehydrated medium is reconstituted by the saline. To provide anaerobic conditions for some of the compartments, it is necessary to add sterile mineral oil to them.

After incubation for 18 to 24 hours, the reactions are recorded, test reagents are added to some compartments, and test results are tabulated. Once the test results are tabulated, a *profile number* (seven or nine

digits) is computed. By finding the profile number in a code book, the *Analytical Profile Index*, or on the company website, one is able to determine the name of the organism. If no *Analytical Profile Index* is available, characterization can be done by using chart III in Appendix D.

Although this system is intended for the identification of nonenterics, as well as the Enterobacteriaceae, only the identification of the latter will be pursued in this experiment. Proceed as follows to use the API 20E System to identify your unknown enteric.

First Period

Two things will be accomplished during this period: (1) the oxidase test will be performed if it has not been previously performed, and (2) the API 20E test strip will be inoculated. All steps are illustrated in figure 40.2. Proceed as follows to use this system:

Materials

- agar slant or plate culture of unknown
- test tube of 5 ml 0.85% sterile saline
- API 20E test strip
- API incubation tray and cover
- squeeze bottle of tap water
- test tube of 5 ml sterile mineral oil
- Pasteur pipettes (5 ml size)
- oxidase test reagent
- Whatman no. 2 filter paper
- empty petri dish
- Vortex mixer

Figure 40.1 Negative and positive test results on API 20E test.
Courtesy of bioMérieux, Inc.

(1) Select one well-isolated colony to make a saline suspension of the unknown organism. Suspension should be well dispersed with a Vortex mixer.

(2) Dispense saline suspension of organisms into cupules of all 20 compartments. Slightly *underfill* ADH, LDC, ODC, H₂S, and URE. *Completely fill* cupules of CIT, VP, and GEL.

(3) After labeling the end tab of a tray with your name and unknown number, dispense approximately 5 ml of tap water into bottom of tray.

(4) Place an API 20E test strip into the bottom of the moistened tray. Be sure to seal the pouch from which the test strip was removed to prevent contamination of remaining strips.

(5) To provide anaerobic conditions for chambers ADH, LDC, ODC, H₂S, and URE, completely fill cupules of these chambers with sterile mineral oil. Use a fresh sterile Pasteur pipette.

(6) After incubation and after adding test reagents to four compartments, record all results and total numbers to arrive at seven-digit code. Consult the *Analytical Profile Index* to find the unknown.

Figure 40.2 Procedure for preparing and inoculating the API 20E test strip.

1. You must do the oxidase test on your unknown and establish that your unknown is definitely oxidase-negative before using this system. Use the method that is described on page 255.

2. Prepare a **saline suspension** of your unknown by transferring organisms from the center of a well-established colony on an agar plate (or from a slant culture) to a tube of 0.85% saline solution. Disperse the organisms well throughout the saline.

3. Label the end strip of the API 20E tray with your name and unknown number (see illustration 2, figure 40.2).

4. Remove an API 20E test strip from the sealed pouch and place it into the tray (see illustration 3). Be sure to reseal the pouch to protect the remaining strips.

5. Vortex mix the saline suspension to get uniform dispersal, and fill a sterile Pasteur pipette with the suspension. *Take care not to spill any of the organisms on the table or yourself. You may have a pathogen!*

6. Inoculate all the tubes on the test strip with the pipette by depositing the suspension into the cupules as you tilt the API tray (see illustration 4, figure 40.2).
   Important: Slightly *underfill* ADH, LDC, ODC, H₂S, and URE. (Note that the labels for these compartments are underlined on the strip.) Underfilling these compartments leaves room for oil to be added and facilitates interpretation of the results.

7. Since the media in |CIT|, |VP|, and |GEL| compartments require oxygen, *completely fill both the cupule and tube* of these compartments. Note that the labels on these three compartments are bracketed as shown here.

8. To provide anaerobic conditions for the ADH, LDC, ODC, H₂S, and URE compartments, dispense sterile **mineral oil** to the cupules of these compartments. Use another sterile Pasteur pipette for this step.

9. Dispense about 5 ml of tap water into the tray with a squeeze bottle. Note that the bottom of the tray has numerous depressions to accept the water.

10. Place the lid on the incubation tray and incubate at 37°C for 18 to 24 hours. Refrigeration after incubation is not recommended.

ⓘ Second Period

(Evaluation of Tests)

During this period, all reactions will be recorded on the Laboratory Report, test reagents will be added to four compartments, and the seven-digit profile number will be determined so that the unknown can be looked up in the *API 20E Analytical Profile Index*. Proceed as follows:

Materials

- incubation tray with API 20E test strip
- 10% ferric chloride
- Barritt's reagents A and B
- Kovacs' reagent
- nitrite test reagents A and B
- zinc dust or 20-mesh granular zinc
- hydrogen peroxide (1.5%)
- *API 20E Analytical Profile Index*
- Pasteur pipettes

1. Before any test reagents are added to any of the compartments, consult chart II, Appendix D, to determine the nature of positive reactions of each test, except TDA, VP, and IND.

2. Refer to chart II, Appendix D, for an explanation of the 20 symbols that are used on the plastic test strip.

3. Record the results of these tests on Laboratory Report 40.

4. **If GLU test is negative (blue or blue-green), and there are fewer than three positive reactions** before adding reagents, do not progress any further with this test as outlined here in this experiment. Organisms that are GLU-negative are nonenterics.

   For nonenterics, additional incubation time is required. If you wish to follow through on an organism of this type, consult your instructor for more information.

5. **If GLU test is positive (yellow), or there are more than three positive reactions,** proceed to add reagents as indicated in the following steps.

6. Add one drop of **10% ferric chloride** to the TDA tube. A positive reaction (brown-red), if it occurs, will occur immediately. A negative reaction color is yellow.

7. Add 1 drop each of **Barritt's A** and **B solutions** to the VP tube. Read the VP tube within 10 minutes. The pale pink color that occurs immediately has no significance. A positive reaction is dark pink or red and may take 10 minutes before it appears.

8. Add 1 drop of **Kovacs' reagent** to the IND tube. Look for a positive (red ring) reaction within 2 minutes.

   After several minutes, the acid in the reagent reacts with the plastic cupule to produce a color change from yellow to brownish-red, which is considered negative.

9. Examine the GLU tube closely for evidence of bubbles. Bubbles indicate the reduction of nitrate and the formation of N_2 gas. Note on the Laboratory Report that there is a place to record the presence of this gas.

10. Add 2 drops of each **nitrite test reagent** to the GLU tube. A positive (red) reaction should show up within 2 to 3 minutes if nitrates are reduced.

    If this test is negative, confirm negativity with **zinc dust** or 20-mesh granular zinc. A pink-orange color after 10 minutes confirms that nitrate reduction did not occur. A yellow color results if N_2 was produced.

11. Add 1 drop of **hydrogen peroxide** to each of the MAN, INO, and SOR cupules. If catalase is produced, gas bubbles will appear within 2 minutes. Best results will be obtained in tubes that have no gas from fermentation.

Final Confirmation

After all test results have been recorded and the seven-digit profile number has been determined, according to the procedures outlined on the Laboratory Report, identify your unknown by looking up the profile number in the *API 20E Analytical Profile Index*.

Cleanup

When finished with the test strip, be sure to place it in a container of disinfectant that has been designated for test strip disposal.

Laboratory Report

Enter the numbers for each test and tabulate the I.D. number for your bacterium. Follow the example in the results section on page 291.

Laboratory Report

Student: _____

Date: _____   Section: _____

40   Enterobacteriaceae Identification: The API 20E System

A. Results

1. Tabulation
    By referring to charts I and II, Appendix D, determine the results of each test and record these results as positive (+) or negative (−) in the table below. Note that the results of the oxidase test must be recorded in the last column on the right side of the table.

ONPG	ADH	LDC	ODC	CIT	H₂S	URE	TDA	IND	VP	GEL	GLU	MAN	INO	SOR	RHA	SAC	MEL	AMY	ARA	OXI
1	2	4	1	2	4	1	2	4	1	2	4	1	2	4	1	2	4	1	2	4

NO₂	N2 GAS	MOT	MAC	OF-O	OF-F
1	2	4	1	2	4

Additional Digits

2. Construction of Seven-Digit Profile
    Note in the above table that each test has a value of 1, 2, or 4. To compute the seven-digit profile for your unknown, total up the positive values for each group.

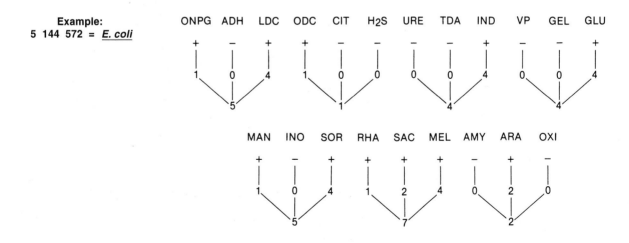

Example:
5 144 572 = *E. coli*

291

3. Using the *API 20E Analytical Index* or the API Characterization Chart
   If the *API 20E Analytical Index* is available on the demonstration table, use it to identify your unknown, using the seven-digit profile number that has been computed. If no *Analytical Index* is available, use characterization chart III in Appendix D.

   Name of Unknown: _____

4. Additional Tabulation Blank
   If you need another form, use the one below:

	api® 20E	Reference Number _____	Patient _____	Date _____
		Source/Site _____	Physician _____	Dept./Service _____

	ONPG 1	ADH 2	LDC 4	ODC 1	CIT 2	H₂S 4	URE 1	TDA 2	IND 4	VP 1	GEL 2	GLU 4	MAN 1	INO 2	SOR 4	RHA 1	SAC 2	MEL 4	AMY 1	ARA 2	OXI 4
5 h																					
24 h																					
48 h																					
Profile Number																					

	NO₂ 1	N₂ GAS 2	MOT 4	MAC 1	OF-O 2	OF-F 4	Additional Information
5 h							
24 h							Identification
48 h							
Additional Digits							00-42-012 E-3 (7/80)

B. Short-Answer Questions

1. What are the advantages and disadvantages of multitest systems for bacterial identification?

   _____

   _____

2. Before using the API 20E System, what test must be performed to confirm the identity of your unknown as a member of the family Enterobacteriaceae? What is the expected result?

   _____

   _____

3. If the seven-digit profile that is tabulated cannot be found in the *API 20E Analytical Profile Index,* what might that indicate about the bacterial culture?

   _____

   _____

Enterobacteriaceae Identification:
The Enterotube II System

Learning Outcomes

After completing this exercise, you should be able to

1. Inoculate an Enterotube II with an unknown bacterium that belongs to the Enterobacteriaceae.

2. Evaluate the test results to generate a five-digit code for the unknown bacterium.

3. Use the five-digit code and *Interpretation Guide* to correctly indentify the unknown bacterium.

The **Enterotube II** miniaturized multitest system was developed by Becton-Dickinson of Cockeysville, Maryland, for rapid identification of Enterobacteriaceae. It incorporates 12 different conventional media and 15 biochemical tests into a single ready-to-use tube that can be simultaneously inoculated with a minimum of equipment.

If you have an unknown gram-negative rod or coccobacillus that appears to be one of the Enterobacteriaceae, you may wish to try this system on it. Before applying this test, however, *make certain that your unknown is oxidase-negative*, since with only a few exceptions, all Enterobacteriaceae are oxidase-negative. If you have a gram-negative rod that is oxidase-positive, you might try the *Oxi/Ferm Tube II* instead, which is featured in Exercise 42.

Figure 41.1 illustrates an uninoculated tube (upper) and a tube with all positive reactions (lower). Figure 41.2 outlines the entire procedure for utilizing this system.

Each of the 12 compartments of an Enterotube II contains a different agar-based medium. Compartments that require aerobic conditions have openings for access to air. Those compartments that require anaerobic conditions have layers of paraffin wax over the media. Extending through all compartments of the entire tube is an inoculating wire. To inoculate

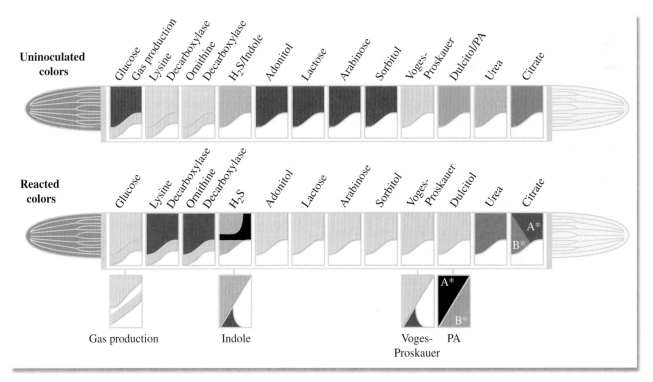

Figure 41.1   Enterotube II color differences between uninoculated and positive tests.
Courtesy and © Becton, Dickinson and Company

(1) Remove organisms from a well-isolated colony. Avoid touching the agar with the wire. To prevent damaging Enterotube II media, do not heat-sterilize the inoculating wire.

(2) Inoculate each compartment by first twisting the wire and then withdrawing it all the way out through the 12 compartments, using a turning movement.

(3) Reinsert the wire (without sterilizing), using a turning motion through all 12 compartments until the notch on the wire is aligned with the opening of the tube.

(4) Break the wire at the notch by bending. The portion of the wire remaining in the tube maintains anaerobic conditions essential for true fermentation.

continued

Figure 41.2 **The Enterotube II procedure.**

the media, one simply picks up some organisms on the end of the wire and pulls the wire through each of the chambers in a single rotating action.

After incubation, the reactions in all the compartments are noted and the indole test is performed. The Voges-Proskauer test may also be performed as a confirmation test. Positive reactions are given numerical values, which are totaled to arrive at a five-digit code. Identification of the unknown is achieved by consulting a coding manual, the *Enterotube II Interpretation Guide*, which lists these numerical codes for the Enterobacteriaceae. Proceed as follows to use an Enterotube II in the identification of your unknown.

⏱ First Period

Inoculation and Incubation

The Enterotube II can be used to identify Enterobacteriaceae from colonies on agar that have been inoculated from urine, blood, sputum, and so on. The culture may be taken from media such as MacConkey, EMB, SS, Hektoen enteric, or trypticase soy agar.

Materials

- culture plate of unknown
- 1 Enterotube II

1. Write your initials or unknown number on the white paper label on the side of the tube.
2. Unscrew both caps from the Enterotube II. The tip of the inoculating end is under the white cap.
3. *Without heat-sterilizing* the exposed inoculating wire, insert it into a well-isolated colony.
4. Inoculate each chamber by first twisting the wire and then withdrawing it through all 12 compartments. Rotate the wire as you pull it through. See illustration 2, figure 41.2.
5. Again, *without sterilizing*, reinsert the wire, and with a turning motion, force it through all

(5) Punch holes with broken-off part of wire through the thin plastic covering over depressions on sides of the last eight compartments (adonitol through citrate). Replace caps and incubate at 35˚C for 18 to 24 hours.

(6) After interpreting and recording positive results on the sides of the tube, perform the indole test by injecting 1 or 2 drops of Kovacs' reagent into the H₂S/indole compartment.

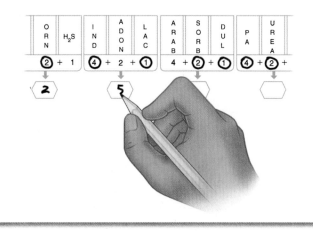

(7) Perform the Voges-Proskauer test, if needed for confirmation, by injecting the reagents into the Voges-Proskauer compartment.

   After encircling the numbers of the positive tests on the Laboratory Report, total up the numbers of each bracketed series to determine the five-digit code number. Refer to the *Enterotube II Interpretation Guide* for identification of the unknown by using the code number.

Figure 41.2   **(continued)**

12 compartments until the notch on the wire is aligned with the opening of the tube. (The notch is about $1\frac{5}{8}''$ from handle end of wire.) The tip of the wire should be visible in the citrate compartment. See illustration 3, figure 41.2.

6. Break the wire at the notch by bending, as shown in step 4, figure 41.2. The portion of the wire remaining in the tube maintains anaerobic conditions essential for fermentation of glucose, production of gas, and decarboxylation of lysine and ornithine.

7. Without touching the end of the wire, use the retained portion of the needle to punch holes through the thin plastic coverings over the small depressions on the sides of the last eight compartments (adonitol, lactose, arabinose, sorbitol, Voges-Proskauer, dulcitol/PA, urea, and citrate). These holes will enable aerobic growth in these eight compartments.

8. Replace the caps at both ends.
9. Incubate at 35–37°C for 18 to 24 hours with the Enterotube II lying on its flat surface. *When incubating several tubes together, leave space between them to allow for air circulation.*

⏱ Second Period

Reading Results

Reading the results on the Enterotube may be done in one of two ways: (1) by simply comparing the results with information on chart IV, Appendix D, or (2) by finding the five-digit code number you compute for your unknown in the *Enterotube II Interpretation Guide*. Of the two methods, the latter is much preferred. The chart in the appendix should be used *only* if the *Interpretation Guide* is not available.

Whether or not the *Interpretation Guide* is available, these three steps will be performed during this period to complete this experiment: (1) positive test results must *first* be recorded on the Laboratory Report; (2) the indole test, a presumptive test, is performed on compartment 4; and (3) confirmatory tests, if needed, are performed. The Voges-Proskauer test falls in the latter category. Proceed as follows:

Materials

- Enterotube II, inoculated and incubated
- Kovacs' reagent
- 10% KOH with 0.3% creatine solution
- 5% alpha-naphthol in absolute ethyl alcohol
- syringes with needles, or disposable Pasteur pipettes
- test-tube rack
- Enterotube II Results Pad (optional)
- coding manual: *Enterotube II Interpretation Guide*

1. Compare the colors of each compartment of your Enterotube II with the lower tube illustrated in figure 41.1.

2. With a pencil, mark a small plus (+) or minus (−) sign near each compartment symbol on the white label on the side of the tube.
3. Consult table 41.1 for information as to the significance of each compartment label.
4. Record the results of the tests on the Laboratory Report. *All results must be recorded before doing the indole test.*
5. **Important:** If at this point you discover that your unknown is GLU-negative, proceed no further with the Enterotube II because your unknown is not one of the Enterobacteriaceae. Your unknown may be *Acinetobacter* sp. or *Pseudomonas maltophilia*. If an Oxi/Ferm tube is available, try it, using the procedure outlined in the next exercise.
6. **Indole Test:** Perform the indole test as follows:
   a. Place the Enterotube II into a test-tube rack with the GLU-GAS compartment pointing upward.
   b. Inject 1 or 2 drops of Kovacs' reagent onto the surface of the medium in the H_2S/Indole compartment. This may be done with a

Table 41.1 Biochemical Reactions of Enterotube II

SYMBOL	UNINOCULATED COLOR	REACTED COLOR	TYPE OF REACTION
GLU-GAS			**Glucose (GLU)** The end products of bacterial fermentation of glucose are either acid or acid and gas. The shift in pH due to the production of acid is indicated by a color change from red (alkaline) to yellow (acidic). Any degree of yellow should be interpreted as a positive reaction; orange should be considered negative.
			Gas Production (GAS) Complete separation of the wax overlay from the surface of the glucose medium occurs when gas is produced. The amount of separation between the medium and overlay will vary with strain of bacteria.
LYS			**Lysine Decarboxylase** Bacterial decarboxylation of lysine, which results in the formation of the alkaline end product cadaverine, is indicated by a change in the color of the indicator from pale yellow (acidic) to purple (alkaline). Any degree of purple should be interpreted as a positive reaction. The medium remains yellow if decarboxylation of lysine does not occur.
ORN			**Ornithine Decarboxylase** Bacterial decarboxylation of ornithine causes the alkaline end product putrescine to be produced. The acidic (yellow) nature of the medium is converted to purple as alkalinity occurs. Any degree of purple should be interpreted as a positive reaction. The medium remains yellow if decarboxylation of ornithine does not occur.
H_2S/IND			**H_2S Production** Hydrogen sulfide, liberated by bacteria that reduce sulfur-containing compounds such as peptones and sodium thiosulfate, reacts with the iron salts in the medium to form a black precipitate of ferric sulfide usually along the line of inoculation. Some *Proteus* and *Providencia* strains may produce a diffuse brown coloration in this medium, which should not be confused with true H_2S production.
			Indole Formation The production of indole from the metabolism of tryptophan by the bacterial enzyme tryptophanase is detected by the development of a pink to red color after the addition of Kovacs' reagent.

Table 41.1   Biochemical Reactions of Enterotube II (continued)

SYMBOL	UNINOCULATED COLOR	POSITIVE REACTIONS	TYPE OF REACTION
ADON			**Adonitol** Bacterial fermentation of adonitol, which results in the formation of acidic end products, is indicated by a change in color of the indicator present in the medium from red (alkaline) to yellow (acidic). Any sign of yellow should be interpreted as a positive reaction; orange should be considered negative.
LAC			**Lactose** Bacterial fermentation of lactose, which results in the formation of acidic end products, is indicated by a change in color of the indicator present in the medium from red (alkaline) to yellow (acidic). Any sign of yellow should be interpreted as a positive reaction; orange should be considered negative.
ARAB			**Arabinose** Bacterial fermentation of arabinose, which results in the formation of acidic end products, is indicated by a change in color from red (alkaline) to yellow (acidic). Any sign of yellow should be interpreted as a positive reaction; orange should be considered negative.
SORB			**Sorbitol** Bacterial fermentation of sorbitol, which results in the formation of acidic end products, is indicated by a change in color from red (alkaline) to yellow (acidic). Any sign of yellow should be interpreted as a positive reaction; orange should be considered negative.
VP			**Voges-Proskauer** Acetylmethylcarbinol (acetoin) is an intermediate in the production of 2,3-butanediol from glucose fermentation. The presence of acetoin is indicated by the development of a red color within 20 minutes. Most positive reactions are evident within 10 minutes.
DUL-PA			**Dulcitol** Bacterial fermentation of dulcitol, which results in the formation of acidic end products, is indicated by a change in color of the indicator present in the medium from green (alkaline) to yellow or pale yellow (acidic). **Phenylalanine Deaminase** This test detects the formation of pyruvic acid from the deamination of phenylalanine. The pyruvic acid formed reacts with a ferric salt in the medium to produce a characteristic black to smoky gray color.
UREA			**Urea** The production of urease by some bacteria hydrolyzes urea in this medium to produce ammonia, which causes a shift in pH from yellow (acidic) to reddish-purple (alkaline). This test is strongly positive for *Proteus* in 6 hours and weakly positive for *Klebsiella* and some *Enterobacter* species in 24 hours.
CIT			**Citrate** Organisms that are able to utilize the citrate in this medium as their sole source of carbon produce alkaline metabolites that change the color of the indicator from green (acidic) to deep blue (alkaline). Any degree of blue should be considered positive.

Courtesy and © Becton, Dickinson and Company

syringe and needle through the thin Mylar plastic film that covers the flat surface, or with a disposable Pasteur pipette through a small hole made in the Mylar film with a hot inoculating needle.

c. A positive test is indicated by the development of a **red color** on the surface of the medium or Mylar film within 10 seconds.

7. **Voges-Proskauer Test:** Since this test is used as a confirmatory test, it should be performed *only*

when called for in the *Enterotube II Interpretation Guide*. If it is called for, perform the test in the following manner:

a. Use a syringe or Pasteur pipette to inject 2 drops of potassium hydroxide containing creatine into the VP section.

b. Inject 3 drops of 5% alpha-naphthol.

c. A positive test is indicated by a **red color** within 10 minutes.

8. Record the indole and VP results on the Laboratory Report.

Laboratory Report

Determine the name of your unknown by following the instructions in Laboratory Report 41. Note that two methods of making the final determination are given.

Student: _____

Date: _____ Section: _____

41 Enterobacteriaceae Identification: The Enterotube II System

A. Results

1. Tabulation

Record the results of each test in the following table with a plus (+) or minus (−) sign.

2. Identification by Chart Method

If no *Interpretation Guide* is available, apply the above results to chart IV, Appendix D, to find the name of your unknown. Note that the spacing of the above table matches the size of the spaces on chart IV. If this page is removed from the manual, folded, and placed on chart IV, the results on the above table can be moved down the chart to make a quick comparison of your results with the expected results for each organism.

3. Using the *Enterotube II Interpretation Guide*

If the *Interpretation Guide* is available, determine the five-digit code number by circling the numbers (4, 2, or 1) under each test that is positive, and then totaling these numbers within each group to form a digit for that group. Note that there are two tally charts in this Laboratory Report for your use.

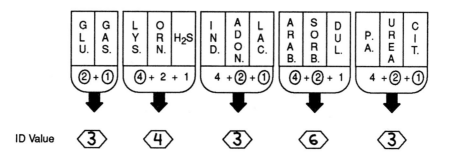

The "ID Value" 34363 can be found by thumbing the pages of the *Interpretation Guide*. The listing is as follows:

ID Value	Organism	Atypical Test Results
34363	*Klebsiella pneumoniae*	None

Conclusion: Organism was correctly identified as *Klebsiella pneumoniae*. In this case, the identification was made independent of the VP test.

4. Additional Tabulation Blanks

ENTEROTUBE® II*

ID Value

Culture Number, Case Number or Patient Name   Date   Organism Identified

*VP utilized as confirmatory test only.

ENTEROTUBE® II*

ID Value

Culture Number, Case Number or Patient Name   Date   Organism Identified

*VP utilized as confirmatory test only.

B. Short-Answer Questions

1. What are the advantages and disadvantages of multitest systems for bacterial identification?

_____

_____

2. Before using the Enterotube II System, what test must be performed to confirm the identity of your unknown as a member of the family Enterobacteriaceae? What is the expected result?

_____

_____

3. For each of the following aspects, compare and contrast the Enterotube II System to the API 20E System (Exercise 40) for Enterobacteriaceae identification.

   a. time requirement

_____

_____

    b. specimen preparation

_____

_____

    c. tests utilized

_____

_____

    d. anaerobic conditions

_____

_____

    e. interpretation of results

_____

_____

4. The five-digit code for all members of the family Enterobacteriaceae starts with the number 2 or 3. What does this indicate about their common biochemistry?

_____

_____

5. If the first number of the five-digit code is 0, what does this indicate? What bacterial genera are likely to give this result? What should you do next?

_____

_____

6. The VP test is a confirmatory test. In what situations would this test be utilized?

_____

_____

7. If the five-digit code that is tabulated cannot be found in the _Enterotube II Interpretation Guide,_ what might that indicate about the bacterial culture?

_____

_____

O/F Gram-Negative Rods Identification:

The Oxi/Ferm Tube II System

Learning Outcomes

After completing this exercise, you should be able to

1. Inoculate an Oxi/Ferm II tube with an oxidase-positive, gram-negative bacterium.

2. Evaluate the test results to generate a five-digit code for the unknown bacterium.

3. Use the five-digit code and the *Biocode Manual* to correctly identify the unknown bacterium.

The Oxi/Ferm Tube II, produced by Becton-Dickinson, takes care of the identification of the oxidase-positive, gram-negative bacteria that cannot be identified by using the Enterotube II System. The two multitest systems were developed to work together. If an unknown gram-negative rod is oxidase-negative, the Enterotube II is used. If the organism is oxidase-positive, the Oxi/Ferm Tube II must be used. Whenever an oxidase-negative gram-negative rod turns out to be glucose-negative on the Enterotube II test, one must move on to use the Oxi/Ferm Tube II.

The Oxi/Ferm Tube II System is intended for the identification of nonfastidious species of oxidative-fermentative gram-negative rods from clinical specimens. This includes the following genera: *Aeromonas, Plesiomonas, Vibrio, Achromobacter, Alcaligenes, Bordetella, Moraxella*, and *Pasteurella*. Some other gram-negative bacteria can also be identified with additional biochemical tests. The system incorporates 12 different conventional media that can be inoculated simultaneously with a minimum of equipment. A total of 14 physiological tests are performed.

Like the Enterotube II System, the Oxi/Ferm Tube II has an inoculating wire that extends through all 12 compartments of the entire tube. To inoculate the media, one simply picks up some organisms on the end of the wire and pulls the wire through each of the chambers in a rotating action.

After incubation, the results are recorded and Kovacs' reagent is injected into one of the compartments to perform the indole test. Positive reactions are given numerical values that are totaled to arrive at a five-digit code. By looking up the code in an *Oxi/Ferm Tube II Biocode Manual*, one can quickly determine the name of the unknown and any tests that might be needed to confirm the identification.

Figure 42.1 illustrates an uninoculated tube and a tube with all positive reactions. Figure 42.2 illustrates the entire procedure for utilizing this system. A minimum of two periods is required to use this system. Proceed as follows:

🕐 First Period

Inoculation and Incubation

The Oxi/Ferm Tube II must be inoculated with a large inoculum from a well-isolated colony. Culture purity is of paramount importance. If there is any doubt of purity, a TSA plate should be inoculated and incubated at 35°C for 24 hours, followed by 24 hours of incubation at room temperature. If no growth occurs on TSA, but growth does occur on blood agar, the organism has special growth requirements. *Such organisms are too fastidious and cannot be identified with the Oxi/Ferm Tube II.*

Materials

- culture plate of unknown
- 1 Oxi/Ferm Tube II
- 1 plate of trypticase soy agar (TSA) (for purity check, if needed)

1. Write your initials or unknown number on the side of the tube.

2. Unscrew both caps from the Oxi/Ferm Tube II. The tip of the inoculating end is under the white cap.

3. *Without heat-sterilizing* the exposed inoculating wire, insert it into a well-isolated colony. Do not puncture the agar.

4. Inoculate each chamber by first twisting the wire and then withdrawing it through all 12 compartments. Rotate the wire as you pull it through. See illustration 2, figure 42.2.

5. If a purity check of the culture is necessary, streak a petri plate of TSA with the inoculating wire that has just been pulled through the tube. **Do not flame.**

6. Again, *without sterilizing*, reinsert the wire, and with a turning motion, force it through all 12 compartments until the notch on the wire is aligned with the opening of the tube. (The notch is about

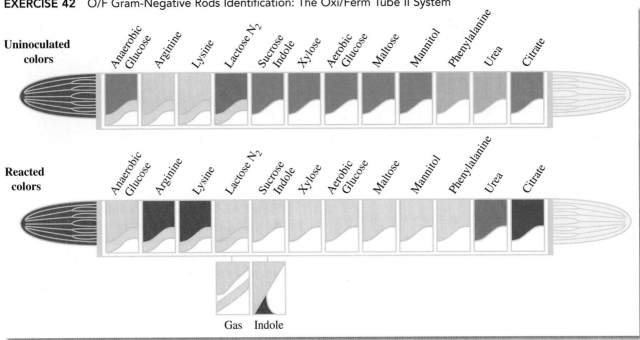

Uninoculated colors

Anaerobic Glucose · Arginine · Lysine · Lactose N₂ · Sucrose Indole · Xylose · Aerobic Glucose · Maltose · Mannitol · Phenylalanine · Urea · Citrate

Reacted colors

Anaerobic Glucose · Arginine · Lysine · Lactose N₂ · Sucrose Indole · Xylose · Aerobic Glucose · Maltose · Mannitol · Phenylalanine · Urea · Citrate

Gas   Indole

Figure 42.1 Oxi/Ferm Tube II color differences between uninoculated and positive tests.
Courtesy and © Becton, Dickinson and Company

(1) Remove organisms from a well-isolated colony. Avoid puncturing the agar with the wire. To prevent damaging Oxi/Ferm II media, do not heat-sterilize the inoculating wire.

(2) Inoculate each compartment by first twisting the wire and then withdrawing it all the way out through the 12 compartments, using a turning movement.

(3) Reinsert the wire (without sterilizing), using a turning motion through all 12 compartments until the notch on the wire is aligned with the opening of the tube.

(4) Break the wire at the notch by bending. The portion of the wire remaining in the tube maintains anaerobic conditions essential for true fermentation.

continued

Figure 42.2 The Oxi/Ferm Tube II procedure.

(5) Punch holes with broken-off part of wire through the thin plastic covering over depressions on sides of the last eight compartments (sucrose/indole through citrate). Replace caps and incubate at 35°C for 18 to 24 hours.

(6) After interpreting and recording positive results on the sides of the tube, perform the indole test by injecting 1 or 2 drops of Kovacs' reagent into the sucrose/indole compartment.

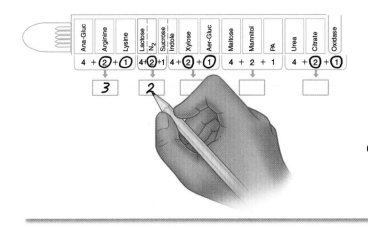

(7) After encircling the numbers of the positive tests on the Laboratory Report, total up the numbers of each bracketed series to determine the five-digit code number. Refer to the *Biocode Manual* for identification of the unknown by using the code number.

Figure 42.2   **(continued)**

$1\frac{5}{8}''$ from the handle end of the wire.) The tip of the wire should be visible in the citrate compartment. See illustration 3, figure 42.2.

7. Break the wire at the notch by bending, as noted in step 4, figure 42.2. The portion of the wire remaining in the tube maintains anaerobic conditions essential for true fermentation.

8. With the retained portion of the needle, punch holes through the thin plastic coverings over the small depressions on the sides of the last eight compartments (sucrose/indole, xylose, aerobic glucose, maltose, mannitol, phenylalanine, urea, and citrate). These holes will enable aerobic growth in these eight compartments.

9. Replace both caps on the tube.

10. Incubate at 35–37°C for 24 hours, with the tube lying on its flat surface or upright. At the end of 24 hours, inspect the tube to check results and continue incubation for another 24 hours. The

24-hour check may be needed for doing confirmatory tests as required in the *Biocode Manual*. Occasionally, an Oxi/Ferm Tube II should be incubated longer than 48 hours.

⏱ Second Period

During this period, you will record the results of the various tests on your Oxi/Ferm Tube II, do an indole test, tabulate your results, use the *Biocode Manual*, and perform any confirmatory tests called for. Proceed as follows:

Materials

- Oxi/Ferm Tube II, inoculated and incubated
- Kovacs' reagent
- syringes with needles, or disposable Pasteur pipettes
- Becton-Dickinson *Biocode Manual* (a booklet)

1. Compare the colors of each compartment of your Oxi/Ferm Tube II with the lower tube illustrated in figure 42.1.
2. With a pencil, mark a small plus (+) or minus (−) sign near each compartment symbol on the white label on the side of the tube.
3. Consult table 42.1 for information as to the significance of each compartment label.
4. Record the results of all the tests on the Laboratory Report. *All results must be recorded before doing the indole test.*

Table 42.1 Biochemical Reactions of the Oxi/Ferm Tube II

REACTION	NEGATIVE	POSITIVE	SPECIAL REMARKS
Anaerobic Glucose			Positive fermentation is shown by change in color from green (neutral) to yellow (acid). Most oxidative-fermentative, gram-negative rods are negative.
Arginine Dihydrolase			Decarboxylation of arginine results in the formation of alkaline end products that changes bromcresol purple from yellow (acid) to purple (alkaline). Gray is negative.
Lysine			Decarboxylation of lysine results in the formation of alkaline end products that changes bromcresol purple from yellow (acid) to purple (alkaline). Gray is negative.
Lactose			Fermentation of lactose changes the color of the medium from red (neutral) to yellow (acid). Most O/F gram-negative rods are negative.
N₂ Gas-production			Gas production causes separation of wax overlay from medium. Occasionally, the gas will also cause separation of the agar from the compartment wall.
Sucrose			Bacterial oxidation of sucrose causes a change in color from green (neutral) to yellow (acid).
Indole			The bacterial enzyme tryptophanase metabolizes tryptophan to produce indole. Detection is by adding Kovacs' reagent to the compartment 48 hours after incubation.
Xylose			Bacterial oxidation of xylose causes a change in color from green (neutral) to yellow (acid).
Aerobic Glucose			Bacterial oxidation of glucose causes a change in color from green (neutral) to yellow (acid).
Maltose			Bacterial oxidation of maltose causes a change in color from green (neutral) to yellow (acid).
Mannitol			Bacterial oxidation of this carbohydrate is evidenced by a change in color from green (neutral) to yellow (acid).
Phenylalanine			Pyruvic acid is formed by deamination of phenylalanine. The pyruvic acid reacts with a ferric salt to produce a brownish tinge.
Urea			The production of ammonia by the action of urease on urea increases the alkalinity of the medium. The phenol red in this medium changes from beige (acid) to pink or purple. Pale pink should be considered negative.
Citrate			Organisms that grow on this medium are able to utilize citrate as their sole source of carbon. Utilization of citrate raises the alkalinity of the medium. The color changes from green (neutral) to blue (alkaline).

5. **Indole Test** (illustration 6, figure 42.2): Do an indole test by injecting 2 or 3 drops of Kovacs' reagent through the flat, plastic surface into the sucrose/indole compartment. Release the reagent onto the inside flat surface and allow it to drop down onto the agar.

If a Pasteur pipette is used instead of a syringe needle, it will be necessary to form a small hole in the Mylar film with a hot inoculating needle to admit the tip of the Pasteur pipette.

A positive test is indicated by the development of a **red color** on the surface of the medium or Mylar film within 10 seconds.

6. Record the results of the indole test on the Laboratory Report.

Laboratory Report

Follow the instructions in Laboratory Report 42 for determining the five-digit code. Use the *Biocode Manual* booklet for identifying your unknown.

42 O/F Gram-Negative Rods Identification: The Oxi/Ferm Tube II System

A. Results

1. Tabulation of Results and Code Determination

Once you have marked the positive reactions on the side of the tube and circled the numbers that are assigned to each of the positive chambers, as indicated in the example below, add the numbers in each bracketed group to get the five-digit code.

The final step is to look up the code number in the *Oxi/Ferm Tube II Biocode Manual* to determine the genus and species. If confirmatory tests are necessary, the manual will tell you which ones to perform.

In the example below, the code number is 32303. If you look up this number in the *Biocode Manual* you will find on page 25 that the organism is *Pseudomonas aeruginosa*.

Use this procedure to identify your unknown by applying your results to the blank diagrams provided.

2. Additional Tabulation Blanks

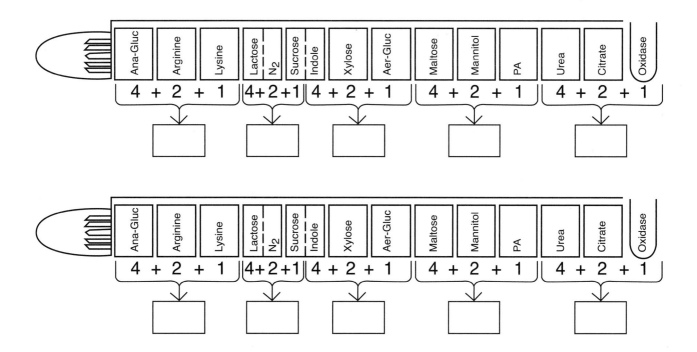

B. Short-Answer Questions

1. What are the advantages and disadvantages of multitest systems for bacterial identification?

   _____

   _____

2. Under what circumstances would you choose the Oxi/Ferm Tube II System over the Enterotube II System (Exercise 41) for the identification of gram-negative bacilli?

   _____

   _____

3. If the five-digit code that is tabulated cannot be found in the *Oxi/Ferm Tube II Biocode Manual,* what might that indicate about the bacterial culture?

   _____

   _____

Staphylococcus Identification:
The API Staph System

Learning Outcomes

After completing this exercise, you should be able to

1. Inoculate an API Staph strip with an unknown staphylococcus.

2. Evaluate the test results to generate a seven-digit profile number for the unknown bacterium.

3. Use the seven-digit profile number and *Staph Profile Register* to correctly identify the unknown bacterium.

The API Staph System, produced by bioMérieux of Raleigh, North Carolina, is a reliable method for identifying 23 species of gram-positive cocci, including 20 clinically important species of staphylococci. This system consists of 19 microampules that contain dehydrated substrates and/or nutrient media. Except for the coagulase test, all the tests are important in the differentiation of *Staphylococcus, Kocuria,* and *Micrococcus.*

Figure 43.1 illustrates two inoculated strips: the upper one just after inoculation and the lower one with positive reactions. Note that the appearance of each microcupule undergoes a pronounced color change when a positive reaction occurs.

Figure 43.2 illustrates the overall procedure. The first step is to make a saline suspension of the organism from an isolated colony. A Staph strip is then placed in a tray that has a small amount of water added to it to provide humidity during incubation. Next, a sterile Pasteur pipette is used to dispense 2–3 drops of the bacterial suspension to each microcupule. The

inoculated tray is covered and incubated aerobically for 18 to 24 hours at 35–37°C. Finally a seven-digit profile number is obtained and used to determine the identity of the organism in Appendix D.

As simple as this system might seem, there are a few limitations that one must keep in mind. Final species determination by a competent microbiologist must take into consideration other factors such as the source of the specimen, the catalase reaction, colony characteristics, and antimicrobial susceptibility pattern. Very often there are confirmatory tests that must also be made.

If you have been working with an unknown that appears to be one of the staphylococci, use this system to confirm your conclusions. If you have already done the coagulase test and have learned that your organism is coagulase-negative, this system will enable you to identify one of the numerous coagulase-negative species that are not identifiable by the procedures in Exercise 51.

🕐 First Period

Inoculations and Coagulase Test

Materials

- API Staph test strip
- API incubation tray and cover
- blood agar plate culture of unknown (must not have been incubated over 30 hours)
- blood agar plate (if needed for purity check)
- serological tube of 2 ml sterile saline
- test-tube rack
- sterile swabs (optional in step 2 on page 313)

Figure 43.1 Negative and positive results on API Staph test strips.
Courtesy of bioMérieux, Inc.

(1) Use several loopfuls of organisms to make saline suspension of unknown. Turbidity of suspension should match McFarland No. 3 barium sulfate standard.

(2) With a Pasteur pipette dispense 2–3 drops of the bacterial suspension into each of the 20 microcupules. Cover the tray with the lid and incubate at 35–37°C for 18 to 24 hours.

(3) Place a STAPH test strip into the bottom of the moistened tray. Take care not to contaminate the microcupules with fingers when handling test strip.

(4) After labeling the end tab of a tray with your name and unknown number, dispense approximately 5 ml of tap water into the bottom of the tray.

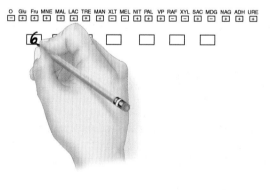

(5) Once all results are recorded on Laboratory Report, total up the positive values in each group to determine the seven-digit profile. Consult chart VII, Appendix D, to find unknown.

Figure 43.2 **The API Staph procedure.**

- squeeze bottle of tap water
- tubes containing McFarland No. 3 (BaSO₄) standard (see Appendix B)
- sterile Pasteur pipette (5 ml size)

1. If the coagulase test has not been performed, refer to Exercise 51, page 363, for the procedure and perform it on your unknown.
2. Prepare a saline suspension of your unknown by transferring organisms to a tube of sterile saline from one or more colonies with a loop or sterile swab. Turbidity of the suspension should match a tube of No. 3 McFarland barium sulfate standard.
   Important: Do not allow the bacterial suspension to go unused for any great length of time. Suspensions older than 15 minutes become less effective.
3. Label the end strip of the tray with your name and unknown number. See illustration 2, figure 43.2.
4. Remove the API test strip from its sealed envelope and place the strip in the bottom of the tray.
5. After shaking the saline suspension to disperse the organisms, fill a sterile Pasteur pipette with the bacterial suspension.
6. Inoculate each of the microcupules with 2 or 3 drops of the suspension. If a purity check is necessary, use the excess suspension to inoculate another blood agar plate.
7. Dispense about 5 ml of tap water into the bottom of the tray with a squeeze bottle. Note that the bottom of the tray has numerous depressions to accept the water.
8. Place the plastic lid on the tray and incubate the strip aerobically for 18 to 24 hours at 35–37°C.

◐ Second Period

(Five Hours Later)

During this period, the results will be recorded on the Laboratory Report, the profile number will be determined, and the unknown will be identified by looking up the number on the *Staph Profile Register* (or chart VI, Appendix D).

Materials

- API Staph test strip (incubated 18 to 24 hours)
- *Staph Profile Register*
- Barritt's reagents A and B
- nitrate reagents A and B
- ZYM reagent A
- ZYM reagent B

1. After 18 to 24 hours of incubation, read and interpret the test results (figure 43.3).
   a. For the Voges-Proskauer test, add 1 drop each of Barritt's A and Barritt's B to the VP ampule. Incubate 10 minutes. A positive test is indicated by a violet-pink color; a negative test shows no color change.
   b. For the nitrate test, add 1 drop each of nitrate A and nitrate B test reagents to the NIT ampule. Incubate 10 minutes. A positive test is a violet-pink color; a negative test shows no color change.
   c. For the alkaline phosphatase (PAL) test, add 1 drop each of ZYM A and ZYM B to the PAL ampule. Incubate 10 minutes. A positive test is violet in color; a negative test is yellow.
2. Record the results in the Laboratory Report.
3. Construct the profile number according to the instructions on the Laboratory Report and determine the name of your unknown. Use chart VII, Appendix D.

Disposal

Once all the information has been recorded be sure to place the entire incubation unit in a receptacle that is to be autoclaved.

Results

Record your results in the chart on page 315. Determine the profile number to find a match for your organism.

POSITIVE TESTS

Figure 43.3 Test results of a positive strip.
Courtesy of bioMérieux, Inc.

43 Staphylococcus Identification: The API Staph System

A. Results

1. Tabulation

   By referring to chart VI in Appendix D, determine the results of each test, and record these results as positive (+) or negative (−) in the Profile Determination Table below. Note there are two more tables below number 3 for tabulation of additional organisms.

	O 1	GLU 2	FRU 4	MNE 1	MAL 2	LAC 4	TRE 1	MAN 2	XLT 4	MEL 1	NIT 2	PAL 4	VP 1	RAF 2	XYL 4	SAC 1	MDG 2	NAG 4	ADH	URE	Lysost*
RESULTS																					
PROFILE NUMBER																					

GRAM STAIN ☐   COAGULASE ☐   Additional Information   Identification

MORPHOLOGY ☐   CATALASE ☐

*Lysostaphin-lysostatin resistance to be done as a separate test as described by the manufacturer. To be included to determine the seven-digit ID number.

2. Construction of Seven-Digit Profile

   Note in the above table that each test has a value of 1, 2, or 4. To compute the seven-digit profile for your unknown, total up the positive values for each group.

3. Final Determination

   Refer to the *Staph Profile Register* (or chart VII, Appendix D) to find the organism that matches your profile number. Write the name of your unknown in the space below and list any additional tests that are needed for final confirmation. If the materials are available for these tests, perform them.

   Name of Unknown: _____

   Additional Tests: _____

	O 1	GLU 2	FRU 4	MNE 1	MAL 2	LAC 4	TRE 1	MAN 2	XLT 4	MEL 1	NIT 2	PAL 4	VP 1	RAF 2	XYL 4	SAC 1	MDG 2	NAG 4	ADH	URE	Lysost*
RESULTS																					
PROFILE NUMBER																					

GRAM STAIN ☐   COAGULASE ☐   Additional Information   Identification

MORPHOLOGY ☐   CATALASE ☐

*Lysostaphin-lysostatin resistance to be done as a separate test as described by the manufacturer. To be included to determine the seven-digit ID number.

	O 1	GLU 2	FRU 4	MNE 1	MAL 2	LAC 4	TRE 1	MAN 2	XLT 4	MEL 1	NIT 2	PAL 4	VP 1	RAF 2	XYL 4	SAC 1	MDG 2	NAG 4	ADH	URE	Lysost*
RESULTS																					

PROFILE NUMBER

GRAM STAIN ☐   COAGULASE ☐

MORPHOLOGY ☐   CATALASE ☐

Additional Information

Identification

*Lysostaphin-lysostatin resistance to be done as a separate test as described by the manufacturer. To be included to determine the seven-digit ID number.

B. Short-Answer Questions

1. What are the advantages and disadvantages of multitest systems for bacterial identification?

_____

_____

2. Before using the API Staph System, what tests should be performed on bacteria to confirm that they are staphylococci?

_____

_____

3. What single test differentiates *Staphylococcus aureus* from other species of staphylococci? What is the expected result?

_____

_____

4. If the seven-digit code that is tabulated cannot be found in the Profile Determination Table, what might that indicate about the bacterial culture?

_____

_____

PART 10

Applied Microbiology

Applied microbiology encompasses many aspects of modern microbiology. We use microorganisms to produce many of the foods we eat such as cheese, yogurt, bread, sauerkraut, and a whole list of fermented beverages. Microorganisms are important in industrial applications where they are involved in producing antibiotics, pharmaceuticals, and even solvents and starting materials for the manufacture of plastics. Their presence and numbers in our foods and drinking water determine if it is safe to consume these substances as they could cause us harm and disease. In the following exercises, you will explore some of the applications of microbiology by determining bacterial numbers and/or kinds in food and water. You will also study the process of alcohol fermentation as an example of food production.

© Brand X Pictures/PunchStock RF

Bacterial Counts of Foods

Learning Outcomes

After completing this exercise, you should be able to

1. Understand the role that microorganisms have in food production and in food spoilage.

2. Perform a standard plate count on a food sample to determine the number of bacteria in the sample.

The presence of microorganisms in food does not necessarily indicate that the food is spoiled or that it has the potential to cause disease. Some foods can have high counts because microorganisms are used in their production. Yogurt, sauerkraut, and summer sausage are examples of foods prepared by microbial fermentation and, therefore, they have high bacterial counts associated with them during production. However, post-production treatments such as pasteurization or smoking will significantly reduce the numbers of bacteria present. During processing and preparation, food can become contaminated with bacteria, which naturally occur in the environment. These bacteria may not be necessarily harmful or pathogenic. Bacteria are naturally associated with some foods when they are harvested. For example, green beans, potatoes, and beets have soil bacteria associated with them when harvested. Even after washing some bacteria can remain and will be preserved with the food when it is frozen. The chalky appearance of grapes is due to yeasts that are naturally associated with grapes and many other fruits. Milk in the udders of healthy cows is sterile, but bacteria such as *Streptococcus* and *Lactobacillus* are introduced during milking and processing because they are part of the bacterial flora associated with the animal, especially the outside of the udder. Pasteurization kills many of the bacteria that are introduced during processing, and any pathogens that may be present, but it does not kill all the bacteria present in milk. Some bacteria in milk can survive pasteurization temperatures and eventually cause spoilage and souring of milk. These are called thermoduric bacteria. Hamburger can also have high counts of bacteria that can be introduced during processing and grinding of the meat. Many bacteria in hamburger are harmless saprophytes (organisms that live on decaying plant and animal material) that come from the environment where processing occurs. For example, endospore-forming bacteria and others can be introduced into ground beef during its preparation.

We must also bear in mind that food can be an important means for the transmission of disease. The Centers for Disease Control estimates that 76 million people per year in the United States become sick; 300,000 are hospitalized; and 5,000 people die from foodborne illnesses. Foodborne illnesses usually result because pathogenic bacteria or their toxins are introduced into food products during processing, handling, or preparation. Food handlers can transmit opportunistic pathogens associated with the human body, like *Staphylococcus aureus* or intestinal bacteria, because of unsanitary practices such as failure to wash their hands before preparing or handling food. Botulism food poisoning results from ingesting a toxin produced by *Clostridium botulinum* when its endospores grow in improperly home-canned foods. The endospores occur in the soil and the environment and contaminate the prepared vegetables. *Salmonella* and *Campylobacter* are associated with poultry and eggs and can cause illness if these foods are not properly prepared. *Escherichia coli* O157:H7 is found in the intestines of cattle and can become associated with meat if fecal material from the animal's intestines contaminates meat during the butchering process. This pathogen is then incorporated into hamburger during grinding and processing. Serious illness results from eating improperly cooked hamburger because cooking temperatures are insufficient to kill the organism. Transmission of this pathogen has also occurred when fecal material of cattle contaminated fruits and vegetables such as lettuce and spinach.

Although high bacterial counts in food do not necessarily mean that the food is spoiled or that it harbors disease-causing organisms, it can suggest the potential for more rapid spoilage of the food. Thus, high counts can be important for this reason. One method to ascertain if food is contaminated with fecal bacteria and, therefore, has the potential to spread disease is to perform coliform counts. Coliforms are organisms such *Escherichia coli* that occur in the intestines of humans and warm-blooded animals. Their presence in food or water indicates that fecal contamination has occurred and that there is the high potential for the spread of serious disease such

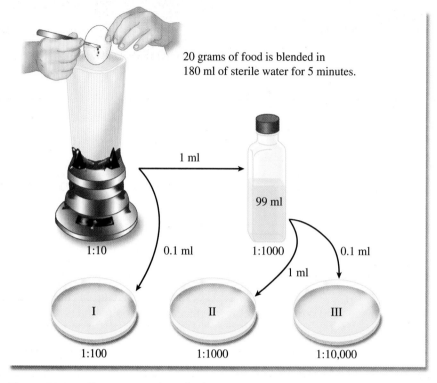

20 grams of food is blended in 180 ml of sterile water for 5 minutes.

1 ml

99 ml

1:10   0.1 ml   1:1000   0.1 ml

1 ml

I   II   III

1:100   1:1000   1:10,000

Figure 44.1   Dilution procedure for bacterial counts of food.

as typhoid fever, bacillary dysentery, cholera, and intestinal viral diseases.

The standard plate count and the coliform count can be used to evaluate foods in much the same manner that they are used on milk and water to determine total bacterial counts and the number of coliforms. However, because most foods are solid in nature, organisms associated with a food sample must be put into suspension by using a food blender before counts can be performed. In this exercise, samples of ground beef, dried fruit, and frozen food will be tested for total numbers of bacteria. This procedure, however, will not determine the numbers of coliforms. Your instructor will indicate the specific kinds of foods to be tested and make individual assignments. Figure 44.1 illustrates the general procedure.

Materials

per student:
- 3 petri plates
- 1 bottle (45 ml) of Plate Count agar or Standard Methods agar
- one 99-ml sterile water blank
- two 1.1-ml dilution pipettes

per class:
- food blender
- sterile blender jars (one for each type of food)
- sterile weighing paper
- 180 ml sterile water blanks (one for each type of food)
- samples of ground meat, dried fruit, and frozen vegetables, thawed 2 hours

1. Using aseptic technique, weigh out on sterile weighing paper 20 grams of food to be tested.
2. Add the food and 180 ml of sterile water to a sterile mechanical blender jar. Blend the mixture for 5 minutes. This suspension will provide a 1:10 dilution.
3. With a 1.1 ml dilution pipette, dispense from the blender 0.1 ml to plate 1 and 1.0 ml to the water blank. See figure 44.1.
4. Shake the water blank 25 times in an arc for 7 seconds with your elbow on the table as done in Exercise 19 (Enumeration of Bacteria).
5. Using a fresh pipette, dispense 0.1 ml to plate III and 1.0 ml to plate II.
6. Pour agar (50°C) into the three plates and incubate them at 35°C for 24 hours.
7. Count the colonies on the best plate and record the results in Laboratory Report 44.

44

44 Bacterial Counts of Foods

A. Results

Record your count and the bacterial counts of various other foods made by other students.

TYPE OF FOOD	PLATE COUNT	DILUTION	ORGANISMS PER ML

B. Short-Answer Questions

1. Which type of food had the highest bacterial count? Explain.

   _____

   _____

2. Which type of food had the lowest bacterial count? Explain.

   _____

   _____

3. Coliform bacteria are common contaminants of meats.

   a. Why might one expect to find coliforms in samples of meat?

   _____

   _____

   b. High coliform counts in food indicate the potential for finding which intestinal pathogens?

   _____

   _____

   c. In terms of food safety, why is it suggested to cook hamburgers medium-well to well-done, whereas steaks can be cooked rare?

   _____

   _____

4. What considerations should be made to safely thaw frozen foods for later consumption?

   _____

   _____

5. Why is refrigeration not always an effective means for preventing food spoilage?

   _____

   _____

6. Why are dried fruits somewhat resistant to spoilage?

   _____

   _____

7. Give two examples of foods that have high bacterial counts but are not spoiled.

   _____

   _____

Bacteriological Examination of Water:

Most Probable Number Determination

Learning Outcomes

After completing this exercise, you should be able to

1. Understand the role of water in transmitting microbial diseases.

2. Define a *coliform* and how coliforms are assayed for in water samples.

3. Perform a coliform count on a water sample and determine the most probable number of coliforms.

Prior to the modern age of public health, water was a major means for the spread of infectious diseases such as cholera, dysentery, and typhoid fever. A physician, John Snow, showed in the 1840s that a cholera epidemic in London was the result of cesspool overflow into the Thames River from a tenement where cholera patients lived. When water for drinking was drawn by inhabitants near the cesspool discharge, the contaminated water and pump became the source for the spread of the disease to people in the area. Snow's solution was simply to remove the handle to the pump, and the epidemic abated. Water safety is still a primary concern of municipalities in today's world, and it has become complex. Because of good public health measures, most of us are confident that the water we draw from our faucets is safe and will not cause us disease.

From a microbiological standpoint, it is not the numbers of bacteria that are present in water that is of primary concern to us but rather the kinds of bacteria. Water found in rivers, lakes, and streams can contain a variety of bacteria that may only be harmless saprophytes, which do not cause disease in humans. However, it is important that water not contain the intestinal pathogens that cause typhoid, cholera, and dysentery. In modern cities, treated sewage is discharged into receiving waters of lakes, rivers, and streams, and this constitutes a major sanitary problem because those same bodies of water are the sources of our drinking water. As a result, we have developed methods to treat water to eliminate the potential for disease, and we do microbiological tests to determine if water is potable and safe for consumption.

At first glance, it might seem reasonable to directly examine water for the presence of the pathogens *Vibrio cholerae, Salmonella typhi,* and *Shigella dysenteriae.* However, this is not the case because it would be tedious and difficult to specifically test for each of the pathogens.

Furthermore, these bacteria are often fastidious, and they might be overgrown by other bacteria in the water if we tried to culture and test for them. It is much easier to demonstrate the presence of some indicator bacterium, such as *Escherichia coli,* which is routinely found in the human intestine but is not found in the soil or water. The presence of these bacteria in water would then indicate the likelihood of fecal contamination and the potential for serious disease.

E. coli is a good indicator of fecal contamination and a good test organism. This is for several reasons: (1) it occurs primarily in the intestines of humans and some warm-blooded animals and it is not found routinely in soil or water; (2) the organism can be easily identified by microbiological tests; (3) it is not as fastidious as the other intestinal pathogens, and hence it survives a little longer in water samples. By definition, organisms such as *E. coli* and *Enterobacter aerogenes* are designated as **coliforms,** which are gram-negative, facultative anaerobic, non-endospore-forming rods that ferment lactose to produce acid and gas in 48 hours at 35°C. Lactose fermentation with the formation of acid and gas provides the basis for determining the total coliform count of water samples in the United States and therefore designates water purity. The presence of other bacteria, such as *Enterococcus faecalis,* which is a gram-positive coccus that inhabits the human intestine, can also indicate fecal contamination, but testing for this bacterium is not routinely done in the United States.

Three different tests are done to determine the coliform count (figure 45.1): presumptive, confirmed, and completed. Each test is based on one or more of the characteristics of a coliform. A description of each test follows.

Presumptive Test   In the presumptive test, 15 tubes of lactose broth are inoculated with measured amounts of water to see if the water contains any lactose-fermenting bacteria that produce gas. If, after incubation, gas is seen in any of the lactose broths, it is *presumed* that coliforms are present in the water sample. This test is also used to determine the **most probable number (MPN)** of coliforms present per 100 ml of water.

Confirmed Test   In this test, plates of Levine EMB agar or Endo agar are inoculated from positive (gas-producing) tubes to see if the organisms that are producing the gas are gram-negative (another coliform characteristic). Both of these media inhibit the growth

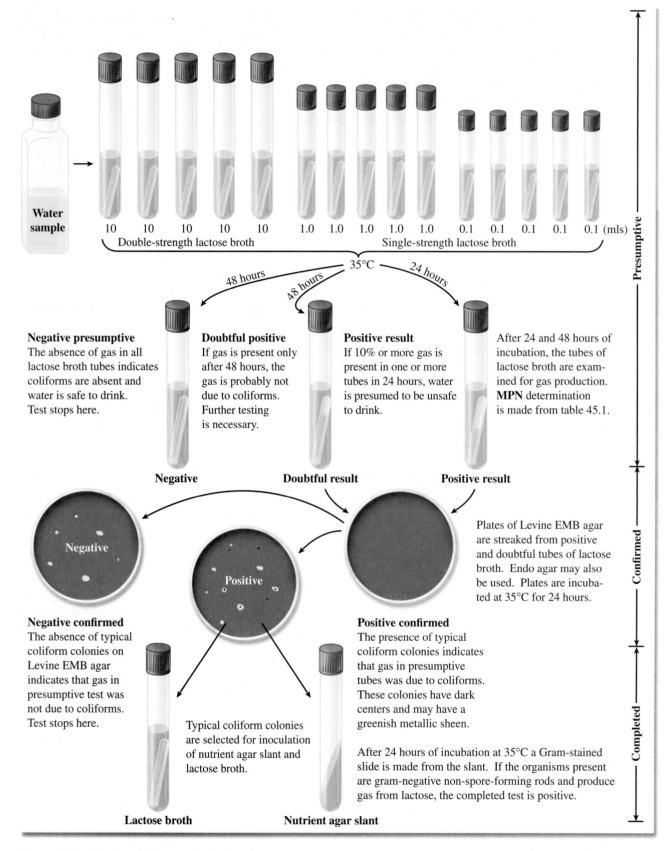

Water sample

10 | 10 | 10 | 10 | 10 | 1.0 | 1.0 | 1.0 | 1.0 | 1.0 | 0.1 | 0.1 | 0.1 | 0.1 | 0.1 (mls)

Double-strength lactose broth | Single-strength lactose broth

Presumptive

35°C

48 hours | 48 hours | 24 hours

Negative presumptive
The absence of gas in all lactose broth tubes indicates coliforms are absent and water is safe to drink. Test stops here.

Doubtful positive
If gas is present only after 48 hours, the gas is probably not due to coliforms. Further testing is necessary.

Positive result
If 10% or more gas is present in one or more tubes in 24 hours, water is presumed to be unsafe to drink.

After 24 and 48 hours of incubation, the tubes of lactose broth are examined for gas production. **MPN** determination is made from table 45.1.

Negative | **Doubtful result** | **Positive result**

Negative

Positive

Plates of Levine EMB agar are streaked from positive and doubtful tubes of lactose broth. Endo agar may also be used. Plates are incubated at 35°C for 24 hours.

Confirmed

Negative confirmed
The absence of typical coliform colonies on Levine EMB agar indicates that gas in presumptive test was not due to coliforms. Test stops here.

Typical coliform colonies are selected for inoculation of nutrient agar slant and lactose broth.

Positive confirmed
The presence of typical coliform colonies indicates that gas in presumptive tubes was due to coliforms. These colonies have dark centers and may have a greenish metallic sheen.

After 24 hours of incubation at 35°C a Gram-stained slide is made from the slant. If the organisms present are gram-negative non-spore-forming rods and produce gas from lactose, the completed test is positive.

Completed

Lactose broth | **Nutrient agar slant**

Figure 45.1 **Bacteriological analysis of water.**

of gram-positive bacteria and cause colonies of coliforms to be distinguishable from noncoliforms. On EMB agar, coliforms produce small colonies with dark centers (nucleated colonies). On Endo agar, coliforms produce reddish colonies. The presence of coliform-like colonies confirms the presence of a lactose-fermenting, gram-negative bacterium.

Completed Test   In the completed test, our concern is to determine if the isolate from the agar plates truly matches our definition of a coliform. The media for this test include a nutrient agar slant and a Durham tube of lactose broth. If gas is produced in the lactose tube and a slide from the agar slant reveals that we have a gram-negative, non-spore-forming rod, we can be certain that we have a coliform.

The completion of these three tests with positive results establishes that coliforms are present; however, there is no certainty that *E. coli* is the coliform present. The organism might be *E. aerogenes*. Of the two, *E. coli* is the better sewage indicator since *E. aerogenes* can be of nonsewage origin. To differentiate these two species, one must perform the **IMViC tests,** which are described on page 271 in Exercise 38.

In this exercise, water will be tested from local ponds, streams, swimming pools, and other sources supplied by students and instructor. Enough known positive samples will be evenly distributed throughout the laboratory so that all students will be able to see positive test results. All three tests in figure 45.1 will be performed. If time permits, the IMViC tests may also be performed.

The Presumptive Test

As stated earlier, the presumptive test is used to determine if gas-producing lactose fermenters are present in a water sample. The American Public Health Association's *Standard Methods for Examination of Water and Waste Water,* 21st edition employs a 15-tube method for the analysis of water samples which is more statistically valid than previous procedures using fewer tubes. This 15-tube procedure is most applicable to clear surface waters but can be used for brackish water and waters with sediments. Figure 45.1 shows the procedure for determining the MPN of a water sample using sample volumes of 10 ml, 1 ml, and 0.1 ml each delivered to five tubes of test medium. The figure also describes the subsequent procedures for performing the confirmed and completed tests for tubes in the presumptive test that show gas production. In addition to detecting the presence or absence of coliforms in the presumptive test, the pattern of positive tubes is used to ascertain the most probable number (MPN) of coliforms in 100 ml of water. See table 45.1 to determine that value from the number of positive tubes.

Set up the test for your water sample using the procedure outlined in figure 45.1. As stated, the method will be used for clear surface waters as well as turbid waters with sediments.

Materials

- 5 Durham tubes of DSLB
- 10 Durham tubes of SSLB
- one 10-ml pipette
- one 1-ml pipette
  Note: DSLB designates double-strength lactose broth. It contains twice as much lactose as SSLB (single-strength lactose broth).

1. Set up 5 DSLB and 10 SSLB tubes as illustrated in figure 45.1. Label each tube according to the amount of water that is to be dispensed to it: *10 ml, 1.0 ml,* and *0.1 ml,* respectively.
2. Mix the bottle of water to be tested by shaking it 25 times.
3. With a 10 ml pipette, transfer 10 ml of water to each of the DSLB tubes.
4. With a 1.0 ml pipette, transfer 1 ml of water to each of the middle set of tubes, and 0.1 ml to each of the last five SSLB tubes.
5. Incubate the tubes at 35°C for 24 hours.
6. Examine the tubes and record the number of tubes in each set that have 10% gas or more.
7. Determine the MPN by referring to table 45.1. Consider the following:
   Example: If gas was present in the first five tubes (10 ml samples), and detected in only one tube of the second series (1 ml samples), but was not present in any of the last five tubes (0.1 ml samples), your test results would read as 5-1-0. Table 45.1 indicates that the MPN for this pattern of tubes is 33. This means that this water sample would have approximately 33 organisms per 100 ml with a 95 percent probability of there being between 10 and 100 organisms. *Keep in mind that the MPN of 33 is only a statistical probability figure.* Please also note that table 45.1 cannot be used to determine the MPN for water samples using less than 15 tubes, for example 9 tubes as was once used by the American Public Health Association. *Standard Methods for Examination of Water and Waste Water* provides tables for interpreting results if other combinations of dilutions are employed. If other dilutions are used for the MPN determination, the MPN can also be estimated using **Thomas' simple formula.**

$$\text{MPN/100 ml} = \frac{\text{no. positive tubes} \times 100}{\sqrt{\frac{\text{ml sample in}}{\text{negative tubes}} \times \frac{\text{ml sample in}}{\text{all tubes}}}}$$

8. Record the data in Laboratory Report 45.

Table 45.1 MPN Index and 95% Confidence Limits for Various Combinations of Positive Results When Five Tubes Are Used for Dilution (10 ml, 1.0 ml, 0.1 ml)*

Combination of Positives	MPN Index/ 100 ml	Confidence Limits		Combination of Positives	MPN Index/ 100 ml	Confidence Limits	
		Low	High			Low	High
0-0-0	<1.8	—	6.8	4-0-3	25	9.8	70
0-0-1	1.8	0.090	6.8	4-1-0	17	6.0	40
0-1-0	1.8	0.090	6.9	4-1-1	21	6.8	42
0-1-1	3.6	0.70	10	4-1-2	26	9.8	70
0-2-0	3.7	0.70	10	4-1-3	31	10	70
0-2-1	5.5	1.8	15	4-2-0	22	6.8	50
0-3-0	5.6	1.8	15	4-2-1	26	9.8	70
1-0-0	2.0	0.10	10	4-2-2	32	10	70
1-0-1	4.0	0.70	10	4-2-3	38	14	100
1-0-2	6.0	1.8	15	4-3-0	27	9.9	70
1-1-0	4.0	0.71	12	4-3-1	33	10	70
1-1-1	6.1	1.8	15	4-3-2	39	14	100
1-1-2	8.1	3.4	22	4-4-0	34	14	100
1-2-0	6.1	1.8	15	4-4-1	40	14	100
1-2-1	8.2	3.4	22	4-4-2	47	15	120
1-3-0	8.3	3.4	22	4-5-0	41	14	100
1-3-1	10	3.5	22	4-5-1	48	15	120
1-4-0	10	3.5	22	5-0-0	23	6.8	70
2-0-0	4.5	0.79	15	5-0-1	31	10	70
2-0-1	6.8	1.8	15	5-0-2	43	14	100
2-0-2	9.1	3.4	22	5-0-3	58	22	150
2-1-0	6.8	1.8	17	5-1-0	33	10	100
2-1-1	9.2	3.4	22	5-1-1	46	14	120
2-1-2	12	4.1	26	5-1-2	63	22	150
2-2-0	9.3	3.4	22	5-1-3	84	34	220
2-2-1	12	4.1	26	5-2-0	49	15	150
2-2-2	14	5.9	36	5-2-1	70	22	170
2-3-0	12	4.1	26	5-2-2	94	34	230
2-3-1	14	5.9	36	5-2-3	120	36	250
2-4-0	15	5.9	36	S-2-4	150	58	400
3-0-0	7.8	2.1	22	5-3-0	79	22	220
3-0-1	11	3.5	23	5-3-1	110	34	250

(continued)

Table 45.1   MPN Index and 95% Confidence Limits for Various Combinations of Positive Results When Five Tubes Are Used for Dilution (continued)   (10 ml, 1.0 ml, 0.1 ml)*

Combination of Positives	MPN Index/ 100 ml	Confidence Limits		Combination of Positives	MPN Index/ 100 ml	Confidence Limits	
		Low	High			Low	High
3-0-2	13	5.6	35	5-3-2	140	52	400
3-1-0	11	3.5	26	5-3-3	170	70	400
3-1-1	14	5.6	36	5-3-4	210	70	400
3-1-2	17	6.0	36	5-4-0	130	36	400
3-2-0	14	5.7	36	5-4-1	170	58	400
3-2-1	17	6.8	40	5-4-2	220	70	440
3-2-2	20	6.8	40	5-4-3	280	100	710
3-3-0	17	6.8	40	5-4-4	350	100	710
3-3-1	21	6.8	40	5-4-5	430	150	1100
3-3-2	24	9.8	70	5-5-0	240	70	710
3-4-0	21	6.8	40	5-5-1	350	100	1100
3-4-1	24	9.8	70	5-5-2	540	150	1700
3-5-0	25	9.8	70	5-5-3	920	220	2600
4-0-0	13	4.1	35	5-5-4	1600	400	4600
4-0-1	17	5.9	36	5-5-5	>1600	700	—
4-0-2	21	6.8	40				

*Results to two significant figures.

From A. D. Eaton, L. S. Clesceri, E. W. Rice, A. E. Greenberg, eds. *Standard Methods for the Examination of Water and Wastewater*, 21st Edition. Washington, DC: APHA, AWWA, WEF; 2005:9-54 (Table 9221:IV).

The Confirmed Test

Once it has been established that gas-producing lactose fermenters are present in the water, it is *presumed* to be unsafe. However, gas formation may be due to noncoliform bacteria. Some of these organisms, such as *Clostridium perfringens*, are gram-positive. To confirm the presence of gram-negative lactose fermenters, the next step is to inoculate media such as Levine eosin-methylene blue (EMB) agar or Endo agar from positive presumptive tubes.

Levine EMB agar contains methylene blue, which inhibits gram-positive bacteria. Gram-negative lactose fermenters (coliforms) that grow on this medium will produce "nucleated colonies" (dark centers). Colonies of *E. coli* and *E. aerogenes* can be differentiated on the basis of size and the presence of a greenish metallic sheen. *E. coli* colonies on this medium are small and have this metallic sheen, whereas *E. aerogenes* colonies usually lack the sheen and are larger. Differentiation in this

manner is not completely reliable, however. It should be remembered that *E. coli* is the more reliable sewage indicator since it is not normally present in soil, while *E. aerogenes* has been isolated from soil and grains.

Endo agar contains a fuchsin sulfite indicator that makes identification of lactose fermenters relatively easy. Coliform colonies and the surrounding medium appear red on Endo agar. Nonfermenters of lactose, on the other hand, are colorless and do not affect the color of the medium.

In addition to these two media, there are several other media that can be used for the confirmed test. Brilliant green bile lactose broth, Eijkman's medium, and EC medium are just a few examples that can be used.

To demonstrate the confirmation of a positive presumptive in this exercise, the class will use Levine EMB agar and Endo agar. One-half of the class will use one medium; the other half will use the other medium. Plates will be exchanged for comparisons.

Materials

- 1 petri plate of Levine EMB agar (odd-numbered students)
- 1 petri plate of Endo agar (even-numbered students)

1. Select one positive lactose broth tube from the presumptive test and streak a plate of medium according to your assignment. Use a streak method that will produce good isolation of colonies. If all your tubes were negative, borrow a positive tube from another student.
2. Incubate the plate for 24 hours at 35°C.
3. Look for typical coliform colonies on both kinds of media. Record your results on Laboratory Report 45. If no coliform colonies are present, the water is considered bacteriologically safe to drink.

   Note: In actual practice, confirmation of all presumptive tubes would be necessary to ensure accuracy of results.

The Completed Test

A final check of the colonies that appear on the confirmatory media is made by inoculating a nutrient agar slant and lactose broth with a Durham tube. After incubation for 24 hours at 35°C, the lactose broth is examined for gas production. A Gram-stained slide is made from the slant, and the slide is examined under oil immersion optics.

If the organism proves to be a gram-negative, non-spore-forming rod that ferments lactose, we know that coliforms were present in the tested water sample. If time permits, complete these last tests and record the results in Laboratory Report 45.

The IMViC Tests

Review the discussion of the IMViC tests on page 271. The significance of these tests should be much more apparent at this time. Your instructor will indicate whether these tests should be performed if you have a positive completed test.

45 Bacteriological Examination of Water: Most Probable Number Determination

A. Results

1. **Presumptive Test (MPN Determination)**

   Record the number of positive tubes on the chalkboard and on the following table. When all students have recorded their results with the various water samples, complete this tabulation. Determine the MPN according to the instructions on page 325 and table 45.1.

WATER SAMPLE (SOURCE)	NUMBER OF POSITIVE TUBES			MPN
	5 Tubes DSLB 10 ml	5 Tubes SSLB 1.0 ml	5 Tubes SSLB 0.1 ml	

2. **Confirmed Test**

   Record the results of the confirmed tests for each water sample that was positive on the presumptive test.

WATER SAMPLE (SOURCE)	POSITIVE	NEGATIVE

3. **Completed Test**

   Record the results of completed tests for each water sample that was positive on the confirmed test.

WATER SAMPLE (SOURCE)	LACTOSE FERMENTATION RESULTS	MORPHOLOGY	EVALUATION

B. Short-Answer Questions

1. Does a positive presumptive test mean that the water is absolutely unsafe to drink? _____

   Explain: _____

2. What might cause a false positive presumptive test? _____

   _____

3. List three characteristics required of a good sewage indicator:

   a. _____ b. _____ c. _____

4. The fermentation of what disaccharide is the basis for determining the presence of coliforms?

   _____

5. Describe the appearance of coliforms on Levine EMB agar. _____

   _____

6. Why don't health departments routinely test for pathogens instead of using a sewage indicator?

   _____

   _____

7. List five characteristics of coliform bacteria.

   _____

   _____

   _____

8. How is each of the following media used for the detection of coliforms?

   a. lactose broth with Durham tube

   _____

   _____

   b. Levine EMB agar

   _____

   _____

   c. nutrient agar slant

   _____

   _____

9. Once the completed test establishes the presence of coliforms in the water sample, why might you perform the IMViC tests on these isolates?

   _____

   _____

Bacteriological Examination of Water:

The Membrane Filter Method

Learning Outcomes

After completing this exercise, you should be able to

1. Understand how coliform counts of water are obtained faster using the membrane filter method.
2. Perform the membrane filter procedure for determining the number of coliforms in a water sample.

The most probable number method for determining coliform bacteria in water samples is complicated and requires several days to complete. Furthermore, more than one kind of culture medium is needed for each phase of the test to finally establish the presence of coliforms in a water sample. A more rapid method is the **membrane filter method**, also recognized by the United States Public Health Service as a reliable procedure for determining coliforms. In this test, known volumes of a water sample are filtered through membrane filters that have pores 0.45 μm in diameter. Most bacteria, including coliforms, are larger than the pore diameters, and hence they are retained on the membrane filter. Once the water sample has been filtered, the filter disk containing bacterial cells is placed in a petri dish with an absorbent pad saturated with Endo broth. The plate is then incubated at 35°C for 22 to 24 hours, during which time individual cells on the filter multiply, forming colonies.

Any coliforms that are present on the filter will ferment the lactose in the Endo broth, producing acids. The acids produced from fermentation interact with basic fuchsin, a dye in the medium, causing coliform colonies to have a characteristic metallic sheen. Non-coliform bacteria will not produce the metallic sheen. Gram-positive bacteria are inhibited from growing because of the presence of bile salts and sodium lauryl sulfate, which inhibit these bacteria. Colonies are easily counted on the filter disk, and the total coliform count is determined based on the volume of water filtered.

Figure 46.1 illustrates the procedure we will use in this experiment.

Materials

- vacuum pump or water faucet aspirators
- membrane filter assemblies (sterile)
- side-arm flask, 1000 ml size, and rubber hose
- sterile graduates (100 ml or 250 ml size)
- sterile, plastic petri dishes, 50 mm dia (Millipore #PD10 047 00)
- sterile membrane filter disks (Millipore #HAWG 047 AO)
- sterile absorbent disks (packed with filters)
- sterile water
- 5 ml pipettes
- bottles of *m* Endo MF broth (50 ml)*
- water samples

1. Prepare a small plastic petri dish as follows:
   a. With a flamed forceps, transfer a sterile absorbent pad to a sterile plastic petri dish.
   b. Using a 5 ml pipette, transfer 2.0 ml of *m* Endo MF broth to the absorbent pad.
2. Assemble a membrane filtering unit as follows:
   a. *Aseptically* insert the filter holder base into the neck of a 1-liter side-arm flask.
   b. With a flamed forceps, place a sterile membrane filter disk, grid side up, on the filter holder base.
   c. Place the filter funnel on top of the membrane filter disk and secure it to the base with the clamp.
3. Attach the rubber hose to a vacuum source (pump or water aspirator) and pour the appropriate amount of water into the funnel.

   The amount of water used will depend on water quality. No less than 50 ml should be used. Water with few bacteria and low turbidity permit samples of 200 ml or more. Your instructor will advise you as to the amount of water that you should use. Use a sterile graduate for measuring the water.
4. Rinse the inner sides of the funnel with 20 ml of sterile water.
5. Disconnect the vacuum source, remove the funnel, and carefully transfer the filter disk with sterile forceps to the petri dish of *m* Endo MF broth. *Keep grid side up.*
6. Incubate at 35°C for 22 to 24 hours. *Don't invert.*
7. After incubation, remove the filter from the dish and dry for 1 hour on absorbent paper.
8. Count the colonies on the disk with low-power magnification, using reflected light. Ignore all colonies that lack the golden metallic sheen. If desired, the disk may be held flat by mounting between two 2″ × 3″ microscope slides after drying. Record your count on the first portion of Laboratory Report 46.

See Appendix C for special preparation method.

(1) Sterile absorbent pad is aseptically placed in the bottom of a sterile plastic petri dish.

(2) Absorbent pad is saturated with 2.0 ml of *m* Endo MF broth.

(3) Sterile membrane filter disk is placed on filter holder base with grid side up.

(4) Water sample is poured into assembled funnel, utilizing vacuum. A rinse of 20 ml of sterile water follows.

(5) Filter disk is carefully removed with sterile forceps after disassembling the funnel.

(6) Membrane filter disk is placed on medium-soaked absorbent pad with grid side up. Incubate at 35°C for 24 hours.

Figure 46.1 **Membrane filter routine.**

Laboratory Report

46

Student: _____

Date: _____ Section: _____

46 Bacteriological Examination of Water: The Membrane Filter Method

A. Results

A table similar to the one below will be provided for you, either on the chalkboard or as a photocopy. Record your coliform count on it. Once all data are available, complete this table.

SAMPLE	SOURCE	COLIFORM COUNT	AMOUNT OF WATER FILTERED	MPN*
A				
B				
C				
D				
E				
F				
G				
H				

$$*MPN = \frac{Coliform\ Count \times 100}{Amount\ of\ Water\ Filtered}$$

B. Short-Answer Questions

1. In what ways is the membrane filter method for coliform detection superior to the most probable number method (Exercise 45)?

   _____

   _____

2. How is the proper amount of water to be filtered determined? Why is this determination critical to the outcome of the testing?

   _____

   _____

3. Why would the water that passed through the filter not be considered sterile even though bacteria were removed? How can sterility be achieved with the use of filtration?

   _____

   _____

4. What is the pore size of the membrane filter used in the membrane filter method for water analysis?

    _____

    _____

5. On what medium is the membrane filter incubated? Describe the appearance of coliforms on this medium.

    _____

    _____

Reductase Test

Learning Outcomes

After completing this exercise, you should be able to

1. Understand the basis for the reductase test for determining the quality of a milk sample.

2. Perform the reductase test and determine the quality of a milk sample.

The reductase test has been used for decades in the dairy industry for estimating the microbiological quality of raw milk. In the United States, the aerobic plate count of Grade A raw milk may not exceed 1×10^5 CFU/ml for milk from an individual supplier or 3×10^5 CFU/ml for comingled milk. Raw milk may be held for a maximum of 5 days at 40°F (4.4°C) before pasteurization. During this time, high initial microbial counts in milk are likely to cause faster spoilage of this product and economic loss.

Milk that contains large numbers of actively growing bacteria will have a lowered oxidation-reduction potential due to the exhaustion of dissolved oxygen by microorganisms. This lowered oxidation-reduction potential can be measured by the addition of the dye methylene blue to a milk sample, which is the basis for the **reductase test** of milk. It is based on the change in the color of methylene blue from blue to clear when the dye is reduced. In this test, 1 ml of methylene blue (1:25,000) is added to 10 ml of milk. The tube is sealed with a rubber stopper and slowly inverted three times to mix. It is placed in a water bath at 35°C and examined at intervals for up to 6 hours. The time it takes for the methylene blue to become colorless is the **methylene blue reduction time (MBRT).** The shorter the MBRT, the lower the quality of the milk. An MBRT of 6 hours is very good, whereas an MBRT of 30 minutes is of very poor quality (figure 47.1). However, microorganisms may vary in the extent to which they reduce the dye. Also, the reductase test is not applicable to all types of foods, because some foods, including raw meats, naturally have reductive substances that change the color of the dye.

The validity of this test is based on the assumption that all bacteria in milk lower the oxidation-reduction potential at 35°C. Large numbers of psychrophiles, thermophiles, and thermodurics (nonpathogenic bacteria that survived pasteurization), which do not grow at this temperature, would not produce a positive test. Raw milk, however, will contain optimally *Streptococcus*

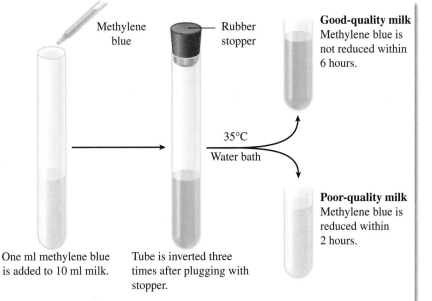

Figure 47.1 Procedure for performing the reductase test on raw milk and interpretation of results.

Methylene blue

Rubber stopper

Good-quality milk
Methylene blue is not reduced within 6 hours.

35°C
Water bath

Poor-quality milk
Methylene blue is reduced within 2 hours.

One ml methylene blue is added to 10 ml milk.

Tube is inverted three times after plugging with stopper.

Figure 47.2 **Procedure for preparing dilutions and plating of milk samples.**

lactis and *Escherichia coli,* which are strong reducers; thus, this test is suitable for screening raw milk at receiving stations. Its principal value is that less technical training of personnel is required for its performance.

In this exercise, samples of low- and high-quality raw milk will be tested.

Materials

- 2 sterile test tubes with rubber stoppers per student
- raw milk samples of low and high quality (samples A and B)
- water bath set at 35°C
- methylene blue (0.004%; 4 µl/100 ml)
- 10 ml pipettes
- 1 ml pipettes
- 2 bottles of nutrient agar (100 ml each)
- 10 petri plates
- 6 sterile 99 ml water blanks
- incubator set at 32°C

1. Label two test tubes with your name and the type of milk. Each student will test a good-quality as well as a poor-quality milk.
2. Using separate 10 ml pipettes for each type of milk, transfer 10 ml to each test tube. To the milk in the tubes add 1 ml of the methylene blue with a 1 ml pipette. Insert the rubber stoppers and gently invert three times to mix. Record your name and the time on the labels and place the tubes in the water bath set at 35°C. Refrigerate the remaining portions of both types of milk for analysis using the standard plate count.
3. After 5 minutes of incubation, remove the tubes from the water bath and invert once to mix. Return the tubes to the water bath. This is the last time they should be mixed.
4. To be performed by groups of four students. While waiting to observe the color changes in the tubes at every 30 minutes, perform a standard plate count (Exercise 19) on the two types of milk that you saved in the refrigerator. For each milk sample, make the following dilutions and plate aliquots as shown in figure 47.2: **1:100; 1:10,000; 1:1,000,000.**
5. Incubate the nutrient agar plates at 32°C for 48 hours. Count the bacterial colonies and record the results in Laboratory Report 47.
6. Carefully remove the tubes from the water bath after 30 minutes of incubation and for every half hour until the end of the laboratory period. *When at least four-fifths of the tube has turned white,* the end point of reduction has occurred. Record this time in Laboratory Report 47. The classification of milk quality is as follows:

Class 1: Excellent, not decolorized in 8 hours.
Class 2: Good, decolorized in less than 8 hours, but not less than 6 hours.
Class 3: Fair, decolorized in less than 6 hours, but not less than 2 hours.
Class 4: Poor, decolorized in less than 2 hours.

Laboratory Report

Student: _____

Date: _____ Section: _____

47

47 Reductase Test

A. Results

How would you grade the two samples of milk that you tested? Give the MBRT for each one.

Sample A: _____

Sample B: _____

B.

Record the standard plate count results (CFU/ml) for each milk sample.

Sample A: _____

Sample B: _____

How do these numbers correlate with MBRT values?

C. Short-Answer Questions

1. Why are the tubes sealed with a rubber stopper rather than left open to the atmosphere?

   _____

   _____

2. Is milk with a short reduction time necessarily unsafe to drink? _____

   Explain: _____

   _____

   _____

3. When oxygen is exhausted in a milk sample, what color is the methylene blue indicator? _____

   _____

   _____

4. What advantage do you see in this method over the direct microscopic count method? _____

   _____

   _____

   _____

   _____

5. What kinds of organisms may be plentiful in a milk sample, yet give a negative reductase test?

   _____

   _____

   _____

   _____

Temperature:
Lethal Effects

Learning Outcomes

After completing this exercise, you should be able to

1. Understand how temperature kills bacterial cells.

2. Define *thermal death point* and *thermal death time* and their importance in canning of foods.

3. Demonstrate how endospore-forming bacteria are more resistant to the lethal effects of temperature than are vegetative cells.

Most microorganisms are killed by elevated temperatures primarily because of the susceptibility of their macromolecules to heat. Elevated temperatures cause proteins to denature and unfold, resulting in a loss of their tertiary structure and biological activity. Because enzymes are proteins, the metabolic capabilities of an organism are irreversibly damaged by heat. Damage to enzymatic systems in the cell, such as those involved in energy production, protein synthesis, and cell division, invariably results in the death of a cell. Nucleic acids can also be damaged by heat, resulting in the loss of DNA and RNA structure. Loss of nucleic acids prevents cell division and protein synthesis and thus causes cell death. Also, small molecules in the cell, such as NAD^+ and other coenzymes, can be damaged or destroyed by elevated temperatures, and loss of these essential factors contributes to cell death. Some organisms are more resistant to heat than others because they form endospores. The endospores of bacteria that belong to the genera *Bacillus* and *Clostridium* are more resistant to heat than are vegetative cells for several reasons. Endospores have a much lower water content than vegetative cells. As a result, their macromolecules are less susceptible to denaturation. Because the endospore is dehydrated, water is unavailable for chemical reactions that can damage macromolecules, coenzymes, and other essential small molecules. Endospores, unlike vegetative cells, contain calcium dipicolinate, which plays a vital role in heat resistance by further preventing thermally induced denaturation of proteins. The calcium associated with dipicolinic acid may likewise provide resistance to oxidizing agents that can irreversibly destroy proteins, nucleic acids, and small molecules in the cell. Additionally, specific proteins found in the endospore can bind to nucleic acids and prevent their denaturation. All of these factors are important for insuring that the endospore remains dormant and undamaged so that it will give rise to viable cells when conditions are reestablished for germination and growth of the organism.

In attempting to compare the susceptibility of different organisms to elevated temperatures, it is necessary to use some metric of comparison. Two methods of comparison are used: the **thermal death point** and the **thermal death time**. The thermal death point (TDP) is the *lowest* temperature at which *a population of a target organism* is killed in 10 minutes. The thermal death time (TDT) is the *shortest* time required to kill a suspension of cells or spores *under defined conditions* at a given temperature. However, various factors such as pH, moisture, medium composition, and age of cells can greatly influence results, and therefore these variables must be clearly stated.

The thermal death time and thermal death point are important in the food industry because canned foods must be heated to temperatures that kill the endospores of *Clostridium botulinum* and *Clostridium perfringens,* two bacteria involved in food-borne illnesses.

In this exercise, you will subject cultures of three different bacteria to temperatures of 60°C, 70°C, 80°C, 90°C, and 100° C. At intervals of 10 minutes, samples of the test bacteria will be removed and plated out to determine the number of survivors. The endospore-former *Bacillus megaterium* will be compared with the non-endospore-formers, *Staphylococcus aureus* and *Escherichia coli*. The overall procedure is illustrated in figure 48.1.

Note in figure 48.1 that *before* the culture is heated, a **control plate** is inoculated with 0.1 ml of the organism. When the culture is placed in the water bath, a tube of nutrient broth with a thermometer inserted into it is placed in the water bath at the same time. Timing of the experiment starts when the thermometer reaches the test temperature.

Due to the large number of plates that have to be inoculated to perform the entire experiment, it will be necessary for each member of the class to be assigned a specific temperature and bacterium. Table 48.1 provides suggested assignments

Every 10 minutes, 0.1 ml of culture is pipetted into one of four plates.

Water bath temperature is held constant for 40 minutes.

0.1 ml

0.1 ml

0.1 ml

0.1 ml

0.1 ml

0.1 ml

0.1 ml

10 minutes

30 minutes

20 minutes

40 minutes

Control

Figure 48.1 **Procedure for determining thermal endurance.**

by student numbers. After the plates have been incubated, each student's results will be tabulated on a Laboratory Report chart at the demonstration table. The lab instructor will have copies made of the chart to give to each student so that everyone will have all the pertinent data needed to draw the essential conclusions.

Although this experiment is not difficult to perform, it can fail to yield the appropriate results because of errors. Common errors are (1) omission of the control (no-heat) plate inoculation, (2) putting the thermometer in the culture tube instead of in a tube of sterile nutrient broth, and (3) not using fresh sterile pipettes when instructed to do so. Other sources of error include not using similar size test tubes and a temperature gradient between the portion of the tube immersed in the water bath and that portion above the water level.

Materials

per student:
- 5 petri plates
- 5 pipettes (1 ml size)
- 1 tube of nutrient broth
- 1 bottle of nutrient agar (60 ml)
- 1 culture of organisms

class equipment:
- water baths set up at 60°, 70°, 80°, 90°, and 100°C

broth cultures:
- *Staphylococcus aureus, Escherichia coli,* and *Bacillus megaterium* (minimum of 5 cultures of each species per lab section)

1. Consult the chart on page 343 to determine what organism and temperature has been assigned to you. If several thermostatically controlled water baths have been provided in the lab, locate the one that you will use. If a bath is not available for your temperature, set up a bath on an electric hot plate or over a tripod and Bunsen burner.

   If your temperature is 100°C, use a hot plate to prepare a beaker of boiling water. When setting up a water bath, use hot tap water to start with to save heating time.

2. Liquefy a bottle containing 60 ml of nutrient agar in a boiling water bath, steamer, etc. When it is completely liquefied, place it in a 50°C water bath to cool.

3. Label 5 petri plates: **control (no heat), 10 min, 20 min, 30 min,** and **40 min.**

4. Shake the culture of organisms and transfer 0.1 ml of culture with a 1 ml pipette to the control (no heat) plate.

5. Place the culture and a tube of sterile nutrient broth into the water bath. Remove the cap from

Chart for Inoculation Assignments

	STUDENT NUMBER				
ORGANISM	60°C	70°C	80°C	90°C	100°C
Staphylococcus aureus	1, 16	4, 19	7, 22	10, 25	13, 28
Escherichia coli	2, 17	5, 20	8, 23	11, 26	14, 29
Bacillus megaterium	3, 18	6, 21	9, 24	12, 27	15, 30

the tube of nutrient broth and insert a thermometer into the tube. *Don't make the mistake of inserting the thermometer into the culture of organisms!*

6. As soon as the temperature of the nutrient broth reaches the desired temperature, record the time here: _____.
Watch the temperature carefully to make sure it does not vary appreciably.

7. After 10 minutes have elapsed, remove the culture from the water bath and mix throughly. Using a fresh pipette, quickly transfer 0.1 ml from the culture to the 10-minute plate and immediately return the culture tube to the water bath. Repeat this procedure at 10-minute intervals until all the plates have been inoculated. *Use fresh pipettes each time and be sure to resuspend the culture before each delivery.*

8. Pour liquefied nutrient agar (50°C) into each plate, rotate, and cool.

9. Incubate at 37°C for 24 to 48 hours. After evaluating your plates, record your results on the chart in Laboratory Report 48 and on the chart on the demonstration table.

Laboratory Report

Complete Laboratory Report 48 once you have a copy of the class results.

48 Temperature: Lethal Effects

A. Results

Examine your five petri plates, looking for evidence of growth. Record on the chalkboard, using a chart similar to the one below, the presence or absence of growth as (+) or (−). When all members of the class have recorded their results, complete this chart.

ORGANISM	60°C					70°C					80°C					90°C					100°C				
	C*	10	20	30	40	C*	10	20	30	40	C*	10	20	30	40	C*	10	20	30	40	C*	10	20	30	40
S. aureus																									
E. coli																									
B. megaterium																									

*control (no-heat) tubes

1. If they can be determined from the above information, record the **thermal death point** for each of the organisms.

   S. aureus: _____   E. coli: _____   B. megaterium: _____

2. From the table *shown above,* determine the thermal death time for each organism at the tabulated temperatures and *record this information in the table below.*

ORGANISM	THERMAL DEATH TIME				
	60°C	70°C	80°C	90°C	100°C
S. aureus					
E. coli					
B. megaterium					

B. Short-Answer Questions

1. What is the importance of inoculating a control plate in this experiment?

   _____

   _____

2. To measure the culture temperature, why is the thermometer placed in a tube separate from the culture?

   _____

   _____

345

3. *Bacillus megaterium* has a high thermal death point and a long thermal death time, but it is not classified as a thermophile. Explain.

   _____

   _____

4. Give three reasons why endospores are much more resistant to heat than are vegetative cells.

   a. _____

   b. _____

   c. _____

5. List four diseases caused by spore-forming bacteria.

   a. _____

   b. _____

   c. _____

   d. _____

6. Would heating the culture in a sealed, small-diameter tube that is totally immersed in a water bath produce more accurate results than the use of tubes that are partially submerged in a water bath? Give reasons for your answer.

   _____

   _____

Microbial Spoilage of Canned Food

Learning Outcomes

After completing this exercise, you should be able to

1. Define how endospore-forming bacteria are involved in the spoilage of canned foods.
2. Explain the types of spoilage that occur in canned foods.
3. Characterize the bacteria in spoiled canned food using the Gram and endospore stains.

Spoilage of inadequately heat-processed, commercially canned foods is confined almost entirely to the action of bacteria that produce heat-resistant endospores. Two genera of endospore-forming bacteria, *Clostridium* and *Bacillus*, are of major concern in canned foods. *Clostridium* species are anaerobic, whereas aerobic to facultatively anaerobic strains occur in the *Bacillus* group. Within both genera are strains that are mesophilic or thermophilic, and the heat resistance of their endospores can vary. Mesophilic anaerobic spore-formers, including *Clostridium* species, may be proteolytic or nonproteolytic. Only the proteolytic strains degrade proteins to produce hydrogen sulfide and gas. Nonproteolytic strains of *Clostridium botulinum* are an insidious problem because they can grow and produce toxin in canned food without the typical signs of spoilage such as gas production, off-flavors, or hydrogen sulfide production. Canning of foods normally involves heat exposure for long periods of time at temperatures that are adequate to kill bacterial endospores. Particular concern is given to the processing of low-acid foods in which *Clostridium botulinum* can thrive to produce botulism toxin, and thereby cause botulism food poisoning.

Spoilage occurs when the heat processing fails to meet accepted standards. This can occur for several reasons: (1) lack of knowledge on the part of the processor (usually the case in home canning); (2) carelessness in handling the raw materials before canning, resulting in an unacceptably high level of contamination that ordinary heat processing may be inadequate to control; (3) equipment malfunction that results in undetected underprocessing; and (4) defective containers that permit the entrance of organisms after the heat process.

Our concern here will be with the most common types of food spoilage caused by heat-resistant, spore-forming bacteria. There are three types: flat sour, T.A. spoilage, and stinker spoilage.

Flat sour pertains to spoilage in which acids are formed with no gas production; result: sour food in cans that have flat ends. **T.A. spoilage** is caused by thermophilic anaerobes that produce acid and gases (CO_2 and H_2, but not H_2S) in low-acid foods. Cans swell to various degrees, sometimes bursting. **Stinker spoilage** is due to spore-formers that produce hydrogen sulfide and blackening of the can and contents. Blackening is due to the reaction of H_2S with the iron in the can to form iron sulfide.

In this experiment, you will have an opportunity to become familiar with some of the morphological and physiological characteristics of organisms that cause canned food spoilage, including both aerobic and anaerobic endospore formers of *Bacillus* and *Clostridium*, as well as a non-spore-forming bacterium.

Working as a single group, the entire class will inoculate 10 cans of vegetables (corn and peas) with five different organisms. Figure 49.1 illustrates the procedure. Note that the cans will be sealed with solder after inoculation and incubated at different temperatures. After incubation the cans will be opened so that stained microscope slides can be made to determine Gram reaction and presence of endospores. Your instructor will assign individual students or groups of students to inoculate one or more of the 10 cans. One can of corn and one can of peas will be inoculated with each of the organisms. Proceed as follows:

🕐 First Period

(Inoculations)

Materials

- 5 small cans of corn
- 5 small cans of peas
- cultures of *Geobacillus stearothermophilus*, *B. coagulans*, *C. sporogenes*, *Thermoanaerobacterium thermosaccharolyticum*, and *E. coli*

(1) Each can of corn or peas is perforated with an awl or ice pick.

(2) To create an air space under the cover, some liquid is poured off.

(3) Contents of each can is inoculated with one of five different organisms.

(4) Hole in each can is sealed by soldering over it.

24–48 hours, incubation

For temperature see text.

Second period

(1) Type of spoilage caused by each organism is noted.

(2) Gram- and spore-stained slides are made from contents of cans.

Figure 49.1 **Canned food inoculation procedure.**

- ice picks or awls
- hammer
- solder and soldering iron
- plastic bags
- gummed labels and rubber bands
- bleach solution (0.01%; 10 µl in 100 ml)
- Bunsen burner

1. Label the can or cans with the name of the organism that has been assigned to you. In addition, label one of the plastic bags to be used after sealing of the cans.
2. Pour the bleach solution onto the tops of all the cans. After 10 minutes, drain off the bleach solution and heat the top of each can over a Bunsen burner to remove excess moisture. Use an autoclaved ice pick or awl to punch a hole through a flat area in the top of the can. This can be done with the heel of your hand or a hammer if available.
3. Pour off about 5 ml of the liquid from the can to leave an air space under the lid.
4. Use an inoculating needle to inoculate each can of corn or peas with the organism indicated on the label.

5. Take the cans up to the demonstration table where the instructor will seal the hole with solder.
6. After sealing, place each can in two plastic bags. Each bag must be closed separately with rubber bands, and the outer bag must have a label on it.
7. Incubation will be as follows until the next period:
   - **45°C; 72 hours**—*T. thermosaccharolyticum*
   - **37°C; 24–48 hrs.**—*C. sporogenes* and *B. coagulans*
   - **37°C; 24–48 hrs.**—*E. coli*
   - **55°C; 24–48 hrs.**—*G. stearothermophilus*

Note: If cans begin to swell during incubation, they should be placed in refrigerator.

⏱ Second Period

(Interpretation)

After incubation, place the cans under a hood to open them. The odors of some of the cans will be very strong due to H_2S production.

Materials

- can opener, punch-type
- small plastic beakers
- Parafilm
- Gram-staining kit
- spore-staining kit

1. Open each can carefully with a punch-type can opener. If the can is swollen, hold an inverted plastic funnel over the can during perforation to minimize the effects of any explosive release of contents.
2. Remove about 10 ml of the liquid through the opening, pouring it into a small plastic beaker. Cover with Parafilm. This fluid will be used for making stained slides.
3. Return the cans of food to the plastic bags, reclose them, and dispose in the biohazard bin.
4. Prepare Gram-stained and endospore-stained slides from your canned food extract as well as from the extracts of all the other cans. Examine under brightfield oil immersion.
5. Record your observations on the report sheet on the demonstration table. It will be duplicated and a copy will be made available to each student.

Laboratory Report

Complete the first portion of Laboratory Report 49.

49 Microbial Spoilage of Canned Food

A. Results

1. Observations

   Record your observations of the effects of each organism on the cans of vegetables. Share results with other students.

ORGANISM	PEAS		CORN	
	Gas Production + or −	Odor	Gas Production + or −	Odor
E. coli				
B. coagulans				
G. stearothermophilus				
C. sporogenes				
T. thermosaccharolyticum				

2. Microscopy

   After making Gram-stained and spore-stained slides of organisms from the canned food extracts, sketch in representatives of each species:

E. coli　　　B. coagulans　　　G. stearothermophilus　　　C. sporogenes　　　T. thermosaccharolyticum

B. Short-Answer Questions

1. Which organisms, if any, caused flat sour spoilage? _____

   _____

2. Which organisms, if any, caused T.A. spoilage? _____

   _____

3. Which organisms, if any, caused stinker spoilage? _____

   _____

4. Does flat sour cause a health problem? _____

   _____

5. Describe how typical spoilage resulting in botulism occurs. _____

   _____

   _____

6. Why is spoilage more likely to occur for individuals who do home canning than in a canning factory?

   _____

   _____

Microbiology of Alcohol Fermentation

Learning Outcomes

After completing this exercise, you should be able to

1. Understand why fermentation is a method to preserve food.
2. Demonstrate how yeast ferments grape juice to produce various end products.

Fermented food and beverages are as old as civilization. Historical evidence indicates that beer and wine making were well-established as long ago as 2000 B.C. An Assyrian tablet states that Noah took beer aboard the ark.

Beer, wine, vinegar, buttermilk, cottage cheese, sauerkraut, pickles, and yogurt are some of the products of fermentation. Most of these foods and beverages are produced by different strains of yeasts (*Saccharomyces*) or bacteria (*Lactobacillus*, *Acetobacter*, etc.).

Fermentation is actually a means of food preservation because the acids and alcohols produced and the anaerobic environment hold back the growth of many spoilage microbes.

Wine is essentially fermented fruit juice in which alcoholic fermentation is carried out by *Saccharomyces cerevisiae* var. *ellipsoideus*. Although we usually associate wine with fermented grape juice, it may also be made from various berries, dandelions, rhubarb, and so on. Three conditions are necessary: simple sugar, yeast, and anaerobic conditions. The reaction is as follows:

$$C_6H_{12}O_6 \xrightarrow{\text{yeast}} 2C_2H_5OH + 2CO_2$$

Commercially, wine is produced in two forms: red and white. To produce red wines, the distillers use red grapes with the skins left on during the initial stage of the fermentation process. For white wines, either red or white grapes can be used, but the skins are discarded. White and red wines are fermented at 13°C (55°F) and 24°C (75°F), respectively.

In this exercise, we will set up a grape juice fermentation experiment to learn about some of the characteristics of sugar fermentation to alcohol. Note in figure 50.1 that a balloon will be attached over the mouth of the fermentation flask to exclude oxygen uptake and to trap gases that might be produced. To detect the

Mouth of flask is sealed with rubber balloon before incubation.

Lead acetate test paper is taped to inside of flask neck.

100 ml of grape juice is inoculated with 3 ml of yeast culture.

15°–17°C
2–5 days

Balloon is removed after incubation. Odor of gas and test strip change are noted.

pH of juice-yeast mixture is determined before incubation.

pH of fermented juice is checked after incubation.

Figure 50.1 **Alcohol fermentation setup.**

presence of hydrogen sulfide production, which is naturally produced by *S. cerevisiae* during fermentation and which can cause off-flavors to wine, we will tape lead acetate test paper inside the neck of the flask. The pH of the substrate will also be monitored before and after the reaction to note any changes that occur.

First Period

Materials

- 100 ml grape juice (no preservative)
- bottle of juice culture of wine yeast
- 125 ml Erlenmeyer flask
- one 10-ml pipette
- balloon
- hydrogen sulfide (lead acetate) test paper
- tape
- pH meter

1. Label an Erlenmeyer flask with your initials and date.
2. Add about 100 ml of grape juice to the flask.
3. Determine the pH of the juice with a pH meter and record the pH in Laboratory Report 50.
4. Agitate the container of yeast and juice to suspend the culture, remove 5 ml with a pipette, and add it to the flask.
5. Attach a short strip of tape to a piece of lead acetate test paper (3 cm long), and attach it to the inside surface of the neck of the flask. Make certain that neither the tape nor the test paper protrudes from the flask.
6. Cover the flask opening with a balloon.
7. Incubate at 15–17°C for 2 to 5 days.

Second Period

Materials

- pH meter

1. Remove the balloon and note the aroma of the flask contents. Describe the odor in Laboratory Report 50.
2. Determine the pH and record it in the Laboratory Report.
3. Record any change in color of the lead acetate test paper in the Laboratory Report. If any H_2S is produced, the paper will darken due to the formation of lead sulfide as hydrogen sulfide reacts with the lead acetate.
4. Wash out the flask and return it to the drain rack.

Laboratory Report

Complete Laboratory Report 50 by answering all the questions.

50   Microbiology of Alcohol Fermentation

A. Results

Record here your observations of the fermented product:

Aroma: _____

pH: _____

H$_2$S production: _____

B. Short-Answer Questions

1. What compound in the grape juice was fermented? What were the major products of this fermentation?

    _____

    _____

2. Why was the flask sealed with a balloon? How would the product be different if the flask was sealed with a rubber stopper?

    _____

    _____

3. Fermented beverages are produced by what groups of organisms?

    _____

    _____

4. Why is hydrogen sulfide production measured during fermentation?

    _____

    _____

5. Fermentation is actually a means of preservation because end products are antagonistic to spoilage organisms. Name several end products that would achieve this purpose.

    _____

    _____

6. How does microbial fermentation contribute to food preservation?

    _____

    _____

Medical Microbiology

Medical microbiology is the study of microorganisms which affect human health. Microorganisms can affect human health either positively or negatively. Although the microbiology laboratory is primarily focused on the negative ways microorganisms affect human health, it must also be appreciated that microorganisms are vital for human existence. The human body plays host to trillions of microbes that make up its own unique microbiota (microscopic life). These microbes help digest carbohydrates, make vitamins, compete for space so that pathogens cannot invade, and aid the immune system.

The microbial population within the human body is in a continual state of flux and is affected by a variety of factors (health, age, diet, hygiene). When the human

Acc.V Spot Magn Det WD ├───────┤ 1 µm
30.0 kV 1.0 20000x SE 8.3

CDC/Janice Haney

body is exposed to an outside organism three results are possible: (1) the organism can pass through the body, (2) the organism can colonize the body (either transiently or permanently) and become part of the microbiota, (3) the organism can produce disease. Organisms that colonize humans don't interfere with normal body functions and may have a positive effect on the body (see above). In contrast, organisms that cause disease damage the human host. It is important to highlight the difference between colonization and disease.

Disease-causing organisms can be divided into opportunistic pathogens and true pathogens. Opportunistic pathogens do not produce disease in their normal environment but can cause disease when they are given access to unprotected sites of the body or if a patient's immune system is not functioning properly. Opportunistic pathogens are usually typical members of the human microbiota. In Exercises 51 and 52 you will isolate and enrich organisms from within your own unique microbiota. True pathogens are always associated with human disease. Exercise 53 employs attenuated (less dangerous) strains of true pathogens.

In the microbiology laboratory we protect ourselves from pathogens (opportunistic and true) by taking extensive safety measures to reduce our exposure to them. The safety measures taken are specific to the pathogen and are based on the primary portal of entry and communicability of the pathogen. All of these exercises call for gowns and gloves, and require special precautions for contaminated equipment and waste. Consult with your instructor to determine whether eye protection is recommended in your setting. Hands should be washed before gloves are put on and after they are removed. Eyes and mouth should not be touched while you are in the microbiology laboratory because they are common portals of entry for bacteria. Nothing should ever be placed into your mouth while you are in the microbiology laboratory. Gloves, swabs, petri dishes, and other disposable items used during these exercises should be placed in the biohazard bag in the laboratory. Supplies that are to be reused should be placed in a container designated for autoclaving. Additionally, many institutions require *Salmonella, Shigella,* and *Staphylococcus aureus* to have their own separate biohazard waste containers.

The following exercises focus on bacteria that are of medical importance and the methods used to isolate and identify these bacteria. Some tests used to characterize these bacteria have been covered in the section on the identification of unknown bacteria. However, other tests will be new and will apply specifically to a particular bacterium. Exercises 51–53 involve the methods used to characterize the staphylococci, the streptococci, and the procedures to differentiate and identify the gram-negative intestinal bacteria. Exercise 54 is a simulated epidemic that will demonstrate how the spread of disease through a population can be tracked.

The Staphylococci:
Isolation and Identification

Learning Outcomes

After completing this exercise, you should be able to

1. Enrich and isolate staphylococci from human sources and from fomites using selective media and culture techniques.

2. Identify unknown staphylococci that you have isolated using differential media and biochemical tests specific for these bacteria.

The name "staphylococcus" is derived from Greek, meaning "bunch of grapes." Staphylococci are gram-positive spherical bacteria that divide in more than one plane to form irregular clusters of cells (figure 51.1). In *Bergey's Manual* the staphylococci are currently grouped in Family VIII Staphylococcaceae, with four other genera. The staphylococci are a coherent phylogenetic group of 40 species with 24 subspecies. The staphylococci are non-motile, non-spore-forming, and able to grow in media containing high salt concentrations. Most are considered facultative anaerobes. Although the staphylococci were originally isolated from pus in wounds, they were later demonstrated to be part of the normal microbiota of nasal membranes, hair follicles, skin, and the perineum in healthy individuals. Infections by staphylococci are initiated

Figure 51.1 Gram stain of *Staphylococcus aureus*.
Centers for Disease Control

when a breach of the skin or mucosa occurs, when a host's ability to resist infection occurs, or when a staphylococcal toxin is ingested.

The Centers for Disease Control and Prevention estimate 20–30% of the U.S. population carries *Staphylococcus aureus*, and this bacterium is responsible for many serious infections. To further complicate matters, *S. aureus* has developed resistance to many antibiotics, including methicillin. MRSA, or methicillin-resistant *S. aureus*, is a major epidemiological problem in hospitals where it is responsible for some health-care-acquired infections (HAIs). More recently, a community form of MRSA has been isolated from infections in individuals who have not been hospitalized. It is estimated that about 1% of the U.S. population now carries MRSA. Although *S. aureus* species are considered to be the most virulent members of the genus, *Staphylococcus epidermidis*, *S. saprophyticus*, *S. haemolyticus*, and *S. lugdunensis* are also associated with human diseases.

S. aureus, the most clinically significant staphylococcal pathogen, can cause skin infections, wound infections, bone tissue infections, scalded skin syndrome, toxic shock syndrome, and food poisoning. It has a wide variety of virulence factors and many unique characteristics. The most notable virulence factor possessed by *S. aureus* is coagulase production. Virtually all strains of *S. aureus* are coagulase positive and will cause serum to form a clot. The role of coagulase in the pathogenesis of disease is unclear, but coagulase may cause a clot to form around the staphylococcal infection thus protecting it from host defenses. Another enzyme associated with *S. aureus* is DNase, a nuclease that digests DNA. *S. aureus* also produces a hemolysin called α-toxin that causes a wide, clear zone of beta-hemolysis on blood agar. This powerful toxin plays a significant role in virulence because it not only lyses red blood cells, but also damages leukocytes, heart muscle, and renal tissue. Additionally, many strains of *S. aureus* produce a pigment that can act as a virulence factor. The pigment (staphyloxanthin) has antioxidant properties which prevent reactive oxygen produced by the host immune system from killing the bacteria. This pigment is responsible for the golden color of *S. aureus* when it is grown on blood agar and staphylococcus 110 plates. Finally, *S. aureus* ferments mannitol to produce acid. This

Figure 51.2 *Staphylococcus epidermidis* (left) and *Staphylococcus aureus* (right) growing on a mannitol salt agar plate. Note fermentation of mannitol by *S. aureus*.
© Kathy Park Talaro

Table 51.1 Differentiation of Three Species of Staphylococci

	S. AUREUS	S. EPIDER- MIDIS	S. SAPRO- PHYTICUS
α-toxin	+	−	−
Mannitol (acid only)	+	−	(+)
Coagulase	+	−	−
DNase	+	−	−
Novobiocin	S	S	R

Note: S = sensitive; R = resistant; + = positive; (+) = mostly positive

metabolic characteristic can be observed when cultures of *S. aureus* are grown on mannitol salt agar (MSA). The production of acid lowers the pH of the medium, causing the phenol red indicator to turn from red to yellow (figure 51.2).

The coagulase-negative staphylococci (CNS), *S. epidermidis* and *S. saprophyticus*, differ from *S. aureus* in many ways. These species of staphylococci, as indicated by their name, do not produce coagulase. They also do not produce DNase or α-toxin. All people have CNS on their skin, and these species were at one time thought of as harmless commensals. However, their clinical significance has greatly increased over the past 20 years, particularly in patients who have compromised immune systems or prosthetic or indwelling devices. *S. epidermidis* is the most common cause of hospital-acquired urinary tract infections. Infections involving *S. epidermidis* have also been documented with catheters, heart valves, and other prosthetic devices. *S. saprophyticus* is the second most common cause of urinary tract infections in sexually active young women. The CNS are unpigmented and appear opaque when grown on blood agar and staphylococcus 110 plates. *S. saprophyticus* is the only clinically important staphylococci species that is resistant to novobiocin. Some strains of *S. saprophyticus* are able to ferment mannitol to acid.

In this experiment, we will attempt to isolate and differentiate staphylococci species from (1) the nose, (2) a **fomite** (inanimate object), and (3) an "unknown control." We will follow the procedure illustrated in figure 51.3. If the nasal membranes and fomite fail to yield a positive isolate, the unknown control will yield a positive isolate provided all the inoculations and tests are performed correctly. The organisms collected will first be cultured in media containing 7.5% sodium chloride. The high salt concentration will inhibit the growth of most bacteria, while allowing species of staphylococci to grow. Once the cultures have been enriched for staphylococci, the exercise will focus on identifying the isolated bacteria as *S. aureus*, *S. epidermidis*, *S. saprophyticus*, or an unidentified CNS. The characteristics we will look for in our isolates will be (1) beta-type hemolysis (α-toxin), (2) mannitol fermentation, and (3) coagulase production. Organisms found to be positive for these three characteristics will be presumed to be *S. aureus* (please see table 51.1).

Note: Please review the safety information concerning human microbiota discussed on page 358 before proceeding with this exercise.

🕐 First Period

(Specimen Collection)

Note in figure 51.3 that swabs have been applied to the nasal membranes, and fomites will be placed in tubes of enrichment medium containing 7.5% NaCl (*m*-staphylococcus broth). Since your unknown control will lack a swab, initial inoculations from this culture will have to be done with a loop.

Materials

- 1 tube containing numbered unknown control
- tubes of *m*-staphylococcus broth
- 2 sterile cotton swabs

1. Label the three tubes of *m*-staphylococcus broth NOSE, FOMITE, and the number of your unknown control.

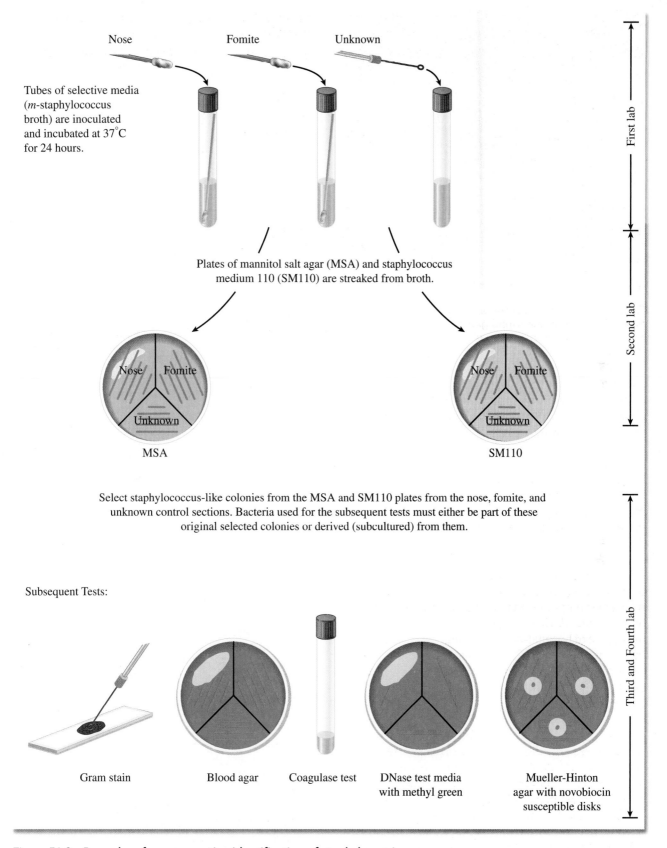

Tubes of selective media (*m*-staphylococcus broth) are inoculated and incubated at 37°C for 24 hours.

Plates of mannitol salt agar (MSA) and staphylococcus medium 110 (SM110) are streaked from broth.

MSA

SM110

Select staphylococcus-like colonies from the MSA and SM110 plates from the nose, fomite, and unknown control sections. Bacteria used for the subsequent tests must either be part of these original selected colonies or derived (subcultured) from them.

Subsequent Tests:

Gram stain

Blood agar

Coagulase test

DNase test media with methyl green

Mueller-Hinton agar with novobiocin susceptible disks

First lab

Second lab

Third and Fourth lab

Figure 51.3 **Procedure for presumptive identification of staphylococci.**

2. After moistening one of the swabs by immersing it partially into the "nose" tube of broth, swab the nasal membrane just inside your nostril. A small amount of moisture on the swab will enhance the pick-up of organisms. Place this swab into the "nose" tube.

3. Swab the surface of a fomite with the other swab that has been similarly moistened and deposit this swab in the "fomite" tube. The fomite you select may be a coin, drinking glass, telephone mouthpiece, or any other item that you might think of.

4. Inoculate the appropriate tube of *m*-staphylococcus broth with one or two loopfuls of your unknown control.

5. Incubate these tubes of broth for 24 hours at 37°C.

🕐 Second Period

(Primary Isolation Procedure)

Two kinds of media will be streaked for primary isolation: mannitol salt agar and staphylococcus medium 110.

Mannitol salt agar (MSA) contains mannitol, 7.5% sodium chloride, and phenol red indicator. The NaCl inhibits organisms other than staphylococci. If the mannitol is fermented to produce acid, the phenol red in the medium changes color from red to yellow.

Staphylococcus medium 110 (SM110) also contains NaCl and mannitol, but it lacks phenol red. Its advantage over MSA is that it favors colony pigmentation by different strains of *S. aureus*. Since this medium lacks phenol red, no color change takes place as mannitol is fermented. **Note:** See Exercise 18 for a discussion of selective and differential media.

Materials

- 3 culture tubes from last period
- 1 petri plate of MSA
- 1 petri plate of SM110

1. Divide the MSA and Staph 110 plates into thirds. Label each third as "nose," "fomite," and "unknown."

2. Remove a loopful of bacteria from the "nose" tube and streak it onto the third of the agar plates labeled **"nose."**

3. Remove a loopful of bacteria from the "fomite" tube and streak it onto the third of the agar plates labeled **"fomite."**

4. Remove a loopful of bacteria from the "unknown" control tube and streak it onto the third of the agar plates labeled **"unknown."**

5. Incubate the plates aerobically at 37°C for 24 to 36 hours.

🕐 Third Period

(Plate Evaluations and Coagulase/DNase/ Novobiocin Tests)

During this period we will evaluate the plates streaked in the previous period, as well as set up a series of experiments that will help you to identify the type of staphylococci you have isolated. Prior to beginning these experiments it is crucial to select an isolated colony in each of the three sections (nose, fomite, and unknown control) of one of the plates.

Materials

- MSA and SM110 plates from previous period
- 1 blood agar plate
- capped serological tubes containing 0.5 ml of 1:4 saline dilution of rabbit or human plasma (one tube for each isolate)
- 1 petri plate of DNase test agar with methyl green
- Gram-staining kit
- 1 Mueller-Hinton agar plate

Evaluation of Plates

1. Examine the mannitol salt agar plate. Has the phenol red in the medium surrounding any of the colonies turned yellow? If this color change exists, it can be presumed that you have isolated a strain of *S. aureus* or *S. saprophyticus*.

2. Examine the plate of SM110. The presence of growth here indicates that the organisms are salt-tolerant. Note color of the colonies. *S. aureus* colonies will appear yellow or orange, while CNS colonies will appear colorless.

3. Record your observations of these plates in Laboratory Report 51.

Blood Agar Inoculations

1. Divide one blood agar plate (BAP) into thirds. Label the sections "nose," "fomite," and "unknown control."

2. Select staphylococcus-like colonies from the MSA and SM110 plates from the nose, fomite, and unknown control sections. If you do not have an isolated colony from each section, consult with your instructor about any modifications you may need to make as you proceed through this lab (for example, if you only have two cultures, divide BAP in half as opposed to thirds).

3. Streak out part of the selected "nose" colony onto the third of the BAP labeled "nose." Streak out part of the selected "fomite" colony onto the third of the BAP labeled "fomite." Streak out part of the selected "unknown" colony onto the third of the BAP labeled "unknown."

4. Incubate the blood agar plates at 37°C for 18 to 24 hours. *Don't leave plates in incubator longer than 24 hours.* Overincubation will cause blood degeneration.

[At this point in the lab you can either proceed to the following tests using the remaining portion of your selected colonies, or incubate your BAP for 24 hours and then proceed with the following tests. Incubating your BAP for 24 hours before proceeding will provide you with a larger amount of pure culture to use in the coagulase, DNase, and novobiocin tests, all of which require a heavy inoculum of bacteria.]

Coagulase Tests

The fact that 97% of the strains of *S. aureus* have proven to be coagulase positive and that the other two species are *always* coagulase negative makes the coagulase test an excellent definitive test for confirming identification of *S. aureus.*

The procedure is simple. It involves inoculating a small tube of plasma with several loopfuls of the organism and incubating it in a 37°C water bath for several hours. If the plasma coagulates, the organism is coagulase positive. Coagulation may occur in 30 minutes or several hours later. *Any degree of coagulation, from a loose clot suspended in plasma to a solid immovable clot, is considered to be a positive result, even if it takes 24 hours to occur.*

It should be emphasized that this test is valid only for gram-positive, staphylococcus-like bacteria because some gram-negative rods, such as *Pseudomonas,* can cause a false-positive reaction. The mechanism of clotting in such organisms is not due to coagulase. Proceed as follows:

1. Label the plasma tubes "nose," "fomite," or "unknown control," depending on which of your plates have staph-like colonies.
2. With a wire loop, inoculate the appropriate tube of plasma with organisms from the colonies on SM110 or MSA. Success is more rapid with a heavy inoculation. If positive colonies are present on both nose and fomite sides, be sure to inoculate a separate tube for each side.
3. Securely cap tubes and place in a 37°C water bath.
4. Check for solidification of the plasma every 30 minutes for the remainder of the period (figure 51.4). Note that solidification may be complete or semisolid.

   Any cultures that are negative at the end of the period will be left in the water bath. At 24 hours your instructor will remove them from the water bath and place them in the refrigerator, so that you can evaluate them in the next laboratory period.
5. Record your results in Laboratory Report 51.

Figure 51.4 Coagulase test: Positive test with negative test shown in inlay.
© McGraw-Hill Education. Lisa Burgess, photo

DNase Test

The fact that coagulase-positive bacteria are also able to hydrolyze DNA makes the DNase test a reliable means of confirming *S. aureus* identification. The following procedure can be used to determine if a staph-like organism can hydrolyze DNA.

1. Heavily streak the organism on a plate of DNase test agar. One plate can be used for several test cultures by making short streaks about 1 inch long.
2. Incubate for 18 to 24 hours at 35°C.

Gram-Stained Slides

While your tubes of plasma are incubating in the water bath, prepare Gram-stained slides from the same colonies that were used for the blood agar plates and coagulase tests.

Examine the slides under an oil immersion lens and draw the organisms in the appropriate areas of Laboratory Report 51.

Novobiocin Susceptibility Test

The two non-coagulase-producing organisms we have discussed in this exercise (*S. saprophyticus* and *S. epidermidis*) could appear indistinguishable thus far in this exercise if you have a strain of *S. saprophyticus* that does not ferment mannitol. Resistance to novobiocin can be used to presumptively identify *S. saprophyticus.*

1. Divide one Mueller-Hinton agar plate into thirds. Label the sections "nose," "fomite," and "unknown control."
2. Using a loop, inoculate each third of the plate with the appropriate test culture. Be sure to inoculate the entire surface of the agar for each section.

3. Aseptically transfer a novobiocin disk (either using forceps or a disk dispenser) to the center of each "third" of the plate.
4. Incubate the plate at 37°C for 24 hours.

⏱ Fourth Period

(Confirmation)

During this period we will make final assessment of all tests and perform any other confirmatory tests that might be available to us.

Materials

- coagulase tubes from previous tests
- blood agar plate from previous period
- DNase test agar plate from previous period

1. Examine any coagulase tubes that were carried over from the last laboratory period that were negative at the end of that period. Record your results in Laboratory Report 51.
2. Examine the colonies on your blood agar plate. Look for clear (beta-type) hemolysis around the colonies. The presence of α-toxin is a definitive characteristic of *S. aureus*. Record your results in the Laboratory Report.
3. Look for zones where the dye has cleared around the streak on the DNase agar plate (figure 51.5).

4. Examine the colonies on your Mueller-Hinton agar plate with novobiocin disks. Susceptibility to novobiocin is indicated by an area of no growth surrounding the novobiocin disk. For a culture to be considered susceptible to novobiocin, the zone of inhibition surrounding the disk must be greater than 16 mm in diameter (figure 51.6).
5. Record your results on the chart on the chalkboard or chart on the demonstration table. If an instructor-supplied tabulation chart is used, the instructor will have copies made of it to be supplied to each student.

Further Testing

In addition to using the API Staph miniaturized test strip system (Exercise 43) to confirm your identification of staphylococci, you may wish to use the latex agglutination slide test described in Exercise 56. Your instructor will inform you as to the availability of these materials and the desirability of proceeding further.

Laboratory Report

After recording your results on the chalkboard (or on the chart on the demonstration table), complete the chart in Laboratory Report 51 and answer all the questions.

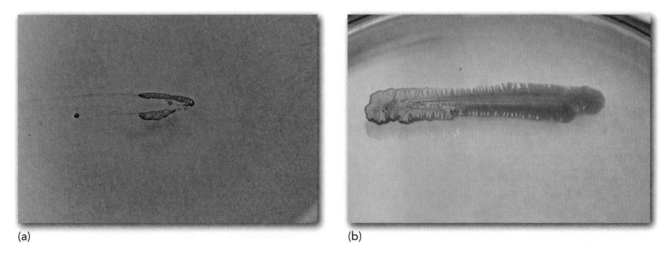

(a)   (b)

Figure 51.5 **(a) DNase test on DNase methyl green agar: clearing indicates DNA breakdown. (b) Negative DNase test: no clearing around the streak.**
© McGraw-Hill Education. Lisa Burgess, photo

Figure 51.6   **Novobiocin test showing sensitivity for *Staphylococcus epidermidis.***
© McGraw-Hill Education. Lisa Burgess, photo

51   The Staphylococci: Isolation and Identification

A. Results

1. **Tabulation**

    At the beginning of the third laboratory period, the instructor will construct a chart similar to this one on the chalkboard. After examining your mannitol salt agar and staphylococcus medium 110 plates, record the presence (+) or absence (−) of staphylococcus growth in the appropriate columns. After performing coagulase tests on the various isolates, record the results also as (+) or (−) in the appropriate columns.

STUDENT INITIALS	FOMITE				UNKNOWN			
	Item	Staph Colonies		Coagulase	Unknown #	Staph Colonies		Coagulase
		MSA	SM110			MSA	SM110	

CULTURE	TOTAL TESTED	TOTAL COAGULASE POSITIVE	PERCENTAGE COAGULASE POSITIVE	TOTAL COAGULASE NEGATIVE	PERCENTAGE COAGULASE NEGATIVE
Fomite					
Unknown					

Microscopy

Provide drawings here of your various isolates as seen under oil immersion (Gram staining).

UNKNOWN CONTROL	NOSE	FOMITE

3. **Record of Culture and Test Results**

	GROWTH ON MSA (+/–)	FERMENTATION OF MANNITOL (+/–)	ALPHA-HEMOLYSIS ON BAP (+/–)	COAGULATION OF PLASMA (+/–)	DNASE ZONE OF CLEARING (+/–)	NOVOBIOCIN (S/R)
Nose						
Fomite						
Unknown						

4. **Identification**

If API Staph miniaturized multitest strips are available, confirm your conclusions by testing each isolate. See Exercise 43.

Nose _____

Fomite _____

Unknown control _____

B. Short-Answer Questions

1. Describe the selective and differential properties of mannitol salt agar (MSA) for the isolation and identification of staphylococci.

   _____

   _____

2. Describe the differential property of blood agar for the isolation and identification of staphylococci.

   _____

   _____

3. Why is the coagulase test considered to be the definitive test for *S. aureus*?

   _____

   _____

4. What is the role of coagulase in the pathogenesis of *S. aureus*?

   _____

   _____

5. What is the role of α-toxin in the pathogenesis of *S. aureus*?

   _____

   _____

6. What are health-care-acquired infections?

   _____

   _____

7. Why are the staphylococci among the leading causes of nosocomial infections?

   _____

   _____

8. Why are staphylococcal infections becoming increasingly difficult to treat?

   _____

   _____

9. Why might hospital patients be tested for nasal carriage of *S. aureus*?

   _____

   _____

10. Describe results from a coagulase, DNase, and novobiocin test that would suggest a mixed culture was used for the tests, as opposed to a pure culture.

    _____

    _____

The Streptococci and Enterococci:
Isolation and Identification

Learning Outcomes

After completing this exercise, you should be able to

1. Isolate streptococci from a mixed culture or from the human throat using selective media and culturing techniques.

2. Identify unknown streptococci that you have isolated using selective media and biochemical tests specific for the streptococci.

The streptococci and enterococci differ from the staphylococci discussed in Exercise 51 in two significant characteristics: (1) Most isolates occur in chains rather than in clusters (figure 52.1), and (2) they lack the enzyme catalase, which degrades hydrogen peroxide to form water and oxygen.

The streptococci and enterococci comprise a large and varied fraction of gram-positive cocci. They are facultative anaerobes and generally considered non-motile. They can occur singly or in pairs, however, they are best known for their characteristic formation of long chains (figure 52.1). At one time, streptococci and enterococci were considered so similar they were placed in the same genus. Current nucleic acid studies have

revealed information on the genetic relationship among these two groups of bacteria, and they are now placed in separate genera in separate families (Streptococcaceae and Enterococcaceae) within the order Lactobacillales in *Bergey's Manual of Systematic Bacteriology.*

There are numerous ways to group and identify the medically important species of these two families. Initial identification of these bacteria is often based on their hemolytic pattern when grown on blood agar. Some species of streptococci and enterococci produce exotoxins that completely destroy red blood cells in blood agar. Complete lysis of red blood cells around a colony is known as **beta-hemolysis** and results in a clear zone surrounding the colonies. Other species of streptococci and enterococci partially break down the hemoglobin inside red blood cells in a blood agar plate, producing a greenish discoloration around the colonies known as **alpha-hemolysis.** Species of bacteria that do not exhibit any hemolysis of blood display **gamma-hemolysis** and have no effect on the red blood cells in a blood agar plate. The three kinds of hemolysis on blood agar are shown in figure 52.2. After patterns of hemolysis have been identified, species of streptococci and enterococci can be further differentiated based on their cell wall carbohydrates. A method developed by Rebecca Lancefield in the 1930s uses the alphabetic system (A, B, C, etc.) to characterize different groups of bacteria based on antigenic differences of these carbohydrates in serological tests. Along with hemolytic patterns and serologic grouping, the clinical laboratory also uses physiologic and biochemical characteristics to identify streptococcal and enterococcal isolates.

Although *Streptococcus* and *Enterococcus* genera contain a large number of species, only a small number of them are human pathogens.

Beta-Hemolytic Groups

The Lancefield serological groups that fall into the beta-hemolytic category are groups A, B, and C. A few species of group D are also beta-hemolytic, however, they will be discussed under the alpha-hemolytic section because they are variably hemolytic.

Group A Streptococci

Streptococcus pyogenes, the main representative of group A streptococci, is by far the most serious

Figure 52.1 Gram stain of *Streptococcus.*
Centers for Disease Control

(a)

(b)

(c)

Zone of hemolysis

Figure 52.2 The three kinds of hemolysis produced by streptococci growing on blood agar plates: (a) beta-hemolysis, (b) alpha-hemolysis, and (c) gamma-hemolysis.

(a) © Kathy Park Talaro; (b) © Evan Roberts; (c) © Fred E. Hossler/Visuals Unlimited

streptococcal pathogen. Humans are the primary reservoir of *S. pyogenes*. Although the pharynx is the most likely place to find this species, it may be isolated from the skin and rectum. Infections from *S. pyogenes* range from pharyngitis and skin infections to scarlet fever, rheumatic fever, and acute glomerulonephritis. When grown on blood agar these colonies are small (0.5 mm diameter), transparent to opaque, and domed. *S. pyogenes* produce hemolysins which

rapidly injure cells and tissues. These hemolysins result in complete lysis of red blood cells around each *S. pyogenes* colony when grown on a blood agar plate, producing a clear zone usually two to four times the diameter of the colony. These bacteria are spherical cocci, arranged in short chains in clinical specimens and longer when grown in broth. In order to differentiate *S. pyogenes* from other streptococci and enterococci, isolates are tested for resistance to bacitracin. If a bacterial isolate is beta-hemolytic and sensitive to bacitracin, it is presumed to be *S. pyogenes*.

Group B Streptococci

S. agalactiae is the only recognized species of Lancefield group B. Like *S. pyogenes*, this pathogen may be found in the pharynx, skin, and rectum; however, it is more likely to be found in the genital and intestinal tracts of healthy adults and infants. This organism is an important cause of serious neonatal infection involving sepsis and meningitis. Colonization of the maternal genital tract is associated with colonization of infants and risk of disease. In the adult population *S. agalactiae* infections consist of abscesses, endocarditis, septicemia, bone and soft tissue infections, and pneumonia. *S. agalactiae* colonies are large, with a narrow zone of beta-hemolysis, in contrast with *S. pyogenes* colonies, which are small with a large zone of hemolysis. *S. agalactiae* cells are spherical to ovoid and occur in short chains in clinical specimens and long chains in culture. Preliminary identification of this species relies heavily on a positive CAMP reaction.

Group C Streptococci

Group C streptococci are uncommon human pathogens but may be involved in zoonoses (infections transmitted from animals to humans). The organism of importance in this group is *S. dysgalactiae*, and infections from this species account for less than 1% of all bacteremias. Members of this group can cause pharyngitis, endocarditis, and meningitis; however, most clinical infections from Group C streptococci occur in patients with underlying illness. *S. dysgalactiae* produce large colonies with a large zone of beta-hemolysis on blood agar. Presumptive differentiation of *S. dysgalactiae* from other beta-hemolytic streptococci (*S. pyogenes* and *S. agalactiae*) is based primarily on resistance to bacitracin and a negative CAMP test.

Alpha-Hemolytic Groups

The grouping of streptococci and enterococci on the basis of alpha-hemolysis is not as clear-cut as it is for beta-hemolytic groups. Note in table 52.1 that some groups exhibit weak alpha-hemolysis, and a few alpha-hemolytic types may exhibit variable

Table 52.1   Physiological Tests for Streptococci and Enterococci Differentiation

	BERGEY'S GROUP	LANCEFIELD GROUP	HEMOLYSIS	BACITRACIN SUSCEPTIBILITY	CAMP REACTION	SXT SENSITIVITY	BILE ESCULIN HYDROLYSIS	TOLERANCE TO 6.5% NaCL	OPTOCHIN SUSCEPTIBILITY
S. pyogenes	Pyogenic	A	β	+	−	R	−	−	−
S. agalactiae	Pyogenic	B	β	−	+	R	−	±	−
S. pneumoniae	Pyogenic	none	α	−	−		−	−	+
S. dysgalactiae	Pyogenic	C	β	−	−	S	−	−	−
E. faecalis	Enterococci	D	β or γ	−	−	R	+	+	−
E. faecium	Enterococci	D	α	−	−	R	+	+	−
S. bovis	Other	D	α[1]	−	−	R/S	+	−	−
S. mitis[2]	Oral (Viridans)	none	α[1]	−	−	S	−	−	−
S. salivarius[2]	Oral (Viridans)	none	α[1]	−	−	S	−	−	−
S. mutans[2]	Oral (Viridans)	none	none	−	−	S	−	−	−

Note: R = resistant; S = sensitive; blank = not significant.
[1]Weakly alpha.
[2]Oral streptococci commonly isolated from throat swabs. Differentiation is based on additional biochemical tests not performed in this exercise.

hemolysis, highlighting the notion that hemolysis can be a misleading characteristic in identification.

Streptococcus pneumoniae

Although *S. pneumoniae* does not possess a Lancefield antigen, it is a significant human pathogen. *S. pneumoniae* is the causative agent of bacterial pneumonia, and can cause meningitis and otitis media (ear infections) in children. It normally colonizes the pharynx, but in certain situations it can spread to the lungs, sinuses, or middle ear. Virulent strains of *S. pneumoniae* are covered with a polysaccharide capsule (avirulent lab strains do not produce the capsule). When this organism is grown on blood agar, its colonies appear smooth, mucoid, and surrounded by a zone of greenish discoloration (alpha-hemolysis). In culture these cells usually grow as diplococci, but they can also occur singly or in short chains. Presumptive identification of *S. pneumoniae* can be made with a positive optochin susceptibility test.

Viridans Streptococci Group

The viridans streptococci are a heterogeneous cluster of alpha-hemolytic and nonhemolytic streptococci. These organisms colonize the oral cavity, pharynx, gastrointestinal tract, and genitourinary tract. Most of the viridans are opportunists, and they are usually regarded as having low pathogenicity. Two species in the viridans group are thought to be the primary cause of dental caries (cavities). Although these organisms have few virulence factors and are constituents of the normal flora, introduction of these species into tissues through dental or surgical means can cause severe infections. The most serious complication of all viridans infections is subacute endocarditis. When grown on blood agar, viridans colonies appear very small, gray to whitish gray, and opaque. In culture they appear rod-like and grow in chains. The viridans group can be differentiated from the pneumococci and enterococci by a negative result in the bile esculin hydrolysis test, the salt-tolerance test, and the optochin susceptibility test.

Group D Enterococci

The enterococci of serological group D are part of the Enterococcaceae family and are considered variably hemolytic. There are two principal species of this enterococcal group, *E. faecalis* and *E. faecium*. They are predominantly inhabitants of the gastrointestinal tract; however, they have also been isolated from the genitourinary tract and oral cavity. Although the enterococci do not possess many virulence factors, they are important pathogens in hospitalized patients where they can cause urinary tract infections, bacteremia, and endocarditis. When grown on blood agar, enterococci form large colonies that can appear nonhemolytic, alpha-hemolytic, or rarely beta-hemolytic. In culture they grow as diplococci in short chains. Colonies of *E. faecalis* appear either nonhemolytic or beta-hemolytic, depending on the strain and the type of blood agar used, while colonies of *E. faecium* generally appear alpha-hemolytic. The enterococci can grow under extreme conditions, and this phenotype can be exploited to help differentiate them from various streptococcal species. Isolates that are able to grow in the presence of 6.5% NaCl and are able to hydrolyze bile esculin are presumed to be enterococci in this exercise.

Group D Nonenterococci

S. bovis is the only clinically relevant nonenterococcal species of group D. It is found in the intestinal tracts of humans, as well as cows, sheep, and other ruminants. Although it is found in many animals, *S. bovis* is a human pathogen and has been implicated as a causative agent of endocarditis and meningitis, and is associated with malignancies of the gastrointestinal tract. Colonies of *S. bovis* appear large, mucoid (many strains have a capsule), and either nonhemolytic or alpha-hemolytic. In culture *S. bovis* grows in pairs and short chains. Key reactions for this group are a positive bile esculin test and negative salt broth test.

The purpose of this exercise is twofold: (1) to learn the standard procedures for isolating streptococci and enterococci and (2) to learn how to differentiate the medically important species of these families. Figure 52.3 illustrates the overall procedure to be followed in the pursuit of these goals. To broaden the application of the tests for identifying streptococci and enterococci, your instructor may supply you with unknown cultures to be identified along with your pharyngeal isolates. Keep in mind as you complete this exercise that these tests result in a presumptive identification of your isolates. Commercial kits are available, such as the RapID Streptococci panel, which can be used to confirm the identification of your isolates.

Note: Please review the safety information concerning human microbiota discussed on page 358 before proceeding with this exercise.

🕐 First Period

(Making a Streak-Stab Agar Plate)

During this period, a plate of blood agar is swabbed and streaked in a special way to determine the type of

hemolytic bacteria that are present in the pharynx and in an unknown mixture.

Since swabbing one's own throat properly can be difficult, it will be necessary for you to work with your laboratory partner to swab each other's throats. Once your throat has been swabbed, you will use the swab and a loop to inoculate a blood agar plate according to a special procedure shown in figure 52.4.

Materials

- 1 tongue depressor
- 1 sterile cotton swab
- inoculating loop
- 2 blood agar plates
- unknown mixture

1. With the subject's head tilted back and the tongue held down with the tongue depressor, rub the back surface of the pharynx up and down with the sterile swab.

    Also, *look for white patches* in the tonsillar area. Avoid touching the cheeks and tongue.
2. Since streptococcal hemolysis is most accurately analyzed when the colonies develop anaerobically beneath the surface of the agar, it will be necessary to use a streak-stab technique as shown in figure 52.4. The essential steps are as follows:

    - Roll the swab over an area approximating one-fifth of the surface. The entire surface of the swab should contact the agar.
    - With a wire loop, streak out three areas, as shown, to thin out the organisms.
    - Stab the loop into the agar to the bottom of the plate at an angle perpendicular to the surface to make a clean cut without ragged edges.
    - Be sure to make the stabs in an unstreaked area so that streptococcal hemolysis will be easier to interpret with a microscope.

Caution

Dispose of swabs and tongue depressors in a beaker of disinfectant or biohazard bag.

3. Repeat the inoculation procedure for the unknown mixture.
4. Incubate the plates aerobically at 37°C for 24 hours. Do not incubate the plates longer than 24 hours.

🕐 Second Period

(Analysis and Subculturing)

During this period, two objectives must be accomplished: first, the type of hemolysis must be correctly

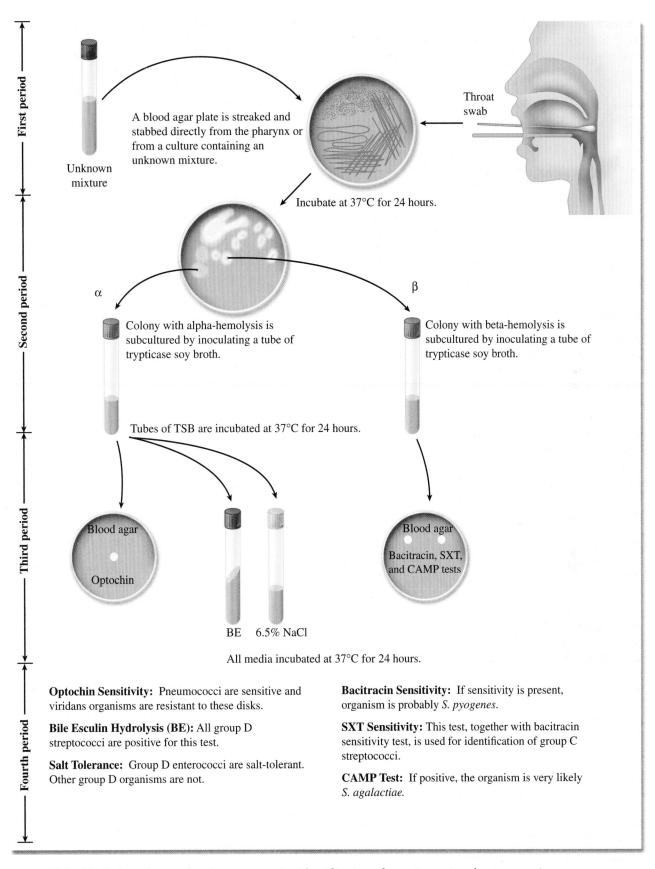

First period

Unknown mixture

A blood agar plate is streaked and stabbed directly from the pharynx or from a culture containing an unknown mixture.

Throat swab

Incubate at 37°C for 24 hours.

Second period

α

Colony with alpha-hemolysis is subcultured by inoculating a tube of trypticase soy broth.

β

Colony with beta-hemolysis is subcultured by inoculating a tube of trypticase soy broth.

Tubes of TSB are incubated at 37°C for 24 hours.

Third period

Blood agar

Optochin

BE   6.5% NaCl

Blood agar

Bacitracin, SXT, and CAMP tests

All media incubated at 37°C for 24 hours.

Fourth period

Optochin Sensitivity: Pneumococci are sensitive and viridans organisms are resistant to these disks.

Bile Esculin Hydrolysis (BE): All group D streptococci are positive for this test.

Salt Tolerance: Group D enterococci are salt-tolerant. Other group D organisms are not.

Bacitracin Sensitivity: If sensitivity is present, organism is probably *S. pyogenes*.

SXT Sensitivity: This test, together with bacitracin sensitivity test, is used for identification of group C streptococci.

CAMP Test: If positive, the organism is very likely *S. agalactiae*.

Figure 52.3 Media inoculations for the presumptive identification of streptococci and enterococci.

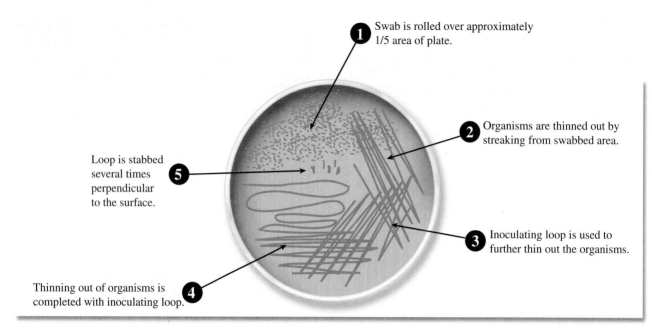

Figure 52.4 Streak-stab procedure for blood agar inoculations.

determined and, second, well-isolated colonies must be selected for making subcultures. The importance of proper subculturing cannot be overemphasized: without a pure culture, future tests are certain to fail. Proceed as follows:

Materials

- blood agar plates from previous period
- tubes of trypticase soy broth (TSB), one for each different type of colony
- dissecting microscope

1. Look for isolated colonies that have hemolysis surrounding them.
2. Do any of the stabs appear to exhibit hemolysis? Examine these hemolytic zones near the stabs under 60× magnification with a dissecting microscope or with the scanning objective (4×) of a light microscope.
3. Hemolysis patterns are divided into beta- and alpha-hemolysis, and alpha-hemolytic patterns can be furthered subdivided into either alpha or alpha-prime (consult figure 52.5). Complete lysis of red blood cells around a colony (a clear area around the colony) is **beta-hemolysis.** Partial lysis of red blood cells around a colony results in greenish discoloration of the area around the colony and is called **alpha-hemolysis.** A small area of intact red blood cells around a colony surrounded by a wider zone of complete hemolysis is **alpha-prime-hemolysis.** Some streptococci from the viridans group display alpha-prime-hemolysis. Alpha-prime-hemolysis is sometimes

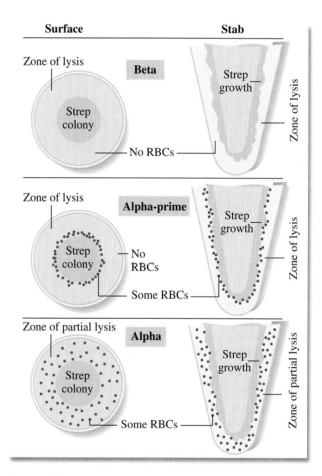

Figure 52.5 Comparison of the types of hemolysis seen on blood agar plates. The illustrations on the left indicate the appearance of red blood cells surrounding a bacterial colony on the surface of a blood agar plate. The illustrations on the right indicate the appearance of red blood cells surrounding stabs into the blood agar.

mistaken for beta-hemolysis, which is why it is recommended the hemolytic colonies/stabs be viewed under either the dissecting scope or the lowest power of a light microscope. Colonies that do not affect the red blood cells surrounding them are said to exhibit nonhemolysis or **gamma-hemolysis.**

4. Record your observations in Laboratory Report 52.
5. Select well-isolated colonies that exhibit hemolysis (alpha, beta) for inoculating tubes of TSB. Be sure to label the tubes ALPHA or BETA. Whether or not the organism is alpha or beta is crucial in identification.

   Since the chances of isolating beta-hemolytic streptococci from the pharynx are usually quite slim, notify your instructor if you think you have isolated one.
6. Incubate the tubes at 37°C for 24 hours.

⏱ Third Period

(Inoculations for Physiological Tests)

Presumptive identification of the various groups of streptococci is based on the physiological tests in table 52.1. Note that Groups A, B, and C are all beta-hemolytic; a few enterococci are also beta-hemolytic. The remainder are all alpha-hemolytic or nonhemolytic.

Since each of the physiological tests is specific for differentiating only two or three groups, it is not desirable to do all the tests on all unknowns. For economy and precision, carefully select which tests you will perform on an isolate or unknown based on the type of hemolysis it exhibits.

Before any inoculations are made, however, it is desirable to do a purity check on each TSB culture from the previous period. To accomplish this, it will be necessary to make a Gram-stained slide of each of the cultures.

Gram-Stained Slides (Purity Check)

Materials

- TSB cultures from previous period
- Gram-staining kit

1. Make a Gram-stained slide from your isolates and examine them under an oil immersion lens. Do they appear to be pure cultures?
2. Draw the organisms in the appropriate circles in Laboratory Report 52.

Beta-Type Inoculations

Use the following procedure to perform tests on each isolate that has beta-type hemolysis:

Materials

for each isolate:
- 1 blood agar plate
- 1 bacitracin differential disk (0.04 units)
- 1 SXT sensitivity disk
- 1 broth culture of *S. aureus*
- dispenser or forceps for transferring disks

1. Label a blood agar plate with the proper identification information of each isolate and unknown to be tested.

3 Bacitracin and SXT differential disks are placed as shown in area streaked by the unknown.

1 Unknown is heavily streaked out over 40% of the area and brought straight downward in a single line.

2 A loopful of *S. aureus* is streaked perpendicular to unknown streak. A gap of 1 cm should separate the two streaks.

Figure 52.6 Blood agar inoculation technique for the CAMP, bacitracin, and SXT tests.

2. Follow the procedure outlined in figure 52.6 to inoculate each blood agar plate with the isolate (or unknown) and *S. aureus*.

Note that a streak of the unknown is brought down perpendicular to the *S. aureus* streak, keeping the two organisms about 1 cm apart.

3. With forceps or dispenser, place one bacitracin differential disk and one SXT disk on the heavily streaked area at points shown in figure 52.6. Press down lightly on each disk.

4. Incubate the blood agar plates at 37°C, aerobically, for 24 hours.

Alpha-Type Inoculations

As shown in figure 52.3, three inoculations will be made for each isolate or unknown that is alpha-hemolytic.

Materials

- 1 blood agar plate (for up to 4 unknowns)
- 1 6.5% sodium chloride broth tube per unknown
- 1 bile esculin (BE) slant per unknown
- 1 optochin (Taxo P) disk per unknown
- candle jar setup or CO_2 incubator

1. Mark the bottom of a blood agar plate to divide it into halves, thirds, or quarters, depending on the number of alpha-hemolytic organisms to be tested. Label each space with the code number of each test organism.

2. Completely streak over each area of the blood agar plate with the appropriate test organism, and place one optochin (Taxo P) disk in the center of each area. Press down slightly on each disk to secure it to the medium.

3. Inoculate one tube of 6.5% sodium chloride broth and streak the surface of one bile esculin slant.

4. Incubate all media at 35–37°C as follows:
   Blood agar plates: 24 hours in a candle jar
   6.5% NaCl broths: 24, 48, and 72 hours
   Bile esculin slants: 48 hours

Note: While the blood agar plates should be incubated in a candle jar or CO_2 incubator, the remaining cultures can be incubated aerobically.

Inoculations of Colonies That Do Not Exhibit Hemolysis (Gamma-Type)

Refer to table 52.1. Isolates of this nature are probably members of group D or viridans. Perform an SXT sensitivity test, a bile esculin hydrolysis test, and a salt-tolerance test as described under Alpha-Type Inoculations.

Note: Optochin susceptibility can be evaluated as well if an optochin disk is placed on the streaked area

of the SXT sensitivity BAP, in the same manner as the SXT disk is placed.

🕐 Fourth Period

(Evaluation of Physiological Tests)

Once all of the inoculated media have been incubated for 24 hours, begin to examine the plates and tubes and add test reagents to some of the cultures. Some of the tests will also have to be checked at 48 and 72 hours.

After the appropriate incubation period, assemble all the plates and tubes from the last period, and examine the blood agar plates first that were double-streaked with the unknowns and *S. aureus*. Note that the CAMP reaction, bacitracin susceptibility test, and SXT sensitivity test can be read from these plates. Proceed as follows:

CAMP Reaction (β test)

If you have an unknown that produces an enlarged arrowhead-shaped hemolytic zone at the juncture where the unknown meets the *Staphylococcus aureus* streak, as seen in figure 52.7, the organism is *Streptococcus agalactiae*. This phenomenon is due to what is called the *CAMP factor*, named for the developers of the test, Christie, Atkins, and Munch-Peterson. The CAMP factor produced by *Streptococcus agalactiae* acts synergistically with Staphylococcal hemolysins causing an enhanced breakdown of red blood cells and therefore producing the arrowhead zone of clearing. The only problem that can arise from this test is that if

Figure 52.7 Note positive SXT disk on right, negative bacitracin disk on left, and positive CAMP reaction (arrowhead). Organism: *S. agalactiae.*

the plate is incubated anaerobically, a positive CAMP reaction can occur on *S. pyogenes* inoculated plates.

Record the CAMP reactions for each of your isolates or unknowns in Laboratory Report 52.

Bacitracin Susceptibility (β test)

Any size zone of inhibition seen around the bacitracin disks should be considered to be a positive test result. Note in table 52.1 that *S. pyogenes* is positive for this characteristic.

This test has two limitations: (1) the disks must be of the *differential type*, not sensitivity type, and (2) the test should not be applied to alpha-hemolytic streptococci. Reasons: Sensitivity disks have too high a concentration of the antibiotic, and many alpha-hemolytic streptococci are sensitive to these disks.

Record the results of this test in the table under number 2 of Laboratory Report 52.

SXT Sensitivity Test (β test)

The disks used in this test contain 1.25 mg of trimethoprim and 27.75 mg of sulfamethoxazole (SXT). The purpose of this test is to distinguish groups A and B from other beta-hemolytic streptococci. Note in table 52.1 that both groups A and B are uniformly resistant to SXT.

If a beta-hemolytic streptococcus proves to be bacitracin resistant and SXT susceptible, it is classified as being a **non–group A or B beta-hemolytic streptococcus.** This means that the organism is probably a species within group C. *Keep in mind that an occasional group A streptococcal strain is susceptible to both bacitracin and SXT disks.* One must always remember that exceptions to most tests do occur; that is why this identification procedure leads us only to *presumptive* conclusions.

Record any zone of inhibition (resistance) as positive for this test.

Note: Some strains of *E. faecalis* are beta-hemolytic. A beta-hemolytic isolate of *E. faecalis* would have a negative CAMP reaction and would appear resistant to bacitracin and SXT. Although this exercise does not outline performing a bile esculin hydrolysis test or salt-tolerance test on a beta-hemolytic colony, these physiological tests would help confirm identification of a beta-hemolytic *E. faecalis* strain.

Bile Esculin (BE) Hydrolysis (α test)

This is the best physiological test that we have for the identification of group D streptococci. Both enterococcal and non-enterococcal species of group D are able to hydrolyze esculin in the agar slant, causing the slant to blacken.

Figure 52.8 **Positive bile esculin hydrolysis on left; negative on right.**

A positive BE test tells us that we have a group D streptococcus; differentiation of the two types of group D streptococci (*Enterococcus* and *S. bovis*) depends on the salt-tolerance test.

Examine the BE agar slants, looking for **blackening of the slant,** as illustrated in figure 52.8. If less than half of the slant is blackened, or if no blackening occurs within 24 to 48 hours, the test is negative.

Salt-Tolerance (6.5% NaCl) (Group D) (α test)

All enterococci of group D produce heavy growth in 6.5% NaCl broth. As indicated in table 52.1, *S. bovis* does not grow in this medium. This test, then, provides us with a good method for differentiating the organisms of group D.

A positive result shows up as turbidity within 72 hours. A color change of **purple to yellow** may also be present. If the tube is negative at 24 hours, incubate it and check it again at 48 and 72 hours. *If the organism is salt tolerant and BE positive, it is considered to be an enterococcus.* Parenthetically, it should be added here that approximately 80% of group B streptococci will grow in this medium.

Optochin Susceptibility (α test)

Optochin susceptibility is used for differentiation of the alpha-hemolytic viridans streptococci from the pneumococci. The pneumococci are sensitive to these disks; the viridans organisms are resistant.

Materials

- blood agar plates with optochin disks
- plastic metric ruler

1. Measure the diameters of zones of inhibition that surround each optochin disk, evaluating whether the zones are large enough to be considered positive. The standards are as follows:
   - For 6 mm diameter disks, the zone must be at least 14 mm diameter to be considered positive.
   - For 10 mm diameter disks, the zone must be at least 16 mm diameter to be considered positive.
2. Record your results in Laboratory Report 52.

Final Confirmation

All the laboratory procedures performed so far lead us to presumptive identification. To confirm these conclusions, it is necessary to perform serological tests on each of the unknowns. If commercial kits are available for such tests, they should be used to complete the identification procedures.

Laboratory Report

Complete Laboratory Report 52.

52 The Streptococci and Enterococci: Isolation and Identification

A. Results

1. **Microscopy**
   Provide drawings here of your isolates and unknowns as seen under oil immersion (Gram staining).

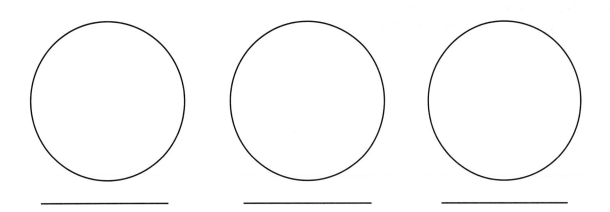

_____   _____   _____

2. **Record of Test Results**
   Record here all information pertaining to the identification of pharyngeal isolates and unknowns.

SOURCE OF UNKNOWN	Hemolysis	Bacitracin Susceptibility	CAMP Reaction	SXT Sensitivity	Bile Esculin Hydrolysis	Tolerance to 6.5% NaCl	Optochin Susceptibility

3. **Final Determination**
   Record here the identities of your various isolates and unknowns:

   Pharyngeal isolates: _____

   _____

   Unknowns: _____

   _____

   _____

B. Short-Answer Questions

1. When bacteria from a throat swab are streaked on blood agar, why is the agar stabbed several times with the loop?

   _____

   _____

2. Differentiate between alpha- and beta-hemolysis.

   _____

   _____

3. What was Rebecca Lancefield's contribution to the study of streptococci?

   _____

   _____

4. In the CAMP reaction, which organism produces the CAMP factor? What substance does the CAMP factor react with to cause enhanced breakdown of red blood cells?

   _____

5. Humans may carry both staphylococci and streptococci as normal microbiota. How might you differentiate between the two genera?

   _____

6. Name two tests that are useful in differentiating *S. pyogenes* and *S. agalactiae*.

   _____

   _____

7. Name two tests that are useful for the differentiation of pneumococci and oral viridans streptococci.

   _____

   _____

8. What test can be performed to differentiate the enterococci from other group D streptococci?

   _____

   _____

9. What test can be performed to differentiate between group A and group C streptococci?

   _____

   _____

10. Describe the appearance of an *S. agalactiae* colony grown on blood agar. Describe how that colony would differ in appearance from a colony of *S. pyogenes*.

   _____

   _____

11. Vaginal swabs are taken from pregnant women in their third trimester. Which streptococcal species is the focus of the investigation?

_____

_____

12. Which streptococci are implicated in the development of dental caries? What is the mechanism of their formation?

_____

_____

MICROBIOLOGY

Gram-Negative Intestinal Pathogens

Learning Outcomes

After completing this exercise, you should be able to

1. Enrich for *Salmonella* and *Shigella* using differential and selective media.

2. Differentiate the coliforms and other lactose fermenters from the non-lactose fermenters, *Salmonella*, *Shigella*, and *Proteus*, using specific biochemical tests and differential media.

The Enterobacteriaceae are a large family of diverse gram-negative rods. They are ubiquitous and can be found in soil, water, vegetation, and the intestinal tracts of most organisms. All members of this family are facultative anaerobes. This heterogeneous collection of organisms is responsible for a variety of human diseases, including bacteremias, urinary tract infections, and numerous intestinal infections. Organisms in this group are divided into potential human pathogens and true intestinal pathogens. The potential human pathogens are part of the normal commensal microbiota of the gastrointestinal tract and can cause opportunistic infections if they are not confined to their natural environment or if a person's immune response is compromised. The true pathogens are not present as commensal microbiota in the gastrointestinal tract of humans, and they are always associated with human disease.

E. coli is an important opportunistic pathogen. Normally, *E. coli* is found in large numbers as a resident of the colon; however, it can grow outside its normal body site and cause urinary tract infections, sepsis, wound infections, and meningitis. Most infections involving *E. coli* are endogenous, meaning the resident *E. coli* of the commensal microbiota established the infection when it grew outside of its natural site. *E. coli* can also cause clinical disease in immune compromised patients. Additionally, this species can acquire virulence factors encoded on plasmids or in bacteriophage DNA, causing some strains to have enhanced virulence. *E. coli* 0157:H7 is a particularly virulent strain that has been associated with various kinds of contaminated food and has caused many deaths. This strain of *E. coli* produces a toxin that damages blood vessels and causes very severe diarrhea. From a nomenclature standpoint *E. coli* are considered **coliforms,** because they are gram-negative organisms that can ferment lactose. Other biochemical characteristics used in the identification of *E. coli* include its motility, its ability to produce indole, and its inability to use citrate as a carbon source (figure 53.1). Additional potential human pathogens that are also coliforms include *Enterobacter, Citrobacter, Klebsiella,* and *Serratia.* These organisms can cause infections in individuals whose defenses are compromised, but they rarely cause infections in immune competent individuals.

Proteus are opportunistic pathogens that are normally found in the intestinal tract and considered harmless, but they can cause urinary tract infections, wound infections, and septicemia when grown outside of the intestinal tract. They are motile and produce many different types of fimbriae, which account for their ability to adhere to the epithelium of the urinary tract and cause urinary tract infections. Most species of *Proteus* produce large quantities of urease, which then raises the urine pH and can cause the formation of crystals. Organisms of the genus *Proteus* are not considered coliforms because they are unable to ferment lactose; however, they are able to ferment glucose.

The true intestinal pathogens of the Enterobacteriaceae family are *Salmonella, Shigella,* and *Yersinia.* These organisms differ from the potential human pathogens by not being part of the normal microbiota of humans and by having well-developed virulence factors. *Salmonella* infections involve gastroenteritis, septicemia, and typhoid fever. Some strains of *Salmonella,* such as the one that causes typhoid fever, can exist in patients for more than a year. This carrier state following *Salmonella* infection represents an important source of human infection. These organisms are also transmitted through poultry and dairy products. When grown on differential or selective media, *Salmonella* produce non-lactose fermenting colonies with black centers if the media contain indicators for hydrogen sulfide production. These pathogens have flagella (are motile) and are negative for indole and urease.

Shigella intestinal pathogens cause dysentery; however, species vary in the severity of the disease they cause. Humans are the only known reservoir of *Shigella,* and a small number of patients develop asymptomatic colonization of the colon, thus creating

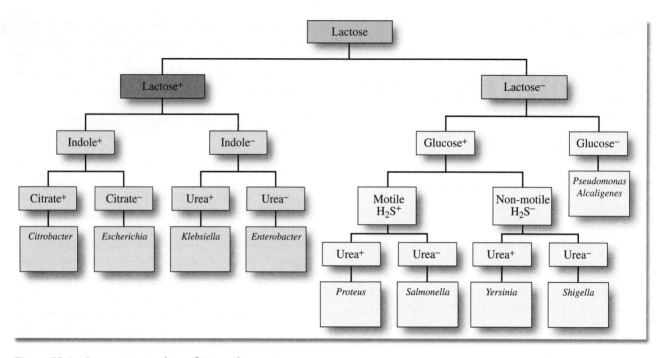

Figure 53.1 Separation outline of Enterobacteriaceae.

a reservoir for infection. *Shigella* species do not ferment lactose (with the exception of *Shigella sonnei,* which slowly ferments lactose) or hydrolyze urea. Unlike *Salmonella, Shigella* are non-motile and do not produce hydrogen sulfide. When outbreaks of *Salmonella* and *Shigella* occur, serological typing (Exercise 55) is useful in tracing epidemics caused by a particular serotype of the respective organism. Serotypes (also called serovars) are strains of bacteria within a species that are biochemically similar but differ in their antigenic composition.

The final clinically relevant primary intestinal pathogen is *Yersinia. Yersinia pestis,* the most famous human pathogen within this genus, causes the highly fatal systemic disease known as the bubonic plague. Humans are accidental hosts of *Yersinia,* and infections occur through ingestion of contaminated animals or the handling of contaminated animal tissues. Similar to *Shigella* and *Salmonella, Yersinia* are non-lactose fermenters when cultured in a laboratory setting. Also, *Yersinia* do not produce hydrogen sulfide and are generally considered non-motile and urease positive.

In this exercise you will be presented with a mixture of organisms that represents a "simulated" GI tract sample. In actual practice, stool samples are used and are plated onto isolation media and grown in enrichment broth. In this experiment you will be given a mixed culture containing a coliform, *Proteus,* and either *Salmonella* or *Shigella.* The pathogens will be attenuated, but their presence will naturally demand utmost caution and handling. Your goal will be to isolate the primary (true) pathogen from the mixed culture and to make a genus identification of the pathogen. Figure 53.1 summarizes the biochemical characteristics that will be used for genus identification.

Note: Please review the safety information concerning human microbiota discussed on page 358 before proceeding with this exercise.

🕐 First Period

(Isolation)

There are several excellent selective differential media that have been developed for the isolation of these pathogens. Various inhibiting agents such as brilliant green, bismuth sulfite, sodium desoxycholate, and sodium citrate are included in them.

Widely used media for the isolation of intestinal pathogens include MacConkey agar, Hektoen Enteric agar (HE), and Eosin Methylene Blue agar (EMB). These media may contain bile salts and/or sodium desoxycholate to inhibit gram-positive bacteria. To inhibit coliforms and other nonenterics, they may contain citrate. All of them contain lactose and a dye so that if an organism is a lactose fermenter, its colony will take on a color characteristic of the dye present.

You will be issued an unknown broth culture with a pathogenic enteric. Your instructor will indicate which selective media will be used. Proceed as follows to inoculate the selective media with your unknown mixture:

Materials

- unknown culture (mixture of a coliform, *Proteus*, and a *Salmonella* or *Shigella*)
- 1 or more petri plates of different selective media: MacConkey, Hektoen Enteric (HE), or Eosin Methylene Blue (EMB) agar

1. Label each plate with your name and unknown number.
2. With a loop, streak each plate with your unknown in a manner that will produce good isolation.
3. Incubate the plates at 37°C for 24 hours.

Second Period

(Fermentation Tests)

As stated above, the fermentation characteristic that separates the *Salmonella* and *Shigella* pathogens from the coliforms is their *inability to ferment lactose*. Once we have isolated colonies on differential media that look like *Salmonella* and *Shigella* (non-lactose fermenters), the next step will be to determine whether the isolates can ferment glucose. Kligler's iron agar is often used for this purpose. It contains two sugars, glucose and lactose, as well as phenol red to indicate when acid is produced (fermentation occurs), and iron salts for the detection of H_2S. The glucose concentration of Kligler's iron agar is only 10% of the lactose concentration. Non-lactose fermenters, such as *Shigella* and *Salmonella*, initially produce a yellow slant due to acid produced by the fermentation of glucose, which lowers the pH, causing the agar to turn yellow. When the glucose supply (remember the agar is only 10% glucose) is depleted by non-lactose fermenters they begin to break down amino acids in the medium, producing ammonia and raising the pH, causing part of the agar to turn red. At the end of 24 hours of incubation a non-lactose fermenter has a red slant and a yellow butt (bottom of the tube). The bottom of the tube is yellow-acidic from the breakdown of glucose. The slant is red because after using up all of the glucose, the bacteria began breaking down amino acids in the media, producing ammonia. The ammonia increased (alkalinized) the pH. Lactose fermenters produce yellow slants and butts. Because these organisms can ferment lactose, and because there is 10 times more lactose in the media than glucose, this sugar is not exhausted over a 24-hour time period, and the pH is kept low (yellow) due to the acids produced during fermentation. If the bacteria being cultured reduce sulfur, hydrogen sulfide will be produced, and iron-containing compounds in the media will react with the hydrogen sulfide and produce a black precipitate.

Proceed as follows to inoculate three slants from colonies on the selective media that look like either *Salmonella* or *Shigella*. The reason for using three slants is that you may have difficulty distinguishing *Proteus* from the *Salmonella-Shigella* pathogens. By inoculating three tubes from different colonies, you will be increasing your chances of success.

Materials

- 3 agar slants (Kligler's iron)
- streak plates from first period

1. Label the three slants with your name and the number of your unknown.
2. Look for isolated colonies that look like *Salmonella* or *Shigella* organisms. The characteristics to look for on each medium are as follows:

- **MacConkey agar**—*Salmonella, Shigella,* and other non-lactose-fermenting species produce smooth, colorless colonies. Coliforms that ferment lactose produce reddish, mucoid, or dark-centered colonies (figure 53.2).
- **Hektoen Enteric (HE) agar**—*Salmonella* and *Shigella* colonies are greenish-blue. Some species of *Salmonella* will have greenish-blue colonies with black centers due to H_2S production. Coliform colonies are salmon to orange and may have a bile precipitate.
- **Eosin Methylene Blue (EMB) agar**—Bacteria that ferment lactose produce blue-black colonies, and strong lactose fermenters (such as *E. coli*) have a characteristic green, metallic sheen. Non-lactose fermenters such as *Shigella* and *Salmonella* produce colorless or amber colonies (figure 53.3).

Figure 53.2 MacConkey agar results: (top) nonlactose fermenter, (left and right) lactose fermenters.
© McGraw-Hill Education. Lisa Burgess, photo

Figure 53.3 EMB agar: (top and left) nonlactose fermenters, (right) lactose fermenter with metallic sheen, (bottom) inhibition of gram-positive bacteria.
© McGraw-Hill Education. Lisa Burgess, photo

3. With a straight wire, inoculate the three agar slants from separate *Salmonella-Shigella* appearing colonies. Use the streak-stab technique. When streaking the surface of the slant before stabbing, move the wire over the entire surface for good coverage.

4. Loosen the caps on the slants and incubate at 37°C for 18 to 24 hours. The timing of the incubation of your slant should be carefully monitored. If your slant is read too early (prior to 24 hours) you may erroneously conclude that the organism is capable of fermenting both glucose and lactose. The glucose supply in the tube will last approximately 12 hours, thus tubes containing non-lactose fermenters will have a yellow slant and butt in the beginning of the incubation. If your slant is older than 24 hours you may observe a condition known as **alkaline reversion,** which produces a false negative result for both glucose and lactose. This phenomenon occurs when all of the lactose in the tube has been depleted (usually around 48 hours). Once the lactose is completely metabolized, the lactose fermenters then begin to break down amino acids in the medium. The breakdown of amino acids forms ammonia, and ammonia increases the pH of the agar, causing the color to change back to red.

🕐 Third Period

(Slant Evaluations and Final Inoculations)

During this period, you will inoculate tubes of SIM medium and urea broth with organisms from the slants of the previous period. Examination of the separation outline in figure 53.1 reveals that the final step in the differentiation of the *Salmonella-Shigella* pathogens is to determine whether a nonlactose fermenter can do three things: (1) exhibit motility, (2) produce hydrogen sulfide, and (3) produce urease. You will also make a Gram-stained slide to perform a purity check. If miniaturized multitest media are available, they can also be inoculated at this time.

Materials

- Kligler's iron agar slants from previous period
- 1 tube of SIM medium for each positive slant
- 1 urea slant for each positive Kligler's iron agar slant
- miniaturized multitest media such as API 20E or Enterotube II (optional)
- Gram-staining kit

1. Make Gram-stained slides from the same slants and confirm the presence of gram-negative rods.
2. Examine the slants from the previous period and **select those tubes that have a yellow butt with a red slant.** These tubes contain organisms that ferment only glucose (non-lactose fermenters). If you used Kligler's iron agar, a black precipitate in the medium will indicate that the organism is a producer of H_2S (refer to figure 38.1).

   Note that slants in figure 53.4 that are completely yellow are able to ferment lactose as well as glucose. Tubes that are completely red are either nonfermenters or examples of alkaline reversion. Ignore those tubes.
3. With a loop, inoculate one urea slant from each slant that has a yellow butt and red slant (non-lactose fermenter).
4. With a straight wire, stab one tube of SIM medium from each of the same agar slants. Stab in the center to two-thirds of the depth of the medium.
5. Incubate these tubes at 37°C for 18 to 24 hours.
6. If miniaturized multitest media are available, such as API 20E or Enterotube II, inoculate and incubate for evaluation in the next period. Consult Exercises 40 and 41 for instructions.
7. Refrigerate the positive Kligler's iron slants for future use as a stock culture, if needed.

🕐 Fourth Period

(Final Evaluation)

During this last period, the tubes of SIM medium, urea broth, and any miniaturized multitest media from the last period will be evaluated. Serotyping can also be performed, if desired.

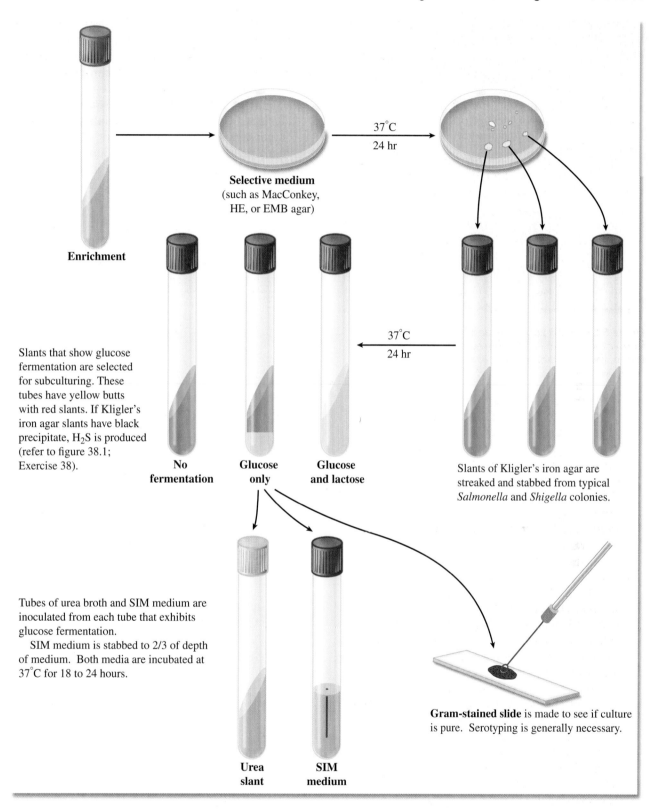

Enrichment

Selective medium
(such as MacConkey,
HE, or EMB agar)

37°C
24 hr

37°C
24 hr

Slants that show glucose
fermentation are selected
for subculturing. These
tubes have yellow butts
with red slants. If Kligler's
iron agar slants have black
precipitate, H₂S is produced
(refer to figure 38.1;
Exercise 38).

**No
fermentation**

**Glucose
only**

**Glucose
and lactose**

Slants of Kligler's iron agar are
streaked and stabbed from typical
Salmonella and *Shigella* colonies.

Tubes of urea broth and SIM medium are
inoculated from each tube that exhibits
glucose fermentation.
  SIM medium is stabbed to 2/3 of depth
of medium. Both media are incubated at
37°C for 18 to 24 hours.

**Urea
slant**

**SIM
medium**

Gram-stained slide is made to see if culture
is pure. Serotyping is generally necessary.

Figure 53.4   Isolation and presumptive identification of *Salmonella* and *Shigella*.

Materials

- urea slant tubes and SIM medium from previous period
- Kovacs' reagent
- 5 ml pipettes
- miniaturized multitest media from previous period
- serological testing materials (optional)

1. Examine the tubes of SIM medium, checking for motility and H$_2$S production. If you see cloudiness spreading from the point of inoculation, the organism is motile. If the bacteria are growing only along the stab line, then motility is absent. A black precipitate will be evidence of H$_2$S production (see Exercise 38, figure 38.2).
2. Test for indole production by adding four drops of Kovacs' reagent to the surface of the media of each SIM tube. A **pink to deep red color** will form on the top of the tube if indole is produced (figure 38.2). *Salmonella* are negative. *Proteus* and some *Shigella* may be positive; there is species variability within these genera. *Citrobacter* and *Escherichia* are positive.
3. Examine the urea slant tubes. If the medium has changed from yellow to **bright pink,** the organism is urease positive (figure 37.4).
4. If a miniaturized multitest medium was inoculated in the last period, complete it now.
5. If time and materials are available, confirm the identification of your unknown with serological typing. Refer to Exercise 55.

Laboratory Report

Record the identity of your unknown in Laboratory Report 53 and answer all the questions.

53 Gram-Negative Intestinal Pathogens

A. Results

1. Record of test results

UNKNOWN NUMBER	FERMENT LACTOSE	FERMENT GLUCOSE	H$_2$S PRODUCED	MOTILITY	INDOLE	UREASE

2. Microscopy

    Provide a drawing of your unknown as seen under oil immersion (Gram staining).

3. What was the genus of your unknown?

    _____   _____

    Genus   No.

4. What problems, if any, did you encounter?

    _____

    _____

    _____

    _____

    _____

    _____

5. Now that you know the genus of your unknown, what steps would you follow to determine the species?

    _____

    _____

    _____

B. **Short-Answer Questions**

1. Name three enteric pathogens of primary medical importance.

_____

_____

2. The ability of *Salmonella* to produce H$_2$S is one characteristic that helps differentiate it from *Shigella.* List the three opportunities you had in this exercise to determine whether or not your unknown produced H$_2$S.

_____

_____

3. What selective agents are added to media to preferentially grow enterobacteria for study? What type of growth is inhibited?

_____

4. What characteristic separates *Salmonella* and *Shigella* from most of the other enterobacteria? What media can be used for this differentiation?

_____

_____

5. What two characteristics separate *Salmonella* from *Shigella?* What media can be used for this differentiation?

_____

_____

6. Which coliform bacteria are the most difficult to distinguish from the *Salmonella-Shigella* pathogens? What is the primary characteristic used to differentiate them?

_____

_____

7. How can acid production by glucose and lactose fermentation be differentiated in the same tube?

_____

_____

8. What is alkaline reversion? Explain why this condition gives a false negative result.

_____

_____

9. In this lab exercise, were the results of the indole test necessary to differentiate between *Salmonella* and *Shigella?* Explain why or why not.

_____

_____

A Synthetic Epidemic

Learning Outcomes

After completing this exercise, you should be able to

1. Define terminology related to the field of epidemiology.

2. Differentiate between common source and propagated epidemics.

3. Perform a simple test, using detectable reagents as a "microbe," to demonstrate how an infectious agent can be passed from person to person.

4. Describe how epidemiology is used to trace the source and spread of communicable diseases.

5. Explain how herd immunity can reduce the spread of communicable diseases through a population and protect susceptible individuals from infection.

A disease caused by microorganisms that enter the body and multiply in the tissues at the expense of the host is said to be an **infectious disease.** Infectious diseases that are transmissible from one person to another are considered to be **communicable.** The transfer of communicable infectious agents between individuals can be accomplished by direct contact, such as in handshaking, kissing, and sexual intercourse, or these agents can be spread indirectly through food, water, objects, animals, and so on.

Epidemiology is the study of how, when, where, what, and who are involved in the spread and distribution of diseases in human populations. An epidemiologist is, in a sense, a medical detective who searches out the sources of infection so that the transmission cycle can be broken.

Whether an epidemic actually exists is determined by comparing the number of new cases with previous records. If the number of newly reported cases in a given period of time in a specific area is excessive, an **epidemic** is considered to be in progress. Notable epidemics in the United States today include chlamydia and pertussis. If the disease spreads to one or more continents, a **pandemic** is occurring. An example of a pandemic disease is HIV/AIDS. According to the World Health Organization, over 34 million people are living with HIV worldwide. Tuberculosis, caused by *Mycobacterium tuberculosis,* is also considered a pandemic, with an estimated one-third of the world's population now

infected with the causative bacterium. An infectious disease that exhibits a steady frequency over a long period of time in a particular region is considered **endemic.** In tropical regions of the world, malaria is endemic.

Epidemics fall into two categories: common source epidemics and host-to-host epidemics (see figure 54.1). Common source epidemics occur rapidly and have many new cases immediately after the initial case. This type of epidemic usually involves a contaminated fomite (inanimate object) or contaminated food or water. After the infected source has been identified and removed in a common source epidemic, the number of new cases of disease drops rapidly, and the epidemic quickly subsides. From August to October 2011, 147 people across the United States were infected with *Listeria monocytogenes* after eating contaminated cantaloupes from a Colorado farm. The identification of this source led to recalls of the fruit and a quick end to the epidemic.

Propagated, or host-to-host, epidemics grow much more slowly and are also slower to subside. These epidemics involve transmission of the infectious agent through direct contact with the infected individual, a carrier, or a vector. A **vector** is a living carrier, such as an insect or rodent, which transmits an infectious agent. Control of propagated epidemics can

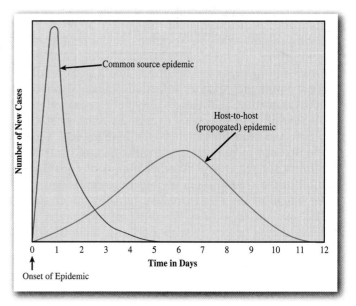

Figure 54.1 Comparison of common source and host-to-host epidemics.

involve education, vaccination, administration of anti-virals or antibiotics, improved hygiene, and voluntary quarantine. The 2009 outbreak of H1N1 influenza was an example of a host-to-host epidemic and was eventually declared a pandemic as it spread from person to person across continents. Many of these control measures were implemented to slow the spread of the infection.

Human history has been shaped by epidemics. Population growth has been curbed, wars have been lost and won, and migrations have occurred all due to past outbreaks of disease. In the Middle Ages, "The Black Death," or bubonic plague caused by *Yersinia pestis*, killed one-third of the European population, redistributing wealth and property. In 1918 the flu pandemic infected one-third of the world population. The rapid spread of the influenza virus during this epidemic is attributed to the living conditions and travel of soldiers during World War I. Smallpox was brought to Mexico by European explorers, and the spread of this disease through the native population of Mexico is thought to be the reason they were defeated by the Spaniards. AIDS has wreaked havoc on the sub-Saharan African population and has slowed economic growth throughout Africa. Continual changes in human behavior and the threat of bioterrorism make pandemics a concern for the future. Today, it is common to travel across continents and expand our presence into environments that historically have been untouched. This creates opportunities for the spread of infection throughout the world as well as interactions with microbes and animals that may transmit diseases to humans.

The ability of infectious diseases to become epidemic and pandemic depends on the transmission cycle of the disease. The transmission cycles of infectious pathogens can be greatly affected by environmental changes. Vocations in the healthcare field, travel, lifestyles, and crowded living arrangements can increase the likelihood that a person will come in contact with an infectious agent. Vaccination, quarantine, and improved hygiene can decrease susceptibility to infectious agents. Extreme weather events can alter contact between humans and vectors and can increase reservoirs, the natural hosts or habitats for pathogens. Another important consideration in the transmission of infectious pathogens is the length of time an individual remains infectious. The span of time from the onset of symptoms to death during infection with Ebola (a hemorrhagic fever virus) is usually 2 to 21 days. Although highly contagious, Ebola is often confined to small regions because the length of time between the onset of symptoms and death is short, thus restricting the amount of time an infected individual can act as a carrier/host. In the case of HIV/AIDS, patients are infectious for

the remainder of their lifetimes. The length of time they can act as carriers is much longer, thus greatly enhancing the number of individuals to whom they can transmit the disease. Finally, some infections, such as chlamydia and hepatitis B, can be asymptomatic. Carriers can unknowingly transmit the infection to others, resulting in significant spread of the causative agent through the population.

Another principle that affects the spread of an epidemic through a population is referred to as community, or herd, immunity. When a certain proportion of a population is vaccinated or already immune to an infectious agent, an outbreak is unlikely in the population. Even individuals who cannot be vaccinated, such as infants, pregnant women, and immunocompromised individuals, are somewhat protected from infection because the spread of the agent is limited in the community (figure 54.2).

The National Institutes of Health has developed an online simulation that you can use to explore the impact of these variables on the transmission of a disease through a population. Visit this activity at the following URL: science.education.nih.gov/supplements/nih1/diseases/activities/activity4.htm.

The microbiology laboratory plays a crucial role in preventing an outbreak from becoming an epidemic. Laboratory support involves, but is not limited to, culturing of infectious agents and environmental sites, isolate identification, and serological typing. Results from the microbiology lab help epidemiologists determine the first incident in a given outbreak, known as the **index case.** Identification of the index case aids in determining the original source of infection in an outbreak. Additionally, the microbiology lab is responsible for reporting suspicion or identification of infectious diseases to health departments, the Centers for Disease Control and Prevention (CDC), and the World Health Organization.

In this experiment, we will have an opportunity to approximate, in several ways, the work of the epidemiologist. Each member of the class will take part in the spread of a "synthetic infection" with simple reagents used to simulate the spread of an imaginary disease.

Two different experiments will be conducted: Procedures A and B. In Procedure A, students transfer a detectable agent by handshaking. The agent used is visible under UV light. Procedure A represents an epidemic in which the infectious agent is transferred from person to person by physical contact, and class data will be analyzed to determine the index case. In Procedure B, students transfer a detectable agent by the exchange of fluid in a test tube. The detectable agent in Procedure B is NaOH, which causes a change in pH that can be detected by the addition of a pH indicator. Procedure B represents an epidemic in which

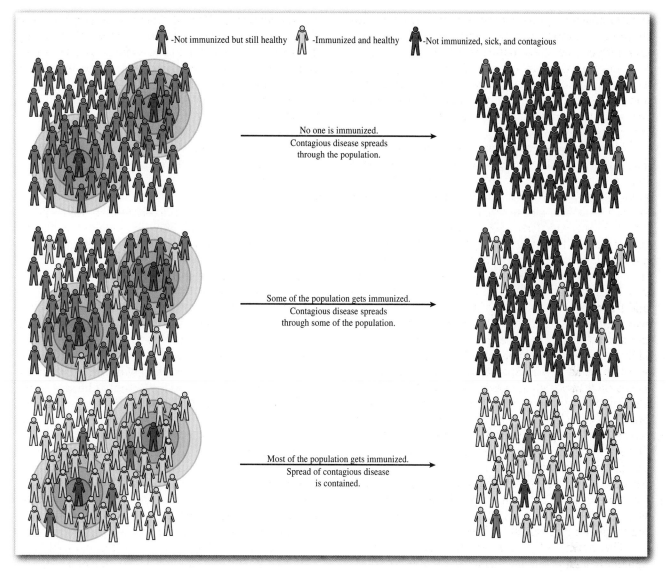

Figure 54.2 Community or herd immunity.
Courtesy: National Institute of Allergy and Infectious Diseases.

the infectious agent is transferred by the exchange of body fluid (saliva, sweat, urine, secretions). This procedure will be slightly modified and repeated three times to demonstrate the concept of community (herd) immunity.

Procedure A

In this experiment, each student will be given a numbered container of white unknown powder. Only one member in the class will be given a detectable agent that is to be considered the infectious agent and is visible under UV light. The other members will be issued a transmissible agent that is considered noninfectious. After each student has spread the powder on his or her hands, all members of the class will engage in two rounds of handshaking, directed by the instructor. A record of the handshaking contacts will be recorded

on a chart similar to the one in the Laboratory Report. After each round of handshaking, the hands will be rubbed on a Kimwipe or tissue that will later be placed under UV light to determine the presence or absence of the infectious agent.

Once all the data are compiled, an attempt will be made to determine two things: (1) the original source of the infection, and (2) who the carriers are. The type of data analysis used in this experiment is similar to the procedure that an epidemiologist would employ. Proceed as follows:

Materials

- 1 numbered petri dish containing an unknown white powder (either powder that is detectable using a UV light, such as GloGerm, or powder that is undetectable under UV light, such as baking soda)
- 2 Kimwipes or tissues

Preliminaries

1. After assembling your materials, label one of your Kimwipes ROUND 1 and one of your Kimwipes ROUND 2.
2. Wash and dry your hands thoroughly.
3. Thoroughly coat your right hand with the powder, focusing on the palm surface.

    IMPORTANT: Once the hand has been prepared, do not rest it on the tabletop or allow it to touch any other object.

Round 1

1. On the cue of the instructor, student number 1 will begin the first round of handshaking. Your instructor will inform you when it is your turn to shake hands with someone. Use your treated right hand to make firm hand contact and be sure that your palms are fully in contact with one another. You may shake with anyone, but it is best not to shake your neighbor's hand. *Be sure to use only your treated hand, and touching anything else with that hand.*
2. In each round of handshaking, you will be selected by the instructor *only once* for handshaking; however, due to the randomness of selection by the handshakers, it is possible that you may be selected as the "shakee" several times. The names of each "shaker" and "shakee" should be recorded on the board as the round progresses.
3. After every member of the class has shaken someone's hand, you need to assess just who might have picked up the "microbe." To accomplish this, rub your right hand thoroughly on the Kimwipe labeled ROUND 1. Set the Kimwipe aside until after the second round of handshaking.

    IMPORTANT: Don't allow your hand to touch any other object. A second round of handshaking follows.

Round 2

1. On cue of the instructor, student number 1 will again select a person at random to shake hands with, proceeding as in Round 1 until everyone has had a turn to initiate a handshake. Avoid contact with any other objects. Be sure to keep a record of each handshake on the board.
2. Once the second handshaking episode is finished, rub the fingers and palm of the contaminated hand on the Kimwipe labeled ROUND 2. Set the Kimwipe aside and thoroughly wash hands.

Chemical Identification

1. The powder that is considered the infectious agent in this lab exercise is comprised of synthetic beads that fluoresce under UV light. Place your Kimwipes under a UV light in a darkened room to determine if they glow. A glowing Kimwipe is interpreted as positive for the infectious agent. Record the results of your ROUND 1 and ROUND 2 Kimwipes.
2. Observe your right hand, which was washed at the conclusion of the handshaking, under the UV light. This illustrates the importance of proper handwashing technique.

Tabulation of Results

1. Tabulate the results on the chalkboard, using a table similar to the one in the Laboratory Report.
2. Once all results have been recorded, proceed to determine the index case in this epidemic. The easiest way to determine this is to put together a flow chart of shaking.
3. Identify those persons that test positive. You will be working backward with the kind of information an epidemiologist has to work with (contacts and infections). Eventually, a pattern will emerge that shows which person or persons may have started the epidemic.
4. Complete Laboratory Report 54.

Procedure A: Results and Analysis

Note to instructors: If GloGerm and a UV light source are unavailable, you can use corn starch or baking powder as safe, inexpensive alternatives. In this case, iodine is used to determine those students who have been "infected." Because student safety takes the highest priority in this manual, microorganisms were not used during this exercise.

Procedure B

In this experiment, you will be investigating the benefit of community immunity for a population. In each round, students will be given a test tube containing water (representing a susceptible individual), 0.1 M NaOH (representing an infected individual), or a pH buffer solution (representing a vaccinated individual). After receiving their tubes, students will exchange a portion of the fluid in the tubes with one another as directed by the instructor. During each round, each student will exchange fluid with three different students. Once all of the exchanges have been completed for each round, a pH indicator will be added to all tubes to determine the presence or absence of the "infectious agent."

Proceed as follows for each round, recording class data in the Laboratory Report.

Materials

- sterile gloves
- 1 numbered test tube with lid containing liquid (either distilled water, a pH buffer solution such as a sodium phosphate buffer pH = 6.8, or 0.1 M NaOH)
- 1 dropper
- phenolphthalein solution, dissolved in alcohol and diluted in water (pH indicator) for instructor

Caution

Sodium hydroxide (NaOH) and phenolphthalein can irritate the eyes and skin. Wear gloves, and alert your instructor if any spills occur.

Round 1: 100% Susceptible Population

For this round, one student will unknowingly be infected with the agent and the rest of the students will be susceptible to infection (sample fluid of water).

1. On cue from the instructor, each student will participate in the first fluid exchange by selecting another student at random to swap fluid with. Each participant will use a dropper to trade a few drops of fluid.
2. After the class has conducted the first exchange, your instructor will announce that each student needs to find a second random student to exchange fluid with. Complete the fluid exchange with this second classmate.
3. Repeat with a third exchange when indicated by your instructor.

4. Add one drop of phenolphthalein to your test tube. If the fluid turns pink, you are positive for the infectious agent.
5. Record the class results in Laboratory Report 54, Procedure B: Results and Analysis.

Round 2: 50% Susceptible Population

For this second round, you will receive a new test tube sample of fluid. One student will unknowingly be infected with the agent, 50% of the students will be susceptible to infection (sample fluid of water), and 50% will represent vaccinated individuals (sample fluid of pH buffer).

1. Repeat the same procedure as above with your new fluid sample, exchanging with three individuals as your instructor leads.
2. Use the indicator to test for infection, and record the class results in Laboratory Report 54, Procedure B: Results and Analysis.

Round 3: 10% Susceptible Population

For this final round, repeat the preceding experiment starting with one infected individual, 90% vaccinated students (sample fluid of pH buffer), and 10% susceptible students (sample fluid of water). Record the class results in Laboratory Report 54, Procedure B: Results and Analysis.

Laboratory Report

Complete Laboratory Report 54.

54   A Synthetic Epidemic

A. Procedure A: Results and Analysis

Record the class information from the board into the table below. The SHAKER is the person designated by the instructor to shake hands with another class member. The SHAKEE is the individual chosen by the shaker. A Kimwipe that glows under UV light is positive; a Kimwipe that does not glow when exposed to UV light is negative.

SHAKER Round 1	RESULT + or –	SHAKEE Round 1	RESULT + or –	SHAKER Round 2	RESULT + or –	SHAKEE Round 2	RESULT + or –
1.				1.			
2.				2.			
3.				3.			
4.				4.			
5.				5.			
6.				6.			
7.				7.			
8.				8.			
9.				9.			
10.				10.			
11.				11.			
12.				12.			
13.				13.			
14.				14.			
15.				15.			
16.				16.			
17.				17.			
18.				18.			
19.				19.			
20.				20.			

1. Was it possible to determine an index case? If so, who was it? If not, explain why. _____

2. What knowledge can be gained from determining the index case? _____

3. How many carriers resulted after Round 1? _____

4. How many carriers resulted after Round 2? _____

5. What percentage of the population was initially infected? What percentage of the population was infected at the end of the experiment?_____

6. What was the mode of transmission in this epidemic?_____

7. Describe an environment that would have a high rate of disease transmission._____

8. If this were a real infectious agent, such as a cold virus or influenza, list some factors that would affect transmission._____

9. How would it have been possible to stop this infection cycle?_____

10. Describe how the incubation period of a disease affects the spread of the disease._____

11. What factors are compared to confirm that an epidemic is occurring?_____

B. **Procedure B: Results and Analysis**

Record the class data for each round of this experiment in the table below. Calculate the percentage of susceptible individuals who contracted the disease during each round of this procedure. To calculate, use the following equation:

$$\frac{\text{Number of New Infected Individuals} \times 100}{\text{Initial Number of Susceptible Individuals}}$$

ROUND	INITIAL NUMBER OF INFECTED INDIVIDUALS	INITIAL NUMBER OF SUSCEPTIBLE INDIVIDUALS	NUMBER OF NEW INFECTED INDIVIDUALS (FINAL NUMBER OF INFECTED INDIVIDUALS MINUS 1 INITIAL)	PERCENTAGE OF SUSCEPTIBLE INDIVIDUALS THAT CONTRACTED INFECTION
1	1	100% of class = _____ individuals		
2	1	50% of class = _____ individuals		
3	1	10% of class = _____ individuals		

1. Explain how vaccines work to protect individuals from infection with pathogenic organisms.

_____

_____

2. Based on your results, describe how the number of vaccinated individuals affects the spread of a disease through a population.

_____

_____

3. Describe community (herd) immunity and its benefits to public health.

_____

_____

Immunology and Serology

When the human body becomes infected by a microbial agent such as a bacterium or virus, the invading microbe will encounter an army of different kinds of cells and host factors that will eventually destroy the microbe and prevent further damage or death to the host. These cells and factors comprise the **immune system,** which is composed of organs such as the liver, the spleen, and the lymph glands. The immune system recognizes the microbe as "foreign," and cells associated with the immune system engulf and remove the invading microbe, thus preventing its multiplication and further spread.

Many of the immune cells are derived from undifferentiated cells in the bone marrow called **stem cells.** Stem cells processed by the thymus gland, a butterfly-shaped gland at the base of the neck, undergo development into **T-lymphocytes (or T-cells).** Other stem cells processed by various lymphatic tissue, such as bone marrow in humans, will develop into

© Liquidlibrary/PictureQuest

B-lymphocytes (or B-cells). Both of these types of lymphocytes have specific roles in various aspects of the immune response.

An invading microorganism is initially encountered by white blood cells such as **macrophages** and **neutrophils.** The microbe is engulfed and destroyed by these cells. However, components of the microbe or virus are preserved and "presented" on the surfaces of phagocytic cells such as the macrophages. These components are usually composed of proteins and carbohydrates and are referred to as **antigens.** Antigens are substances such as proteins, carbohydrates, nucleic acids, and even some lipids that are foreign to the host and therefore elicit the immune response. A unique property of the immune system is that it can differentiate the substances that comprise the host from those associated with the invader. That is, the immune system of an animal or human normally does not respond to its own protein and carbohydrate makeup. A breakdown in this recognition of self can result in a an immune response to its own antigenic makeup and the development of an **autoimmune disease** such as some forms of arthritis.

Microbial antigens that are presented on the surfaces of macrophages stimulate a variety of other immune cells. For example, some T-lymphocytes are stimulated to seek out and destroy any cells infected with the microbe or virus. These killer T-lymphocytes recognize infected cells by the microbial antigens presented on their surfaces. Destruction of these infected cells ensures that the microbe or virus cannot reproduce to infect new cells. Other T-cells, called helper cells, are also stimulated and they send chemical messages in the form of leukotrienes and cytokines that interact with **B-lymphocytes.** The B-lymphocytes differentiate into plasma cells that synthesize host proteins called **antibodies,** which specifically react with antigens and immobilize them so that phagocytic cells can engulf the complexes of antigens and antibodies that are formed by this interaction. Antibodies are usually specific for a portion of an antigen molecule called an **epitope,** which is a surface feature of the antigen. Hence, there is a tremendous diversity in the numbers of antibodies directed against a single protein molecule. Antibody molecules are Y-shaped, and the prongs of the antibody molecule are what specifically attach to the epitope of the antigen. Because the antibody molecule has two prongs (binding sites), it can simultaneously bind to the same epitope on different antigens, thereby forming a lattice complex of antigens and antibody molecules. When these complexes are destroyed by phagocytic cells, it ensures that no free antigen is present to cause further damage to the host.

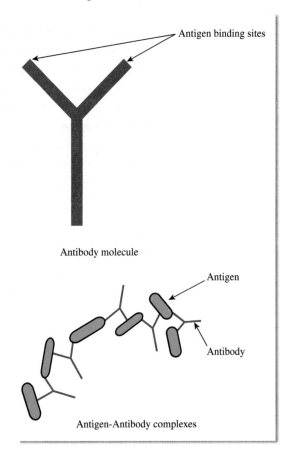

Antigen binding sites

Antibody molecule

Antigen

Antibody

Antigen-Antibody complexes

A host produces different classes of antibodies, also called immunoglobulins. Immunoglobulin D (IgD) is associated with the surface of B-cells and aids in the recognition of antigens "presented" by phagocytic cells. IgA occurs on the surface of cells lining the respiratory and intestinal tracts, and it is thought to play a protective role in the respiratory and digestive systems. IgE is found on the surface of mast cells and binds to antigens called allergens. When this occurs, the chemical histamine is released by mast cells, and it is histamine that is responsible for many of the symptoms associated with allergic reactions. IgM and IgG are found primarily in the serum, which is the clear liquid in which blood cells are suspended. These antibodies are important in immobilizing free antigens for destruction by phagocytic cells. IgM is made in the primary response, that is, the first time an antigen is encountered by a host, whereas IgG is made in the secondary or amnestic (memory) response to an antigen. The following is a list of the various classes of antibodies and their functions.

ANTIBODY CLASSES AND THEIR FUNCTIONS

ANTIBODY CLASS	FUNCTION
IgM	Primary immune response
IgG	Secondary immune response
IgE	Release of histamine from mast cells
IgA	Protection of respiratory and intestinal tracts
IgD	Recognition of antigens found on B-cells

When antibodies react with an antigen, the result will differ depending on the nature of the antigen. If an antigen is particulate such as cells or a virus, antibodies specific for the antigen will cause it to **agglutinate,** or form visible clumps. However, if the antigen is soluble, as is the case for many proteins and polysaccharides, the antibodies will cause the antigen to **precipitate.** These types of reactions are useful for diagnosing disease in a patient. During the course of a disease, the amount of antibodies in a patient's serum, or the **titer** of the antibodies, will increase. The presence of an increasing titer of an antibody can therefore be used as an indicator of the course of a disease. Blood is collected from a patient, and diluted serum is reacted with the antigen. The lowest dilution that still demonstrates agglutination or precipitation is the serum titer. An increasing titer over time indicates an active infection.

In the following exercises you will study antigen-antibody reactions and how they are used to monitor and detect disease.

Slide Agglutination Test:
Serological Typing

Learning Outcomes

After completing this exercise, you should be able to

1. Define the terms *antigen, antibody, antiserum, serotype,* and *agglutination.*

2. Use an agglutination reaction to differentiate between *Salmonella* and a coliform.

Organisms of different species differ not only in their morphology and physiology but also in the various components that make up their molecular structure. The proteins, polysaccharides, nucleic acids, and lipids (known as macromolecules) define the molecular structure of an organism. Some of these macromolecules of a bacterial cell can act as **antigens** because when introduced into an animal, they stimulate the immune system to form antibodies that are specific for these substances. The production of antibodies occurs because the antigens are foreign to the animal and different from the animal's own unique makeup. The antigenic structure of each species of bacteria is unique to that species, and much like a fingerprint of a human, its antigenic makeup can be used to identify the bacterium. The antigens that comprise lipopolysaccharide (O-antigens), capsules (K-antigens), and flagellar antigens (H-antigens) of the *Enterobacteriaceae* can be used to differentiate these bacteria into **serotypes.** Serotypes are more specific than species because they can be identical physiologically but differ significantly in their antigenic makeup. For example, *Escherichia coli* O157:H7, a serotype of *E coli,* is a serious pathogen that can be transmitted in contaminated food to cause disease and death. It can be differentiated by its unique antigenic makeup from strains of *E. coli* that normally inhabit the intestinal tract of humans.

When a human is challenged by a microbe, **B-cells** differentiate into plasma cells that are responsible for producing immunoglobulins or antibodies that are specific for the antigens of the invading microorganism. The immunoglobulins occur in abundance in circulating blood in an immunized individual. If the blood cells are separated from the serum, the clear liquid that remains contains the immunoglobulins or antibodies and is called **antiserum.**

The antigens of a microorganism can be determined by a procedure called **serological typing** (serotyping). It consists of adding a suspension of microorganism to antiserum that contains antibodies specific for antigens associated with the microorganism. If antigens are present, the antibodies in the antiserum will react with the antigens on the bacterial cell and cause the cells to **agglutinate,** or form visible clumps. Serotyping is particularly useful in the identification of *Salmonella* and *Shigella,* which cause infections in humans such as typhoid fever and bacillary dysentery. For example, *Salmonella* can be differentiated into more than 2500 different serotypes based on antigenic differences associated with the cell. Serotyping of *Salmonella* or *Shigella* is useful in tracing epidemics caused by a particular strain or serotype of the respective organism. In the identification of these two genera, biochemical tests are first used to identify the organism as either *Salmonella* or *Shigella* (Exercise 53), followed by serotyping to identify specific strains.

Figure 55.1   Slide agglutination technique.

In this exercise, you will be issued two unknown organisms, one of which is a *Salmonella*. By following the procedure shown in figure 55.1, you will determine which one of the unknowns is *Salmonella*. Note that you will use two test controls. A **negative test control** will be set up in depression A on the slide to see what the absence of agglutination looks like. The negative control is a mixture of antigen and saline (antibody is lacking). A **positive test control** will be performed in depression C with standardized antigen and antiserum to give you a typical reaction of agglutination.

Materials

- 2 numbered unknowns per student (slant cultures of a *salmonella* and a *coliform*)
- *salmonella* O-antigen, group B (Difco #2840-56)
- *salmonella* O antiserum, poly A-I (Difco #2264-47)
- depression slides or spot plates
- dropping bottle of phenolized saline solution (0.85% sodium chloride, 0.5% phenol)
- 2 serological tubes per student
- 1 ml pipettes

Caution

Keep in mind that *Salmonella typhimurium* is a pathogen and can cause gastroenteritis. Be careful!

1. Label three depressions on a spot plate or depression slide **A, B,** and **C,** as shown in figure 55.1.
2. Make a phenolized saline suspension of each unknown in separate serological tubes by suspending one or more loopfuls of organisms in 1 ml of phenolized saline. Mix the organisms sufficiently to ensure complete dispersion of clumps of bacteria. The mixture should be very turbid.
3. Transfer 1 loopful (0.05 ml) from the phenolized saline suspension of one tube to depressions A and B.
4. To depressions B and C, add 1 drop of *salmonella* O polyvalent antiserum. To depression A, add 1 drop of phenolized saline, and to depression C, add 1 drop of *salmonella* O-antigen, group B.
5. Mix the organisms in each depression with a clean wire loop. Do not go from one depression to the other without washing the loop first.
6. Compare the three mixtures. Agglutination should occur in depression C (positive control), but not in depression A (negative control). If agglutination occurs in depression B, the organism is *Salmonella*.
7. Repeat this process on another slide for the other organism.

Caution

Deposit all slides and serological tubes in a container of disinfectant provided by the instructor.

Laboratory Report

Record your results on the first portion of Laboratory Report 55–56.

Slide Agglutination Test for *S. aureus*

Learning Outcomes

After completing this exercise, you should be able to

1. Understand how antibodies can be adsorbed to latex spheres and then used to detect soluble antigens by agglutination reactions.

2. Perform serological tests using antibodies adsorbed to latex spheres to detect the coagulase or protein A of *Staphylococcus*.

When antibodies react with soluble antigens such as proteins or polysaccharides, the result is usually a precipitation reaction in which a visible precipitate is formed. The reaction, however, is best seen when a precise ratio of antibody to antigen occurs, called the **equivalence point.** At the equivalence point a visible precipitate forms between the antigen and antibody. If excess antibody or antigen is present, soluble complexes can form, and no visible precipitation will occur. Hence, unless the precise ratio of antigen to antibody is achieved in a precipitin reaction, a result cannot be readily seen. However, soluble antigens such as proteins can also be detected using a modified procedure involving an agglutination reaction. In these tests antibodies that have been produced against soluble antigens are adsorbed, or chemically linked, to polystyrene latex particles

(figure 56.1). The polystyrene particles act as carriers of the antibodies, and when they react with their soluble antigen an agglutination reaction occurs between the antibodies carried on the polystyrene particles and the antigen (figure 56.2).

Many manufacturers have devised such tests to detect specific pathogens or their products. In this exercise, you will use reagents manufactured by Difco Laboratories to determine if a suspected staphylococcus organism is producing the enzyme **coagulase** and/or **protein A.** Both protein A and coagulase are thought to be important virulence factors for *Staphylococcus aureus.* Coagulase may cause clot formation in capillaries at infection sites, thus preventing leukocytes, which are critical in the inflammatory response, from reaching areas of the infection. Protein A is a cell wall component of *S. aureus* that reacts with antibody molecules and coats the surface of the bacterial cell with host immunoglobulins. This immunoglobulin coat protects the bacterial cell by inhibiting phagocytosis by white blood cells and destruction of the bacterial cell. In the test for coagulase/protein A, the test reagent is a suspension of yellow latex particles to which antibodies against coagulase and protein A have been adsorbed. If the test strain of staphylococcus is producing coagulase and/or protein A, these

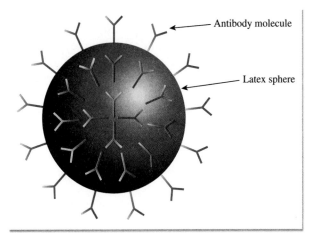

Figure 56.1 **Antibody molecules adsorbed to a latex sphere.**

Figure 56.2 **Reaction between antibodies adsorbed to latex spheres and soluble antigens resulting in agglutination.**

Figure 56.3 **Slide agglutination test (direct method) for the presence of coagulase and/or protein A.**

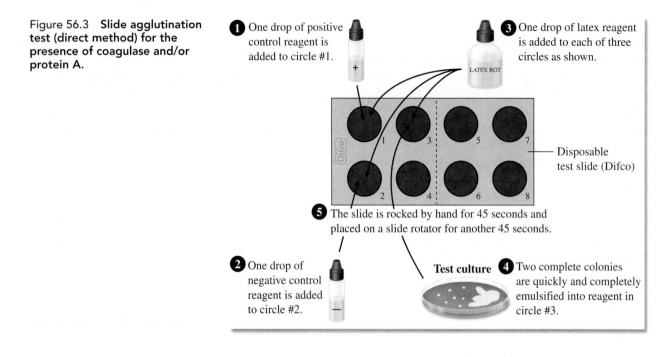

① One drop of positive control reagent is added to circle #1.

③ One drop of latex reagent is added to each of three circles as shown.

LATEX RGT

Disposable test slide (Difco)

⑤ The slide is rocked by hand for 45 seconds and placed on a slide rotator for another 45 seconds.

② One drop of negative control reagent is added to circle #2.

Test culture

④ Two complete colonies are quickly and completely emulsified into reagent in circle #3.

antigens will react with the latex particles and antibodies to cause agglutination of the latex spheres. No agglutination means that the test organism does not produce the virulence factors. This test is specific enough that it is used instead of culturing.

Reagents are included in the test kit to perform both positive and negative controls. The test is performed on disposable cards with eight black circles. As indicated in figure 56.3, three circles are used to test one unknown. The remaining circles are used for testing five additional unknowns at the same time. The black background facilitates the observation of the agglutination reaction by providing good contrast for the reaction.

There are two versions of this test: direct and indirect. The procedure for the direct method is illustrated in figure 56.3. The indirect method differs in that saline is used to suspend the organism being tested.

It should be pointed out that the reliability correlation between this test for coagulase and the tube test (figure 51.4) is very high. Studies reveal that a reliability correlation of over 97% exists. Proceed as follows to perform this test.

Materials

- plate culture of staphylococcus-like organism (trypticase soy agar plus blood)
- Difco Staph Latex Test kit #3850-32-7, which consists of:
    bottle of Bacto Staph Latex Reagent
    bottle of Bacto Staph Positive Control
    bottle of Bacto Staph Negative Control
    bottle of Bacto Normal Saline Reagent

    disposable test slides (black circle cards)
    mixing sticks (minimum of 3)
- slide rotator

Direct Method

If the direct method is to be used, as illustrated in figure 56.3, follow this procedure:

1. Place 1 drop of Bacto Staph Positive Control reagent onto circle #1.
2. Place 1 drop of Bacto Staph Negative Control reagent on circle #2.
3. Place 1 drop of Bacto Staph Latex Reagent onto circles #1, #2, and #3.
4. Using a sterile inoculating needle or loop, quickly and completely emulsify *two isolated colonies* from the culture to be tested into the drop of Staph Latex Reagent in circle #3.

    Also, emulsify the Staph Latex Reagent in the positive and negative controls in circles #1 and #2 using separate mixing sticks supplied in the kit.

    All mixing in these three circles should be done quickly to minimize drying of the latex on the slide and to avoid extended reaction times for the first cultures emulsified.
5. Rock the slide by hand for 45 seconds.
6. Place the slide on a slide rotator capable of providing 110 to 120 rpm and rotate it for another 45 seconds.
7. Read the results immediately, according to the descriptions provided in table 56.1. If agglutination occurs before 45 seconds, the results may be read at that time. *The slide should be read at normal reading distance under ambient light.*

Indirect Method

The only differences between the direct and indirect methods pertain to the amount of inoculum and the use of saline to emulsify the unknown being tested. Proceed as follows:

1. Place 1 drop of Bacto Staph Positive Control reagent onto test circle #1.
2. Place 1 drop of Bacto Staph Negative Control onto circle #2.
3. Place 1 drop of Bacto Normal Saline Reagent onto circle #3.
4. Using a sterile inoculating needle or loop, completely emulsify *four isolated colonies* from the culture to be tested into the circle containing the drop of saline (circle #3).
5. Add 1 drop of Bacto Staph Latex Reagent to each of the three circles.
6. Quickly mix the contents of each circle, using individual mixing sticks.
7. Rock the slide by hand for 45 seconds.
8. Place the slide on a slide rotator capable of providing 110 to 120 rpm and rotate it for another 45 seconds.
9. Read the results immediately according to the descriptions provided in table 56.1. If agglutination occurs before 45 seconds, the results may be read at that time (figure 56.4). *The slide should*

Table 56.1

POSITIVE REACTIONS	
4 +	Large to small clumps of aggregated yellow latex beads; clear background
3 +	Large to small clumps of aggregated yellow latex beads; slightly cloudy background
2 +	Medium to small but clearly visible clumps of aggregated yellow latex beads; moderately cloudy background
1 +	Fine clumps of aggregated yellow latex beads; cloudy background

NEGATIVE REACTIONS	
—	Smooth, cloudy suspension; particulate grainy appearance that cannot be identified as agglutination
—	Smooth, cloudy suspension; free of agglutination or particles

be read at normal reading distance under ambient light.

Laboratory Report

Record your results on the last portion of Laboratory Report 55–56.

(a)

(b)

Figure 56.4   **C-reactive protein agglutination test (a) positive; (b) negative.**
© McGraw-Hill Education. Lisa Burgess, photo

55 Slide Agglutination Test: Serological Typing

A. Results

1. Describe the appearance of the mixtures in each of the wells.

   _____

   _____

2. Which unknown number proved to be *Salmonella*?

   _____

   _____

B. Short-Answer Questions

1. What types of compounds in bacterial cells can serve as antigens?

   _____

   _____

2. What are immunoglobulins?

   _____

   _____

3. What is a serotype of an organism?

   _____

   _____

4. What is agglutination?

   _____

   _____

5. What types of controls are used for the slide agglutination?

   _____

   _____

6. What is the importance of doing controls?

   _____

   _____

A. Results

1. Describe the appearance of the mixtures in each of the wells.

_____

_____

2. Was there a positive result for *S. aureus*? If so, what was the degree of the positive reaction?

_____

_____

B. Short-Answer Questions

1. What two *S. aureus* antigens are being detected with the use of this test kit?

_____

_____

2. What definitive test for *S. aureus* is highly correlated with this agglutination test?

_____

_____

3. What advantages does the agglutination test have over the definitive *S. aureus* test?

_____

_____

4. What is the advantage of the latex test as compared to the ordinary precipitin-like reaction?

_____

_____

Slide Agglutination Test for *Streptococcus*

Learning Outcomes

After completing this exercise, you should be able to

1. Understand the importance of carbohydrate antigens in the Lancefield classification of streptococci.

2. Identify an unknown streptococcus using carbohydrate antigens extracted from cells and reacting them with latex beads coated with antibodies developed against the carbohydrate antigens.

The Lancefield classification of streptococci divides this group of bacteria into immunological groups based on carbohydrate antigens associated with the cells (see Exercise 52). These groups are designated A through V. The majority of pathogenic streptococci, however, occur in groups A, B, C, D, F, and G, and these bacteria usually also display beta-hemolysis when grown on blood agar. The presence of these antigenic carbohydrate groups in streptococcal cells has been used to develop an agglutination test for streptococci belonging to these Lancefield groups. In the test, the antigenic carbohydrates are first enzymatically extracted from cells and then reacted with latex beads which have been coated with antibodies developed against the specific carbohydrate antigens. If the carbohydrate antigens are present on the cells and extracted by the enzymatic treatment, they will cause the latex beads to agglutinate when they react with the adsorbed antibodies.

In practice, suspected streptococci are isolated from clinical samples onto blood agar plates and grown for 18 to 24 hours at 37°C. Several colonies (2–5) showing beta-hemolysis are selected and treated with an extraction enzyme that cleaves the antigens from the cells. The extracts are then mixed with the antibody-coated latex beads and judged for agglutination (figure 57.1). In the following exercise, you will use this procedure to test for pathogenic streptococci.

Materials

- cultures of *Streptococcus pyogenes, S. agalactiae, S. bovis, Enterococcus faecalis,* and *E. faecium* grown on blood agar plates at 37°C for 24 hours.

- unknown cultures of the above organisms (per group)
- Oxoid Diagnostic Streptococcal grouping kit, which consists of:
    latex grouping reagents A, B, C, D, F, and G
    polyvalent positive control
    extraction enzyme
    disposable reaction cards
- test tubes (serological type)
- pipettes (1 ml)
- Pasteur pipettes
- pipette aids
- wooden sticks

Procedure

Students will work in pairs. Each pair will test one known culture and an unknown culture assigned by the instructor.

ORGANISMS TO BE TESTED				
S. PYOGENES	*S. AGALACTIAE*	*S. BOVIS*	*E. FAECALIS*	*E. FAECIUM*
Group No. 1	2	3	4	5
6	7	8	9	10
11	12	13	14	15

1. Reconstitute the extraction enzyme with distilled water using the amount indicated on the label of the reagent bottle.

2. Label one test tube with the known organism that is to be tested and the second test tube with your unknown organism. Dispense 0.4 ml of extraction enzyme to each test tube.

3. Using a loop, select 2–5 colonies (2–3 mm) of your known organism and transfer them to the appropriate extraction enzyme tube.

4. Repeat this procedure for your unknown culture.

5. Incubate the enzyme extraction tubes at 37°C for 5 minutes. After 5 minutes, shake each tube vigorously and continue incubation of the tubes for an additional 5 minutes.

6. Carefully mix the antibody-latex suspension corresponding to the group to which your known belongs (i.e., A, B, C, D, F, or G). Dispense one

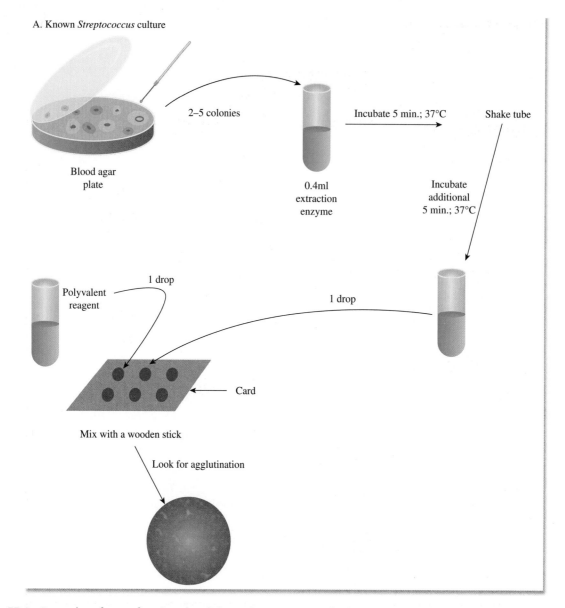

A. Known *Streptococcus* culture

2–5 colonies

Incubate 5 min.; 37°C

Shake tube

Blood agar
plate

0.4ml
extraction
enzyme

Incubate
additional
5 min.; 37°C

Polyvalent
reagent

1 drop

1 drop

Card

Mix with a wooden stick

Look for agglutination

Figure 57.1   Procedure for performing the slide agglutination test for known and unknown streptococcus colonies.

drop of the suspension to one of the rings on the reaction card.

7. Mix and dispense the polyvalent reagent to a separate ring on the card. This will be a positive control.

8. Using a Pasteur pipette, add one drop of extraction enzyme from your known culture to the ring on the card containing the latex suspension corresponding to the group of your known culture. Add a second drop of the extraction enzyme for your known culture to the ring with the polyvalent reagent.

9. Using a wooden stick, spread the extraction enzyme and latex suspension over the entire surface of the ring. Using a separate wooden stick, repeat the procedure for the polyvalent reagent and known enzyme extract.

10. Gently rock the cards back and forth. Agglutination will usually occur in 30 seconds.

11. To test your unknown culture, dispense one drop of each of the individual latex suspensions, A, B, C, D, F, and G to six separate circles on a reaction card.

B. Unknown *Streptococcus* culture

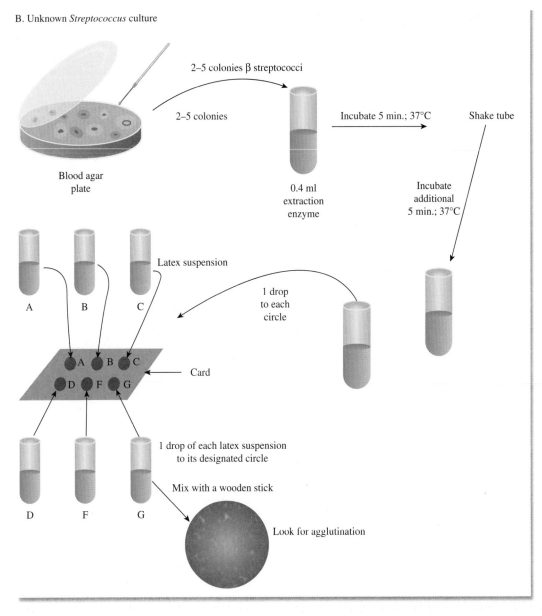

Figure 57.1 (continued)

12. Add one drop of the extraction enzyme from your unknown organism to each of the circles with the various latex suspensions.

13. Using a separate stick, mix the extraction enzyme and latex reagents.

14. Gently rock the card back and forth. Record which latex suspension caused the agglutination of the enzyme extract from your unknown organism.

415

57 Slide Agglutination Test for *Streptococcus*

A. Results

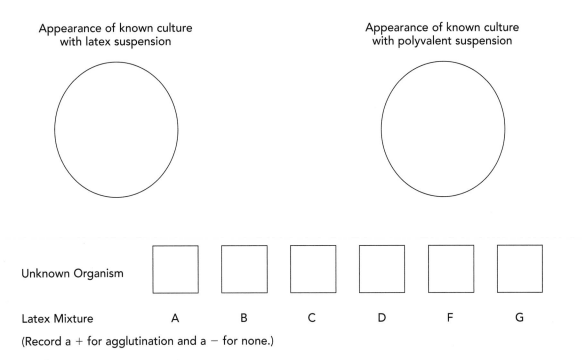

Appearance of known culture with latex suspension

Appearance of known culture with polyvalent suspension

Unknown Organism

Latex Mixture   A   B   C   D   F   G

(Record a + for agglutination and a − for none.)

B. Short-Answer Questions

1. The Lancefield classification of streptococci is based on what property of these cells?

   _____

   _____

2. Why can this test not be used to test for *Streptococcus pneumoniae*?

   _____

   _____

3. Streptococcal pathogens belonging to the groups tested also display what other important characteristic?

   _____

   _____

The Heterophile Antibody Test

Learning Outcomes

After completing this exercise, you should be able to

1. Define heterophile antigens and antibodies.

2. Understand how an agglutination reaction of sheep red blood cells by heterophile antibodies is used to diagnose infectious mononucleosis in a patient.

3. Perform an agglutination assay with serum and sheep red blood cells to detect heterophile antibodies.

In the United States, infectious mononucleosis (mono) is a benign disease that occurs principally in individuals 13 to 25 years old. It is caused by the **Epstein-Barr virus (EBV),** a member of the herpes group of viruses. The virus infects most humans early in childhood after maternal, protecting antibodies disappear, and it becomes established as a lysogenic virus in throat cells. Studies have shown that in the United States, 95% of adults 35 to 40 years of age have been infected with the virus, and infected individuals can shed the virus in their saliva even though they are asymptomatic for the disease. The virus can be spread by intimate contact such as kissing. Because the virus is so ubiquitous in the human population and because it is easily transmitted, it is virtually impossible to prevent.

Infectious mononucleosis is thought to occur in individuals who were not infected by the virus in childhood but who become infected during adolescence. In these individuals, there is a sudden onset of fever, sore throat, and a pronounced enlargement of cervical lymph nodes. In some cases there can also be liver and spleen involvement. Initially the virus infects epithelial cells in the mouth, but eventually it spreads to lymph nodes, where it infects B-lymphocytes that differentiate into antibody-producing cells. The virus can kill these cells, but it usually establishes a latent infection in B-cells whereby viral DNA is incorporated into host DNA or can exist as a plasmid-like genetic element. The infected B-cells begin to multiply and produce IgM, one of the immunoglobulins. In response to the infected B-cells, cytotoxic T-lymphocytes then target and kill the infected B-lymphocytes. There is a 50–90% increase in the number of lymphocytes, which accounts for the increase in white blood cells (leukocytosis) associated with the disease.

Diagnosis of infectious mononucleosis is based on symptoms and a relatively simple, serological screening test called the **heterophile antibody test.** This test takes advantage of the fact that the IgM antibodies produced by infected B-cells will agglutinate cells from other animal species. As the disease progresses, the level or **titer** of heterophile antibodies in a patient's serum will increase and will become coincidentally high in acute cases of infectious mononucleosis. The heterophile antibodies are not specifically directed toward the Epstein-Barr virus and do not react specifically with antigens associated with the Epstein-Barr virus. However, the heterophile antibody test is a useful screening method for infectious mononucleosis because the heterophile antibodies in a patient's serum coincidentally increase in the disease and can be assayed for by an agglutination reaction using sheep red blood cells. Importantly, the heterophile antibodies associated with infectious mononucleosis are normally absent in healthy individuals. Heterophile antibodies are also produced in other diseases such as serum sickness, and in response to the Forssman antigen, a naturally occurring antigen found in other animal species such as horses, cattle, and sheep and in some bacteria. Individuals who are positive for the heterophile antibody test for infectious mononucleosis can be confirmed for the disease by a more complicated procedure that employs antigens of the Epstein-Barr virus and which specifically assays for antibodies developed against these viral antigens. The heterophile antibody test for infectious mononucleosis can be performed using a test tube procedure or using a rapid qualitative test procedure in which a color result on a test card indicates a positive test. In the test tube method a specific serum titer is determined, which may be useful in evaluating the disease.

The test tube procedure for determining a serum titer is performed by adding a suspension of sheep red blood cells to dilutions of inactivated serum from a patient. The serum must first be inactivated by heating at 56°C for 30 minutes, which destroys a protein complex in serum called **complement.** Complement binds to red blood cells and causes them to undergo lysis, thus interfering with the test. The inactivated serum is diluted and mixed with sheep red blood cells, and the tubes are incubated overnight in a refrigerator. The lowest dilution of serum that still causes agglutination of the sheep red blood cells is reported as the titer of the serum. A titer of 1:28 is considered positive for the disease. Thus, a very high dilution of the serum still causing agglutination would indicate a high level of heterophile antibodies in the patient's serum. As the disease progresses, the level of heterophile antibodies will increase and then begin to decrease as the patient recovers. Figure 58.1 illustrates the overall procedure.

First Period

Materials

- test-tube rack (Wasserman type) with 10 clean serological tubes
- bottle of saline solution (0.85% NaCl), clear or filtered
- 1 ml pipettes
- 5 ml pipettes
- 2% suspension of sheep red blood cells
- patient's serum (known to be positive)

1. Place the test serum in a 56°C water bath for 30 minutes to inactivate the complement.
2. Set up a row of 10 serological tubes in the front row of a test-tube rack and number them from 1 to 10 (left to right) with a marking pencil.
3. Into tube 1, pipette 0.8 ml of physiological saline.
4. Dispense 0.5 ml of physiological saline to tubes 2 through 10. Use a 5 ml pipette.
5. With a 1 ml pipette add 0.2 ml of the inactivated serum to tube 1. Mix the contents of this tube.
6. Transfer 0.5 ml from tube 1 to tube 2, mix, and transfer 0.5 ml from tube 2 to tube 3, and so on, through the ninth tube. *Discard 0.5 ml from the ninth tube after mixing.* Tube 10 is the **control.**
7. Add 0.2 ml of 2% sheep red blood cells to all tubes (1 through 10) and shake the tubes. Final dilutions of the serum are shown in figure 58.1.
8. Allow the rack of tubes to stand at room temperature for 1 hour, then transfer the tubes to the refrigerator to remain overnight.

Second Period

Set up the tubes in a tube rack in order of dilution and compare each tube with the control by holding the tubes overhead and looking up at the bottoms of the tubes. Nonagglutinated cells will tumble to the bottom of the tube and form a small button (as in control

Figure 58.1   **Procedure for setting up heterophile antibody test.**

Eyepoint

Black surface

Figure 58.2 **Agglutination is more readily seen when the tube is examined against a black surface.**

tube). Agglutinated cells will form a more-amorphous "blanket" (figure 58.2).

The titer should be recorded as the reciprocal of the last tube in the series that shows positive agglutination.

Laboratory Report

Complete Laboratory Report 58.

58 The Heterophile Antibody Test

A. Results

1. What was the titer of the serum tested?

   _____

   _____

2. Is the result considered significant? Explain.

   _____

   _____

   _____

B. Short-Answer Questions

1. For what disease is this diagnostic test used?

   _____

   _____

2. What is the name of the virus that causes this disease?

   _____

   _____

3. What percentage of all adults possess antibodies for the virus?

   _____

   _____

4. What are heterophile antibodies? Why are they useful for this test?

   _____

   _____

5. What additional blood examination can be performed to confirm an infection?

   _____

   _____

   _____

Blood Grouping

Learning Outcomes

After completing this exercise, you should be able to

1. Understand the basis of blood grouping in the ABO and Rh systems.

2. Determine the ABO and Rh types of human blood.

Exercises 55 through 58 illustrate four uses of agglutination tests as related to (1) the identification of serological types (*Salmonella*), (2) species identification (*S. aureus*), (3) species identification (*Streptococcus*), and (4) disease identification (infectious mononucleosis). The typing of blood is another example of a medical procedure that relies on this useful phenomenon.

The procedure for blood typing was developed by Karl Landsteiner around 1900. He is credited with having discovered that human blood types can be separated into four groups on the basis of two antigens that are present on the surface of red blood cells. These antigens are designated as A and B. The four groups (types) are A, B, AB, and O. The last group, type O, which is characterized by the absence of A or B antigens, is the most common type in the United States (45% of the population). Type A is next in frequency, found in 39% of the population. The incidences of types B and AB are 12% and 4%, respectively. Type O has been called the "universal" donor because cells lack the AB antigens. Type AB has been called the universal recipient because serum lacks the AB antibodies. However, blood is always specifically typed to avoid errors.

Blood typing is performed with antisera containing high titers of anti-A and anti-B antibodies. The test may be performed by either slide or tube methods. In both instances, a drop of each kind of antiserum is added to separate samples of saline suspension of red blood cells. Figure 59.1 illustrates the slide technique. If agglutination occurs only in the suspension to which the anti-A serum was added, the blood is type A. If agglutination occurs only in the anti-B mixture, the blood is type B. Agglutination in both samples indicates that the blood is type AB. The absence of agglutination indicates that the blood is type O.

Between 1900 and 1940, a great deal of research was done to uncover the presence of other antigens in human red blood cells. Finally, in 1940, Landsteiner and Wiener reported that rabbit sera containing antibodies against the red blood cells of the rhesus monkey would agglutinate the red blood cells of 5% of white humans. This antigen in humans, which was first designated as the **Rh factor** (in due respect to the rhesus monkey), was later found to exist as six antigens: C, c, D, d, E, and e. Of these six antigens, the D factor is responsible for the Rh-positive condition and is found in 85% of Whites, 94% of Blacks, and 99% of Asians. Rh typing is also important to prevent incompatibilities and serious illness.

Typing blood for the Rh factor can also be performed by both tube and slide methods, but there are certain differences in the two techniques. First of all, the antibodies in the typing sera are of the incomplete

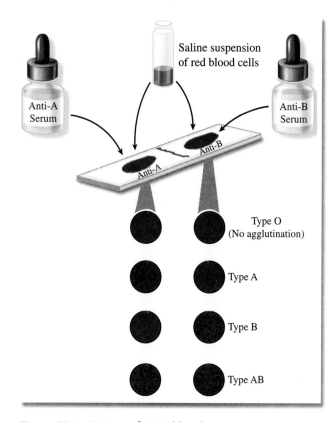

Saline suspension of red blood cells

Anti-A Serum

Anti-B Serum

Anti-A

Anti-B

Type O (No agglutination)

Type A

Type B

Type AB

Figure 59.1 **Typing of ABO blood groups.**

albumin variety, which *will not agglutinate human red cells when they are diluted with saline.* Therefore, it is necessary to use whole blood or dilute the cells with plasma. Another difference is that the test *must be performed at higher temperatures:* 37°C for tube test; 45°C for the slide test.

In this exercise, two separate slide methods are presented for typing blood. If only the Landsteiner ABO groups are to be determined, the first method may be preferable. If Rh typing is to be included, the second method, which utilizes a slide warmer, will be followed. The availability of materials will determine which method is to be used.

Precautions

When working with blood, observe the following precautions:

1. Always disinfect the finger with alcohol prior to piercing it.
2. Use sterile disposable lancets only one time.
3. Dispose of used lancets by placing them into a beaker of disinfectant.
4. Avoid skin contact with blood of other students. Wear disposable latex gloves.
5. Disinfect finger with alcohol after blood has been taken.

ABO Blood Typing

Materials

- small vial (10 mm dia × 50 mm long)
- disposable lancets (B–D Microlance, Serasharp, etc.)
- 70% alcohol and cotton
- china marking pencil
- microscope slides
- typing sera (anti-A and anti-B)
- applicators or toothpicks
- saline solution (0.85% NaCl)
- 1 ml pipettes
- disposable latex gloves

Note: sterilized blood can be purchased from biological supply companies and substituted in this experiment.

1. Mark a slide down the middle with a marking pencil, dividing the slide into two halves, as shown in figure 59.1. Write "anti-A" on the left side and "anti-B" on the right side.
2. Pipette 1 ml of saline solution into a small vial or test tube.
3. Scrub the middle finger with a piece of cotton saturated with 70% alcohol and pierce it with a sterile disposable lancet. Allow 2 or 3 drops of blood to mix with the saline by holding the finger

over the end of the vial and washing it with the saline by inverting the tube several times.
4. Place a drop of this red cell suspension on each side of the slide.
5. Add a drop of anti-A serum to the left side of the slide and a drop of anti-B serum to the right side.

   Do not contaminate the tips of the serum pipettes with the material on the slide.
6. After mixing each side of the slide with separate applicators or toothpicks, look for agglutination. The slide should be held about 6″ above an illuminated white background and rocked gently for 2 or 3 minutes. Record your results in Laboratory Report 59 as of 3 minutes.

Combined ABO and Rh Typing

As stated, Rh typing must be performed with heat on blood that has not been diluted with saline. A warming box such as the one in figure 59.2 is essential in this procedure. In performing this test, two factors are of considerable importance: first, only a small amount of blood must be used (a drop of about 3 mm diameter on the slide) and, second, proper agitation must be executed. The agglutination that occurs in this antibody-antigen reaction results in finer clumps; therefore, closer examination is essential. If the agitation is not properly performed, agglutination may not be as apparent as it should be.

In this combined method, we will use whole blood for the ABO typing as well as for the Rh typing. Although this method works satisfactorily as a classroom demonstration for the ABO groups, it is *not as reliable* as the previous method in which saline and room temperature are used. *This method is not recommended for clinical situations.*

Materials

- slide warming box with a special marked slide
- anti-A, anti-B, and anti-D typing sera
- applicators or toothpicks
- 70% alcohol and cotton
- disposable sterile lancets

1. Scrub the middle finger with a piece of cotton saturated with 70% alcohol and pierce it with a sterile disposable lancet. Place a small drop of blood in each of three squares on the marked slides on the warming box.

   To get the proper proportion of serum to blood, do not use a drop larger than 3 mm diameter on the slide.
2. Add a drop of anti-D serum to the blood in the anti-D square, mix with a toothpick, and note

Figure 59.2   **Blood typing with warming box.**

One drop of each antiserum is sufficient.

Anti-D serum

Anti-B serum

Anti-A serum

Whole blood or plasma-diluted blood must be used for Rh typing. Saline-diluted blood is preferred for the ABO typing.

Agitation is achieved by slowly rocking box back and forth for 2 minutes.

the time. **Only 2 minutes should be allowed for agglutination.**

3. Add a drop of anti-B serum to the anti-B square and a drop of anti-A serum to the anti-A square. Mix the sera and blood in both squares with *separate* fresh toothpicks.

4. Agitate the mixtures on the slide by slowly rocking the box back and forth on its pivot. At the end of 2 minutes, examine the anti-D square carefully for agglutination. If no agglutination is apparent, consider the blood to be Rh-negative. By this time the ABO type can also be determined.

5. Record your results in Laboratory Report 59.

59 Blood Grouping

A. Results

1. Describe each of the mixtures on the blood-typing slide.

   _____

   _____

2. Based upon these results, what is your blood type?

   _____

   _____

3. What antigens are present on your blood cells?

   _____

   _____

4. If you needed a transfusion, what blood types could you accept?

   _____

   _____

5. To which recipient blood types can you donate blood?

   _____

   _____

B. Short-Answer Questions

1. What happens when an individual receives an incompatible blood type in the course of a transfusion?

   _____

   _____

2. Which blood type is known as the universal recipient? Explain.

   _____

   _____

3. Which blood type is known as the universal donor? Explain.

   _____

   _____

4. Which Rh antigen defines an Rh-positive individual?

   _____

   _____

Table I   International Atomic Weights

Element	Symbol	Atomic Number	Atomic Weight
Aluminum	Al	13	26.97
Antimony	Sb	51	121.76
Arsenic	As	33	74.91
Barium	Ba	56	137.36
Beryllium	Be	4	9.013
Bismuth	Bi	83	209.00
Boron	B	5	10.82
Bromine	Br	35	79.916
Cadmium	Cd	48	112.41
Calcium	Ca	20	40.08
Carbon	C	6	12.010
Chlorine	Cl	17	35.457
Chromium	Cr	24	52.01
Cobalt	Co	27	58.94
Copper	Cu	29	63.54
Fluorine	F	9	19.00
Gold	Au	79	197.2
Hydrogen	H	1	1.0080
Iodine	I	53	126.92
Iron	Fe	26	55.85
Lead	Pb	82	207.21
Magnesium	Mg	12	24.32
Manganese	Mn	25	54.93
Mercury	Hg	80	200.61
Nickel	Ni	28	58.69
Nitrogen	N	7	14.008
Oxygen	O	8	16.0000
Palladium	Pd	46	106.7
Phosphorus	P	15	30.98
Platinum	Pt	78	195.23
Potassium	K	19	39.096
Radium	Ra	88	226.05
Selenium	Se	34	78.96
Silicon	Si	14	28.06
Silver	Ag	47	107.880
Sodium	Na	11	22.997
Strontium	Sr	38	87.63
Sulfur	S	16	32.066
Tin	Sn	50	118.70
Titanium	Ti	22	47.90
Tungsten	W	74	183.92
Uranium	U	92	238.07
Vanadium	V	23	50.95
Zinc	Zn	30	65.38
Zirconium	Zr	40	91.22

Table II   Autoclave Steam Pressures and Corresponding Temperatures

Steam Pressure lb /sq in.	Temperature °C	Temperature °F	Steam Pressure lb/sq in.	Temperature °C	Temperature °F	Steam Pressure lb/sq in.	Temperature °C	Temperature °F
0	100.0	212.0						
1	101.9	215.4	11	116.4	241.5	21	126.9	260.4
2	103.6	218.5	12	117.6	243.7	22	127.8	262.0
3	105.3	221.5	13	118.8	245.8	23	128.7	263.7
4	106.9	224.4	14	119.9	247.8	24	129.6	265.3
5	108.4	227.1	15	121.0	249.8	25	130.4	266.7
6	109.8	229.6	16	122.0	251.6	26	131.3	268.3
7	111.3	232.3	17	123.0	253.4	27	132.1	269.8
8	112.6	234.7	18	124.1	255.4	28	132.9	271.2
9	113.9	237.0	19	125.0	257.0	29	133.7	272.7
10	115.2	239.4	20	126.0	258.8	30	134.5	274.1

Figures are for steam pressure only, and the presence of any air in the autoclave invalidates temperature readings from the above table.

Table III   Autoclave Temperatures as Related to the Presence of Air

Gauge Pressure, lb	Pure steam, complete air discharge °C	Pure steam, complete air discharge °F	Two-thirds air discharge, 20-in. vacuum °C	Two-thirds air discharge, 20-in. vacuum °F	One-half air discharge, 15-in. vacuum °C	One-half air discharge, 15-in. vacuum °F	One-third air discharge, 10-in. vacuum °C	One-third air discharge, 10-in. vacuum °F	No air discharge °C	No air discharge °F
5	109	228	100	212	94	202	90	193	72	162
10	115	240	109	228	105	220	100	212	90	193
15	121	250	115	240	112	234	109	228	100	212
20	126	259	121	250	118	245	115	240	109	228
25	130	267	126	259	124	254	121	250	115	240
30	135	275	130	267	128	263	126	259	121	250

Table IV Antibiotic Susceptibility Test Discs

ANTIBIOTIC AGENT	CONCENTRATION	INDIVIDUAL/10 PACK
Amikacin	30 μg	231596/231597
Amoxicillin/Clavulanic Acid	30 μg	231628/23629
Ampicillin	10 μg	230705/231264
Ampicillin/Subactam	10/10 μg	231659/231660
Azlocillin	75 μg	231624/231625
Bacitracin	2 units	230717/231267
Carbenicillin	100 μg	231235/231555
Cefaclor	30 μg	231652/231653
Cefazolin	30 μg	231592/231593
Cefixime	5 μg	231663/NA
Cefoperazone	75 μg	231612/231613
Cefotaxime	30 μg	231606/231607
Cefotetan	30 μg	231655/231656
Cefoxitin	30 μg	231590/231591
Ceftazidime	30 μg	231632/231633
Ceftriaxone	30 μg	231634/231635
Cefuroxime	30 μg	231620/231621
Cephalothin	30 μg	230725/231271
Chloramphenicol	30 μg	230733/231274
Clindamycin	2 μg	231213/231275
Doxycycline	30 μg	230777/231286
Erythromycin	15 μg	230793/231290
Gentamicin	10 μg	231227/231299
Imipenem	10 μg	231644/231645
Kanamycin	30 μg	230825/230829
Mezlocillin	75 μg	231614/231615
Minocycline	30 μg	231250/231251

Courtesy and © Becton, Dickinson and Company

(continued)

Table IV   Antibiotic Susceptibility Test Discs   *(continued)*

ANTIBIOTIC AGENT	CONCENTRATION	INDIVIDUAL/10 PACK
Moxalactam	30 μg	231610/231611
Nafcillin	1 μg	230866/231309
Nalidixic Acid	30 μg	230870/230874
Netilimicin	30 μg	231602/231603
Nitrofurantoin	100 μg	230801/231292
Penicillin	2 units	230914/231320
Piperacillin	100 μg	231608/231609
Rifampin	5 μg	231541/231544
Streptomycin	10 μg	230942/231328
Sulfisoxazole	0.25 mg	230813/231296
Tetracycline	5 μg	230994/231343
Ticarcillin	75 μg	231618/231619
Tobramycin	10 μg	231568/213569
Trimethoprim	5 μg	231600/231601
Vancomycin	30 μg	231034/231353

Table V   Indicators of Hydrogen Ion Concentration

Many of the following indicators are used in the media of certain exercises in this manual. This table indicates the pH range of each indicator and the color changes that occur. To determine the exact pH within a particular range, one should use a set of standard colorimetric tubes that are available from the prep room. Consult with your lab instructor.

Indicator	Full Acid Color	Full Alkaline Color	pH Range
Cresol Red	red	yellow	0.2–1.8
Metacresol Purple (acid range)	red	yellow	1.2–2.8
Thymol Blue	red	yellow	1.2–2.8
Bromphenol Blue	yellow	blue	3.0–4.6
Bromcresol Green	yellow	blue	3.8–5.4
Chlorcresol Green	yellow	blue	4.0–5.6
Methyl Red	red	yellow	4.4–6.4
Chlorphenol Red	yellow	red	4.8–6.4
Bromcresol Purple	yellow	purple	5.2–6.8
Bromtheymol Blue	yellow	blue	6.0–7.6
Neutral Red	red	amber	6.8–8.0
Phenol Red	yellow	red	6.8–8.4
Cresol Red	yellow	red	7.2–8.8
Metacresol Purple (alkaline range)	yellow	purple	7.4–9.0
Thymol Blue (alkaline range)	yellow	blue	8.0–9.6
Cresolphthalein	colorless	red	8.2–9.8
Phenolphthalein	colorless	red	8.3–10.0

Appendix B—Indicators, Stains, Reagents

Indicators

All the indicators used in this manual can be made by (1) dissolving a measured amount of the indicator in 95% ethanol, (2) adding a measured amount of water, and (3) filtering with filter paper. The following chart provides the correct amounts of indicator, alcohol, and water for various indicator solutions.

Indicator Solution	Indicator (gm)	95% Ethanol (ml)	Distilled H$_2$O (ml)
Bromcresol Green	0.4	500	500
Bromcresol Purple	0.4	500	500
Bromthymol Blue	0.4	500	500
Cresol Red	0.4	500	500
Methyl Red	0.4	500	500
Phenolphthalien	1.0	50	50
Phenol Red	0.2	500	500
Thymol Blue	0.4	500	500

Stains and Reagents

Acid-alcohol (for Kinyoun stain)
Ethanol (95%) . . . . .97 ml
Concentrated HCl .3 ml

Alcohol, 70% (from 95%)
Alcohol, 95% . . . . 368.0 ml
Distilled water . . . . 132.0 ml

Barritt's Reagent (Voges-Proskauer test)
Solution A: 6 g alpha-naphthol in 100 ml 95% ethyl alcohol.
Solution B: 16 g potassium hydroxide in 100 ml water.
Note that no creatine is used in these reagents as is used in O'Meara's reagent for the VP test.

Carbolfuchsin Stain (Kinyoun stain)
Basic fuchsin . . . . 4 gm
Phenol . . . . .8 ml
Alcohol (95%) . . . . .20 ml
Distilled/deionized water . . . . .100 ml
Dissolve the basic fuchsin in the alcohol, and add the water while slowly shaking. Melt the phenol in a 56°C water bath and carefully add 8 ml to the stain.

Note: To facilitate staining of acid-fast bacteria, 1 drop of Tergitol No. 7 (Sigma Chemical Co.) can be added to 30–40 ml of the Kinyoun carbolfuchsin stain.

Crystal Violet Stain (Hucker modification)
Solution A: Dissolve 2.0 g of crystal violet (85% dye content) in 20 ml of 95% ethyl alcohol.
Solution B: Dissolve 0.8 g ammonium oxalate in 80.0 ml distilled water.
Mix solutions A and B.

Diphenylamine Reagent (nitrate test)
Dissolve 0.7 g diphenylamine in a mixture of 60 ml of concentrated sulfuric acid and 28.8 ml of distilled water.

Cool and add slowly 11.3 ml of concentrated hydrochloric acid. After the solution has stood for 12 hours, some of the base separates, showing that the reagent is saturated.

Gram's Iodine (Lugol's)

Dissolve 2.0 g of potassium iodide in 300 ml of distilled water and then add 1.0 g iodine crystals.

Iodine, 5% Aqueous Solution (Ex. 32)

Dissolve 4 g of potassium iodide in 300 ml of distilled water and then add 2.0 g iodine crystals.

Kovacs' Reagent (indole test)

n-amyl alcohol	75.0 ml
Hydrochloric acid (conc.)	25.0 ml
ρ-dimethylamine-benzaldehyde	5.0 g

Lactophenol Cotton Blue Stain

Phenol crystals	20 g
Lactic acid	20 ml
Glycerol	40 ml
Cotton blue	0.05 g

Dissolve the phenol crystals in the other ingredients by heating the mixture gently under a hot water tap.

Malachite Green Solution (spore stain)

Dissolve 5.0 g malachite green oxalate in 100 ml distilled water.

McFarland Nephelometer Barium Sulfate Standards (Ex. 43)

Prepare 1% aqueous barium chloride and 1% aqueous sulfuric acid solutions.

Add the amounts indicated in table 1 to clean, dry ampoules. Ampoules should have the same diameter as the test tube to be used in subsequent density determinations.

Seal the ampoules and label them.

Methylene Blue (Loeffler's)

Solution A: Dissolve 0.3 g of methylene blue (90% dye content) in 30.0 ml ethyl alcohol (95%).

Solution B: Dissolve 0.01 g potassium hydroxide in 100.0 ml distilled water. Mix solutions A and B.

Table 1   Amounts for Standards

Tube	Barium Chloride 1% (ml)	Sulfuric Acid 1% (ml)	Corresponding Approx. Density of Bacteria (million/ml)
1	0.1	9.9	300
2	0.2	9.8	600
3	0.3	9.7	900
4	0.4	9.6	1200
5	0.5	9.5	1500
6	0.6	9.4	1800
7	0.7	9.3	2100
8	0.8	9.2	2400
9	0.9	9.1	2700
10	1.0	9.0	3000

Naphthol, alpha

5% alpha-naphthol in 95% ethyl alcohol

Caution: Avoid all contact with human tissues. Alpha-naphthol is considered to be carcinogenic.

Nessler's Reagent (ammonia test)

Dissolve about 50 g of potassium iodide in 35 ml of cold ammonia-free distilled water. Add a saturated solution of mercuric chloride until a slight precipitate persists. Add 400 ml of a 50% solution of potassium hydroxide. Dilute to 1 liter, allow to settle, and decant the supernatant for use.

Nigrosin Solution (Dorner's)

Nigrosin, water soluble	10 g
Distilled water	100 ml

Boil for 30 minutes. Add as a preservative 0.5 ml formaldehyde (40%). Filter twice through double filter paper and store under aseptic conditions.

Nitrate Test Reagent
(see Diphenylamine)

Nitrite Test Reagents

Solution A: Dissolve 8 g sulfanilic acid in 1000 ml 5N acetic acid (1 part glacial acetic acid to 2.5 parts water).

Solution B: Dissolve 5 g dimethyl-alpha-naphthylamine in 1000 ml 5N acetic acid. Do not mix solutions.

Caution: Although at this time it is not known for sure, there is a possibility that dimethyl-alpha-naphthylamine in solution B may be carcinogenic. For reasons of safety, avoid all contact with tissues.

Oxidase Test Reagent

Mix 1.0 g of dimethyl-p-phenylenediamine hydrochloride in 100 ml of distilled water.

Preferably, the reagent should be made up fresh, daily. It should not be stored longer than one week in the refrigerator. Tetramethyl-p-phenylenediamine dihydrochloride (1%) is even more sensitive, but is considerably more expensive and more difficult to obtain.

Phenolized Saline

Dissolve 8.5 g sodium chloride and 5.0 g phenol in 1 liter distilled water.

Physiological Saline

Dissolve 8.5 g sodium chloride in 1 liter distilled water.

Potassium permanganate
(for fluorochrome staining)

KMnO$_4$ . . . . 2.5 g
Distilled water . . . . 500.0 ml

Safranin (for Gram staining)

Safranin O (2.5% sol'n in 95% ethyl alcohol) . . . . 10.0 ml
Distilled water . . . . 100.0 ml

Trommsdorf's Reagent (nitrite test)

Add slowly, with constant stirring, 100 ml of a 20% aqueous zinc chloride solution to a mixture of 4.0 g of starch in water. Continue heating until the starch is dissolved as much as possible, and the solution is nearly clear. Dilute with water and add 2 g of potassium iodide. Dilute to 1 liter, filter, and store in amber bottle.

Vaspar

Melt together 1 pound of Vaseline and 1 pound of paraffin. Store in small bottles for student use.

Voges-Proskauer Test Reagent
(see Barritt's)

White Blood Cell (WBC) Diluting Fluid

Hydrochloric acid . . . . 5 ml
Distilled water . . . . 495 ml
Add 2 small crystals of thymol as a preservative.

Conventional Media   The following media are used in the experiments of this manual. All of these media are available in dehydrated form from either Difco Laboratories, Detroit, Michigan; or Baltimore Biological Laboratory (BBL), a division of Becton, Dickinson & Co., Cockeysville, Maryland. Compositions, methods of preparation, and usage will be found in their manuals, which are supplied upon request at no cost. The source of each medium is designated as (B) for BBL and (D) for Difco.

Bile esculin (D)
Brewer's anaerobic agar (D)
Desoxycholate citrate agar (B,D)
Desoxycholate lactose agar (B,D)
DNase test agar (B,D)
Endo agar (B,D)
Eugon agar (B,D)
Fluid thioglycollate medium (B,D)
Heart infusion agar (D)
Hektoen enteric agar (B,D)
Kligler iron agar (B,D)
Lead acetate agar (D)
Levine EMB agar (B,D)
Lipase reagent (D)
Litmus milk (B,D)
Lowenstein-Jensen medium (B,D)
MacConkey agar (B,D)
Mannitol salt agar (B,D)
MR-VP medium (D)
m-Staphylococcus broth (D)
Mueller-Hinton medium (B,D)
Nitrate broth (D)
Nutrient agar (B,D)

Nutrient broth (B,D)
Nutrient gelatin (B,D)
Phenol red sucrose broth (B,D)
Phenylalanine agar (D)
Phenylethyl alcohol medium (B)
Russell double sugar agar (B,D)
Sabouraud's glucose (dextrose) agar (D)
Semisolid medium (B)
Simmons citrate agar (B,D)
Snyder test agar (D)
Sodium hippurate (D)
Spirit blue agar (D)
SS agar (B,D)
Staphylococcus medium 110 (D)
Starch agar (D)
Trypticase soy agar (B)
Trypticase soy broth (B)
Tryptone glucose extract agar (B,D)
Urea (urease test) broth (B,D)
Veal infusion agar (B,D)
Xylose lysine desoxycholate agar (B,D)

Special Media   The following media are not included in the manuals that are supplied by Difco and BBL; therefore, methods of preparation are presented here.

Bile Esculin Slants (Ex. 52)

Heart infusion agar . . . . 40.0 g
Esculin . . . . 1.0 g
Ferric chloride . . . . 0.5 g
Distilled water . . . . 1000.0 ml

Dispense into sterile 15 × 125 mm screw-capped tubes, sterilize in autoclave at 121°C for 15 minutes, and slant during cooling.

Blood Agar

Trypticase soy agar powder . . . . 40 g
Distilled water . . . . 1000 ml
  Final pH of 7.3
Defibrinated sheep or rabbit blood . . . . . . 50 ml

Liquefy and sterilize 1000 ml of trypticase soy agar in a large Erlenmeyer flask. While the TSA is being sterilized, warm up 50 ml of defibrinated blood to

50°C. After cooling the TSA to 50°C, aseptically transfer the blood to the flask and mix by gently rotating the flask (cold blood may cause lumpiness).

Pour 10–12 ml of the mixture into sterile petri plates. If bubbles form on the surface of the medium, flame the surface gently with a Bunsen burner before the medium solidifies. It is best to have an assistant to lift off the petri plate lids while pouring the medium into the plates. A full flask of blood agar is somewhat cumbersome to handle with one hand.

Bromthymol Blue Carbohydrate Broths

Make up stock indicator solution:

Bromthymol blue. .8 g
95% ethyl alcohol . . . . .250 ml
Distilled water . . . . .250 ml

Indicator is dissolved first in alcohol and then water is added.

Make up broth:

Sugar base (lactose, sucrose, glucose, etc.). .5 g
Tryptone . . . . .10 g
Yeast extract . . . . .5 g
Indicator solution. . . . . .2 ml
Distilled water . . . . .1000 ml
Final pH 7.0

Emmons' Culture Medium for Fungi

C. W. Emmons developed the following recipe as an improvement over Sabouraud's glucose agar for the cultivation of fungi. Its principal advantage is that a neutral pH does not inhibit certain molds that have difficulty growing on Sabouraud's agar (pH 5.6). Instead of relying on a low pH to inhibit bacteria, it contains chloramphenicol, which does not adversely affect the fungi.

Glucose . . . . .20 g
Neopeptone . . . . .10 g
Agar. . . . . .20 g
Chloramphenicol. . . . . 40 mg
Distilled water . . . . .1000 ml

After the glucose, peptone, and agar are dissolved, heat to boiling, add the chloramphenicol, which has been suspended in 10 ml of 95% alcohol, and remove quickly from the heat. Autoclave for only 10 minutes.

Freshwater Enrichment Medium (*Desulfovibrio*)

Sodium lactate. . . . . .5 g
CaSO$_4$ . . . . .1 g
MgSO$_4$ · H$_2$O. . . . . .2 g

NH$_4$Cl . . . . .1 g
K$_2$HPO$_4$ . . . . 0.5 g
Water . . . . .1000 ml

Autoclave at standard conditions. Just prior to inoculation, boil and cool the medium.

Glucose Peptone Acid Agar

Glucose . . . . .10 g
Peptone . . . . .5 g
Monopotassium phosphate . . . . .1 g
Magnesium sulfate (MgSO$_4$ · 7H$_2$O). . . . . 0.5 g
Agar. . . . . .15 g
Water . . . . .1000 ml

While still liquid after sterilization, add sufficient sulfuric acid to bring the pH down to 4.0.

Glycerol Yeast Extract Agar

Glycerol. . . . . .5 ml
Yeast extract . . . . .2 g
Dipotassium phosphate . . . . .1 g
Agar. . . . . .15 g
Water . . . . .1000 ml

LB Broth

Tryptone . . . . .10 g
Yeast extract . . . . .5 g
NaCl. . . . . .5 g
Distilled water .1000 ml

Add 1 M NaOH to adjust the pH to 7. Autoclave at standard conditions.

m Endo MF Broth (Ex. 46)

This medium is extremely hygroscopic in the dehydrated form and oxidizes quickly to cause deterioration of the medium after the bottle has been opened. Once a bottle has been opened it should be dated and discarded after one year. If the medium becomes hardened within that time it should be discarded. Storage of the bottle inside a larger bottle that contains silica gel will extend shelf life.

Failure of Exercise 46 can often be attributed to faulty preparation of the medium. It is best to make up the medium the day it is to be used. It should not be stored over 96 hours prior to use. The Millipore Corporation recommends the following method for preparing this medium. (These steps are not exactly as stated in the Millipore Application Manual AM302.)

1. Into a 250 ml screw-cap Erlenmeyer flask place the following:

   Distilled water . . . . .50 ml
   95% ethyl alcohol . . . . .2 ml

Dehydrated medium (*m* Endo MF broth) . . 4.8 g

Shake the above mixture by swirling the flask until the medium is dissolved and then add another 50 ml of distilled water.

2. Cap the flask loosely and immerse it into a pan of boiling water. As soon as the medium begins to simmer, remove the flask from the water bath. Do not boil the medium any further.
3. Cool the medium to 45°C, and adjust the pH to between 7.1 and 7.3.
4. If the medium must be stored for a few days, place it in the refrigerator at 2–10°C, with screw-cap tightened securely.

Milk Salt Agar (15% NaCl)

Prepare three separate beakers of the following ingredients:

1. Beaker containing 200 grams of sodium chloride.
2. Large beaker (2000 ml size) containing 50 grams of skim milk powder in 500 ml of distilled water.
3. Glycerol-peptone agar medium:

$MgSO_4 \cdot 7H_2O$ . . . . 5.0 g
$MgNO_3 \cdot 6H_2O$ . . . . 1.0 g
$FeCl_3 \cdot 7H_2O$ . . . . 0.025 g
Difco proteose-peptone #3 . . . . 5.0 g
Glycerol . . . . 10.0 g
Agar . . . . 30.0 g
Distilled water . . . . 500.0 ml

Sterilize the above three beakers separately. The milk solution should be sterilized at 113–115°C (8 lb pressure) in autoclave for 20 minutes. The salt and glycerol-peptone agar can be sterilized at conventional pressure and temperature. After the milk solution has cooled to 55°C, add the sterile salt, which should also be cooled down to a moderate temperature. If the salt is too hot, coagulation may occur. Combine the milk-salt and glycerol-peptone agar solutions by gently swirling with a glass rod. Dispense aseptically into petri plates.

Nitrate Agar

Beef extract . . . . 3 g
Peptone . . . . 5 g
Potassium nitrate . . . . 1 g
Agar . . . . 12 g
Distilled water . . . . 1000 ml
  Final pH 6.8 at 25°C

Nitrate Broth

Beef extract . . . . 3 g
Peptone . . . . 5 g
Potassium nitrate . . . . 1 g
Distilled water . . . . 1000 ml
  Final pH 7.0 at 25°C

Nitrate Succinate–Mineral Salts Agar

Add 15 g agar to 1000 ml of the above complete medium. Dispense into petri plates and sterilize in the autoclave.

Nitrate Succinate–Mineral Salts Broth

This medium is used as an enrichment medium for isolating denitrifying bacteria from soil. Note that two stock solutions (A and B) should be made up before attempting to put together the complete medium.

Solution A (Trace Mineral Salts)

$FeSO_4 \cdot 7H_2O$ . . . . 300 mg
$MnCl_2 \cdot 4H_2O$ . . . . 180 mg
$Co(NO_3)_2 \cdot 7H_2O$ . . . . 130 mg
$ZnSO_4 \cdot 7H_2O$ . . . . 40 mg
H_2MoO_4 . . . . 20 mg
$CuSO_4 \cdot 5H_2O$ . . . . 1 mg
$CaCl_2$ . . . . 1000 mg

This solution should be stored at 4°C until used.

Solution B

NH_4Cl . . . . 1 g
Na_2HPO_4 . . . . 2.14 g
KH_2PO_4 . . . . 1.09 g
$MgSO_4 \cdot 7H_2O$ . . . . 0.2 g
Trace mineral salts (Sol A) . . . . 10 ml
Water . . . . 1000 ml

Complete Medium

Solution B . . . . 1000 ml
Sodium succinate . . . . 2 g
Potassium nitrate . . . . 3 g

Adjust the pH to 6.8, dispense into bottles, and autoclave at standard conditions.

Nitrogen-Free Base for *Azotobacter*

K$_2$HPO$_4$	1 g
KH$_2$PO$_4$	0.2 g
MgSO$_4$ · 7H$_2$O	0.2 g
FeSO$_4$ · 7H$_2$O	0.05 g
CaCl$_2$ · 2H$_2$O	0.1 g
Na$_2$MoO$_4$ · 2H$_2$O	0.05 g
Water	1000 ml
*Carbohydrate	10 g

Adjust pH to 7.2.

Note: If the medium is to be used immediately, sterilization is not necessary. However, if it is to be stored it should be autoclaved.

Phage Growth Medium (Ex. 22)

KH$_2$PO$_4$	1.5 g
Na$_2$HPO$_4$	3.0 g
NH$_4$Cl	1.0 g
MgSO$_4$ · 7H$_2$O	0.2 g
Glycerol	10.0 g
Acid-hydrolyzed casein	5.0 g
dl-Tryptophan	0.01 g
Gelatin	0.02 g
Tween-80	0.2 g
Distilled water	1000.0 ml

Sterilize in autoclave at 121°C for 20 minutes.

Phage Lysing Medium (Ex. 22)

Add sufficient sodium cyanide (NaCN) to the above growth medium to bring the concentration up to 0.02 M. For 1 liter of lysing medium this will amount to about 1 gram (actually 0.98 g) of NaCN. When an equal amount of this lysing medium is added to the growth medium during the last 6 hours of incubation, the concentration of NaCN in the combined medium is 0.01 M.

Purple Nonsulfur Medium

This culture medium is used for the enrichment and culture of anaerobic phototrophic bacteria. To make up this medium you need to first prepare three stock solutions (A, B, and C) before putting together the entire batch.

---

*Carbohydrate: Either glucose or mannitol can be used to as a carbon source to isolate *Azotobacter*. Depending on the carbohydrate, different species of *Azotobacter* can be obtained. The glucose or mannitol should be sterilized separate from the salts. Dissolve the sugar in 100 ml of water and sterilize at 121°C for 15 minutes. Sterilize the salts in the same manner. After sterilization, the two solutions are mixed aseptically and dispensed into sterile 8 oz, flat-sided bottles (50 ml per bottle).

A. Iron Citrate Solution

Ammonium ferrous sulfate	748 mg
Sodium citrate	1180 mg
Water	500 ml

Store this stock solution at 4°C until needed.

B. Vitamin B$_{12}$ Solution

Certain strains require this vitamin. To make up 100 ml of this solution, add 1 mg to 100 ml of water. Store at 4°C until needed.

C. Trace Metals Solution

H$_3$BO$_4$	2.86 g
MnCl$_2$ · H$_2$O	1.81 g
ZnSO$_4$ · 7H$_2$O	0.222 g
Na$_2$MoO$_4$ · 2H$_2$O	0.390 g
CuSO$_4$ · 5H$_2$O	0.079 g
Co(NO$_3$)$_2$ · 6H$_2$O	0.0494 g
Water	1000 ml

Store at 4°C until needed.

Complete Enrichment Medium

The final batch of this medium has the following ingredients. The succinate provides the organic carbon and the yeast extract provides essential vitamins for certain strains.

KH$_2$PO$_4$	0.5 g
MgSO$_4$ · 7H$_2$O	0.2 g
NaCl	0.4 g
NH$_4$Cl	0.4 g
CaCl$_2$ · 2H$_2$O	0.05 g
Sodium succinate	1.0 g
Yeast extract	0.2 g
Iron citrate solution (A)	5 ml
Vitamin B$_{12}$ solution (B)	0.1 ml
Trace elements solution (C)	1 ml
Water	1000 ml

Adjust the pH to 6.8, dispense into bottles, and autoclave at standard conditions.

Russell Double Sugar Agar (Ex. 53)

Beef extract	1 g
Proteose Peptone No. 3 (Difco)	12 g
Lactose	10 g
Dextrose	1 g
Sodium chloride	5 g
Agar	15 g
Phenol red (Difco)	0.025 g

Distilled water . . . . 1000 ml
   Final pH 7.5 at 25°C

Dissolve ingredients in water, and bring to boiling. Cool to 50–60°C, and dispense about 8 ml per tube (16 mm dia tubes). Slant tubes to cool. Butt depth should be about $\frac{1}{2}$".

Skim Milk Agar

Skim milk powder . . . . 100 g
Agar. . . . . 15 g
Distilled water . . . . 1000 ml

Dissolve the 15 g of agar into 700 ml of distilled water by boiling. Pour into a large flask and sterilize at 121°C, 15 lb pressure.

In a separate container, dissolve the 100 g of skim milk powder into 300 ml of water heated to 50°C. Sterilize this milk solution at 113–115°C (8 lb pressure) for 20 minutes.

After the two solutions have been sterilized, cool to 55°C and combine in one flask, swirling gently to avoid bubbles. Dispense into sterile petri plates.

SOC Medium

Tryptone . . . . 20 g
Yeast extract . . . . 5 g
NaCl. . . . . 0.6 g
KCl . . . . 0.18 g
$MgCl_2$ . . . . 2 g
$MgSO_4$. . . . . 2.5 g
Glucose . . . . 3.6 g

Add 950 ml of distilled water to dissolve the solutes. Adjust the pH to 7 with 1 M NaOH and bring the volume to 1000 ml. Autoclave at standard conditions.

Sodium Chloride (6.5%) Tolerance Broth (Ex. 52)

Heart infusion broth. . . . . 25 g
NaCl. . . . . 60 g
Indicator (1.6 g bromcresol purple in 100 ml
   95% ethanol). . . . . 1 ml
Dextrose. . . . . 1 g
Distilled water . . . . 1000 ml

Add all reagents together up to 1000 ml (final volume). Dispense in 15 × 125 mm screw-capped tubes and sterilize in an autoclave 15 minutes at 121°C.

A positive reaction is recorded when the indicator changes from purple to yellow or when growth is obvious even though the indicator does not change.

Sodium Hippurate Broth

Heart infusion broth. . . . . 25 g
Sodium hippurate . . . . 10 g
Distilled water . . . . 1000 ml

Sterilize in autoclave at 121°C for 15 minutes after dispensing in 15 × 125 mm screw-capped tubes. Tighten caps to prevent evaporation.

Soft Nutrient Agar (for bacteriophage)

Dehydrated nutrient broth . . . . 8 g
Agar. . . . . 7 g
Distilled water . . . . 1000 ml

Sterilize in autoclave at 121°C for 20 minutes.

Spirit Blue Agar (Ex. 37)

This medium is used to detect lipase production by bacteria. Lipolytic bacteria cause the medium to change from pale lavender to deep blue.

Spirit blue agar (Difco) . . . . 35 g
Lipase reagent (Difco). . . . . 35 ml
Distilled water . . . . 1000 ml

Dissolve the spirit blue agar in 1000 ml of water by boiling. Sterilize in autoclave for 15 minutes at 15 psi (121°C). Cool to 55°C and slowly add the 35 ml of lipase reagent, agitating to obtain even distribution. Dispense into sterile petri plates.

Tryptone Agar

Tryptone . . . . 10 g
Agar. . . . . 15 g
Distilled water . . . . 1000 ml

Tryptone Broth

Tryptone . . . . 10 g
Distilled water . . . . 1000 ml

Tryptone Yeast Extract Agar

Tryptone . . . . 10 g
Yeast extract . . . . 5 g
Dipotassium phosphate . . . . 3 g
Sucrose . . . . 50 g
Agar. . . . . 15 g
Water . . . . 1000 ml
   pH 7.4

Appendix D—Identification Charts

Chart I Interpretation of Test Results of API 20E System

Courtesy of bioMérieux, Inc.

Chart II   **Symbol Interpretation of API 20E System**

READING TABLE

TESTS	ACTIVE INGREDIENTS	QTY (mg/cup.)	REACTIONS/ENZYMES	RESULTS	
				NEGATIVE	POSITIVE
ONPG	2-nitrophenyl-ßD-galactopyranoside	0.223	ß-galactosidase (Ortho NitroPhenyl-ßD-Galactopyranosidase)	colorless	yellow (1)
ADH	L-arginine	1.9	Arginine DiHydrolase	yellow	red / orange (2)
LDC	L-lysine	1.9	Lysine DeCarboxylase	yellow	red / orange (2)
ODC	L-ornithine	1.9	Ornithine DeCarboxylase	yellow	red / orange (2)
CIT	trisodium citrate	0.756	CITrate utilization	pale green / yellow	blue-green / blue (3)
H2S	sodium thiosulfate	0.075	H2S production	colorless / greyish	black deposit / thin line
URE	urea	0.76	UREase	yellow	red / orange (2)
TDA	L-tryptophane	0.38	Tryptophane DeAminase	TDA / immediate	
				yellow	reddish brown
IND	L-tryptophane	0.19	INDole production	JAMES / immediate	
				colorless pale green / yellow	pink
VP	sodium pyruvate	1.9	acetoin production (Voges-Proskauer)	VP 1 + VP 2 / 10 min	
				colorless	pink / red (5)
GEL	Gelatin (bovine origin)	0.6	GELatinase	no diffusion	diffusion of black pigment
GLU	D-glucose	1.9	fermentation / oxidation (GLUcose) (4)	blue / blue-green	yellow / greyish yellow
MAN	D-mannitol	1.9	fermentation / oxidation (MANnitol) (4)	blue / blue-green	yellow
INO	inositol	1.9	fermentation / oxidation (INOsitol) (4)	blue / blue-green	yellow
SOR	D-sorbitol	1.9	fermentation / oxidation (SORbitol) (4)	blue / blue-green	yellow
RHA	L-rhamnose	1.9	fermentation / oxidation (RHAmnose) (4)	blue / blue-green	yellow
SAC	D-sucrose	1.9	fermentation / oxidation (SACcharose) (4)	blue / blue-green	yellow
MEL	D-melibiose	1.9	fermentation / oxidation (MELibiose) (4)	blue / blue-green	yellow
AMY	amygdalin	0.57	fermentation / oxidation (AMYgdalin) (4)	blue / blue-green	yellow
ARA	L-arabinose	1.9	fermentation / oxidation (ARAbinose) (4)	blue / blue-green	yellow
OX	(see oxidase test package insert)		cytochrome-OXidase	(see oxidase test package insert)	

SUPPLEMENTARY TESTS

TESTS	ACTIVE INGREDIENTS	QTY (mg/cup.)	REACTIONS/ENZYMES	RESULTS	
				NEGATIVE	POSITIVE
Nitrate reduction GLU tube	potassium nitrate	0.076	NO2 production	NIT 1 + NIT 2 / 2–5 min	
				yellow	red
			reduction to N2 gas	Zn / 5 min	
				orange-red	yellow
MOB	API M Medium or microscope		motility	nonmotile	motile
McC	MacConkey medium		growth	absence	presence
OF-F	glucose (API OF Medium)		fermentation : under mineral oil	green	yellow
OF-O			oxidation : exposed to the air	green	yellow

(1) A very pale yellow should also be considered positive.
(2) An orange color after 36-48 hours incubation must be considered negative.
(3) Reading made in the cupule (aerobic).
(4) Fermentation begins in the lower portion of the tubes, oxidation begins in the cupule.
(5) A slightly pink color after 10 minutes should be considered negative.
• The quantities indicated may be adjusted depending on the titer of the raw materials used.
• Certain cupules contain products of animal origin, notably peptones.

Courtesy of bioMérieux, Inc.

CHART III   Characterization of Gram-Negative Rods—The API 20E System (continued)

IDENTIFICATION TABLE

% of positive reactions after 18–24/48 hrs. at 36°C ± 2°C

API 20E V4.1	ONPG	ADH	LDC	ODC	CIT	H2S	URE	TDA	IND	VP	GEL	GLU	MAN	INO	SOR	RHA	SAC	MEL	AMY	ARA	OX	NO2	N2	MOB	McC	OF/O	OF/F
Buttiauxella agrestis	100	0	0	85	25	0	0	0	0	0	0	100	100	10	1	99	0	92	99	100	0	100	0	100	100	100	100
Cedecea davisae	99	89	0	99	75	0	0	0	0	89	0	100	100	0	1	0	0	0	100	1	0	99	0	87	100	100	100
Cedecea lapagei	99	99	0	0	75	0	0	0	0	90	0	100	99	0	0	0	1	1	100	1	0	99	0	87	100	100	100
Citrobacter braakii	50	45	0	99	75	81	1	0	4	0	0	100	100	1	100	100	1	91	99	99	0	100	0	95	100	100	100
Citrobacter freundii	90	24	0	75	75	75	1	0	1	0	0	100	99	25	99	99	1	82	40	99	0	98	0	95	100	100	100
Citrobacter koseri/amalonaticus	99	75	0	97	75	0	1	0	99	0	0	100	100	25	99	99	1	80	98	99	0	100	0	95	100	100	100
Citrobacter koseri/farmeri	99	2	0	100	25	0	1	0	99	0	0	100	100	1	95	100	99	1	25	100	0	85	0	95	100	100	100
Citrobacter youngae	100	50	0	1	80	80	1	0	1	0	0	100	100	1	95	100	1	100	1	100	0	100	0	100	100	100	100
Edwardsiella hoshinae	0	0	100	99	50	0	0	0	99	0	0	100	100	0	0	1	100	0	0	1	0	100	0	98	100	100	100
Edwardsiella tarda	0	0	100	99	1	94	1	0	99	0	0	100	0	0	0	1	0	0	0	1	0	100	0	97	100	100	100
Enterobacter aerogenes	99	0	99	98	82	0	1	0	0	85	0	99	99	99	99	99	99	99	99	99	0	100	0	92	100	100	100
Enterobacter amnigenus 1	99	25	0	99	40	0	0	0	0	75	0	100	100	0	1	100	1	100	99	99	0	100	0	95	100	100	100
Enterobacter amnigenus 2	99	80	0	99	80	0	0	0	0	75	0	100	100	25	99	100	99	99	99	99	0	95	0	95	100	100	100
Enterobacter asburiae	100	25	0	99	80	0	0	0	0	10	0	100	100	25	100	1	99	1	100	100	0	99	0	99	100	100	100
Enterobacter cancerogenus	100	75	0	99	90	0	0	0	0	89	0	100	100	0	1	100	1	90	99	100	0	95	0	95	100	100	100
Enterobacter cloacae	98	82	1	92	90	0	0	0	0	85	0	99	99	12	90	85	96	90	99	99	0	100	0	95	100	100	100
Enterobacter gergoviae	99	1	32	100	75	0	99	0	0	90	0	100	99	23	1	100	40	100	99	100	0	90	0	90	100	100	100
Enterobacter intermedius	99	0	0	91	1	0	0	0	0	2	0	100	97	88	88	99	99	100	99	99	0	100	0	92	100	100	100
Enterobacter sakazakii	100	96	0	91	94	0	1	0	25	91	10	100	100	75	8	99	99	99	99	99	0	100	0	96	100	100	100
Escherichia coli 1	90	1	74	70	0	1	1	0	89	0	0	99	98	1	91	82	36	75	3	99	0	100	0	95	100	100	100
Escherichia coli 2	26	45	20	0	0	1	1	0	50	0	0	99	90	1	42	30	3	3	1	70	0	98	0	5	100	100	100
Escherichia fergusonii	96	1	99	100	1	0	0	0	99	0	0	100	99	1	0	87	0	3	99	99	0	100	0	93	100	100	100
Escherichia hermannii	100	1	1	100	1	0	0	0	99	0	0	100	100	0	1	99	25	95	99	99	0	100	0	99	100	100	100
Escherichia vulneris	100	30	50	0	0	0	0	0	0	95	0	99	100	0	1	95	7	95	95	99	0	100	0	100	100	100	100
Ewingella americana	98	0	0	0	75	0	0	0	0	95	0	99	99	0	0	1	1	0	50	1	0	100	0	60	100	100	100
Hafnia alvei 1	75	0	99	98	50	0	0	0	0	50	0	99	99	0	0	99	1	1	25	99	0	100	0	85	100	100	100
Hafnia alvei 2	50	0	99	1	50	0	0	0	0	10	0	99	98	0	0	99	1	0	0	99	0	100	0	0	100	100	100
Klebsiella oxytoca	99	0	80	0	89	0	78	0	99	80	0	100	99	99	100	99	99	100	100	100	0	100	0	0	100	100	100
Klebsiella pneumoniae ssp ozaenae	94	18	25	1	18	0	1	0	0	1	0	99	96	57	66	58	20	80	97	85	0	92	0	0	100	100	100
Klebsiella pneumoniae ssp pneumoniae	99	1	73	0	86	0	75	0	0	90	0	99	99	99	99	99	99	99	99	99	0	99	0	0	100	100	100
Klebsiella pneumoniae ssp rhinoscleromatis	1	0	0	0	0	0	0	0	0	0	0	99	100	90	90	75	75	75	99	10	0	100	0	0	100	100	100
Kluyvera spp	95	0	25	99	50	0	1	0	80	0	0	100	99	0	25	93	89	99	99	99	0	95	0	94	100	100	100
Leclercia adecarboxylata	99	1	0	0	12	0	1	0	99	0	1	99	99	2	2	99	66	99	99	100	0	100	0	100	100	100	100
Moellerella wisconsensis	97	0	0	98	40	0	0	0	15	0	0	100	99	0	0	0	0	99	0	0	0	90	0	0	100	100	100
Morganella morganii	1	0	0	98	13	0	99	93	99	0	0	99	0	0	0	0	0	0	0	0	0	88	0	95	100	100	100
Pantoea spp 1	85	1	0	0	13	0	1	0	0	9	1	99	99	36	26	90	98	26	59	61	0	85	1	85	100	100	100
Pantoea spp 2	99	1	0	0	99	0	1	0	0	62	4	99	99	34	82	90	98	81	59	99	0	85	1	85	100	100	100
Pantoea spp 3	99	1	0	0	21	0	1	0	86	86	15	100	100	10	32	97	93	23	65	97	0	85	1	85	100	100	100
Pantoea spp 4	86	1	0	0	29	0	1	0	59	1	0	98	100	0	0	99	72	89	99	99	0	93	0	95	100	100	100
Proteus mirabilis	1	0	0	99	50	75	99	98	1	1	50	98	0	0	0	0	89	0	0	0	0	99	0	85	100	100	100
Proteus penneri	1	0	0	1	1	20	100	99	0	0	74	99	0	0	0	0	100	0	66	1	0	99	0	85	100	100	100
Proteus vulgaris group	1	0	0	0	12	83	99	99	92	0	74	99	1	1	0	1	89	1	1	1	0	94	1	94	100	100	100
Providencia alcalifaciens/rustigianii	0	0	0	0	80	0	0	100	99	0	0	99	1	1	0	0	0	0	0	1	0	100	0	96	100	100	100
Providencia rettgeri	1	0	0	0	74	0	99	99	90	0	0	98	82	78	1	50	25	0	40	1	0	98	0	94	100	100	100
Providencia stuartii	1	0	0	0	85	0	30	98	95	0	0	98	3	80	0	0	15	0	0	0	0	98	0	85	100	100	100
Rahnella aquatilis	100	0	0	0	50	0	0	0	0	99	0	100	100	0	98	100	100	97	0	98	0	99	0	6	100	100	100
Raoultella ornithinolytica	100	0	99	99	52	0	85	0	100	65	0	100	100	99	100	99	100	100	100	100	0	100	0	0	100	100	100
Raoultella terrigena	100	0	97	6	75	0	0	0	1	75	0	100	99	99	99	99	78	100	100	99	0	100	0	0	100	100	100
Salmonella choleraesuis ssp arizonae	98	75	98	99	75	99	0	0	1	0	0	100	99	0	99	99	1	1	0	99	0	100	0	99	100	100	100
Salmonella choleraesuis ssp choleraesuis	0	15	99	99	3	64	0	0	0	0	0	100	99	0	98	99	0	20	0	99	0	100	0	95	100	100	100
Salmonella ser.Gallinarum	0	1	100	1	0	25	0	0	0	0	0	100	100	0	98	1	0	20	0	100	0	100	0	0	100	100	100

CHART III   Characterization of Gram-Negative Rods—The API 20E System (continued)

API 20 E V4.1	ONPG	ADH	LDC	ODC	CIT	H2S	URE	TDA	IND	VP	GEL	GLU	MAN	INO	SOR	RHA	SAC	MEL	AMY	ARA	OX	NO2	N2	MOB	McC	OF/O	OF/F
Salmonella ser. Paratyphi A	0	5	0	99	0	1	0	0	0	0	0	100	99	0	99	98	0	96	0	99	0	100	0	95	100	100	100
Salmonella ser. Pullorum	0	1	75	100	0	85	0	0	0	0	0	100	100	0	0	100	0	99	0	75	0	100	0	0	100	100	100
Salmonella typhi	0	0	99	0	0	8	0	0	0	0	0	100	99	40	99	86	1	99	0	0	0	100	0	97	100	100	100
Salmonella spp	1	56	82	93	65	83	0	0	1	0	1	99	100	50	99	74	1	90	100	99	1	95	0	95	100	100	100
Serratia ficaria	99	0	0	99	100	0	0	0	0	40	90	100	100	97	100	99	99	99	99	99	0	92	0	100	100	100	100
Serratia fonticola	99	0	73	99	75	0	0	0	0	0	0	100	99	80	98	99	30	99	99	99	0	99	0	91	100	100	100
Serratia liquefaciens	95	1	78	98	80	0	2	0	0	59	65	100	99	85	98	2	99	72	97	97	0	100	0	95	100	100	100
Serratia marcescens	94	0	95	95	96	0	25	0	0	70	87	100	99	99	98	1	99	68	97	25	0	95	0	97	100	100	100
Serratia odorifera 1	95	0	95	99	95	0	0	0	99	50	99	100	99	99	70	99	99	99	99	99	0	99	0	100	100	100	99
Serratia odorifera 2	95	0	96	0	95	0	0	0	99	50	99	100	99	99	70	1	99	99	99	95	0	99	0	100	100	100	99
Serratia plymuthica	99	0	0	0	65	0	0	0	0	65	50	100	90	70	70	1	99	85	99	98	0	99	0	50	100	100	99
Serratia rubidaea	99	0	30	0	92	0	1	0	0	71	82	99	99	75	1	3	99	95	99	99	0	100	0	85	100	100	99
Shigella spp	96	0	0	1	0	0	0	0	29	0	0	99	63	0	7	7	1	20	0	50	0	100	0	0	100	100	100
Shigella sonnei	80	0	0	93	0	0	0	0	50	0	0	99	99	0	1	1	1	4	75	75	0	98	0	2	100	100	99
Yersinia enterocolitica	99	0	0	90	0	0	98	0	50	5	0	99	99	25	98	1	99	1	99	99	0	98	0	5	99	100	99
Yersinia frederiksenii/intermedia	99	0	0	75	60	0	99	0	99	1	0	100	99	25	99	99	99	1	99	99	0	98	0	5	91	100	99
Yersinia kristensenii	80	0	0	80	0	0	99	0	97	1	0	100	99	10	99	1	0	0	0	99	0	47	0	0	100	100	100
Yersinia pestis	68	0	0	0	0	0	0	0	0	0	0	99	99	75	70	0	0	30	30	30	0	95	0	0	99	99	99
Yersinia pseudotuberculosis	98	0	0	0	0	0	99	0	0	0	0	99	97	0	0	75	0	50	25	50	0	97	0	95	99	99	99
Aeromonas hydrophila gr. 1	99	90	25	1	25	0	0	0	85	25	90	99	99	1	3	5	97	1	75	75	100	97	0	95	99	99	99
Aeromonas hydrophila gr. 2	99	97	80	1	80	0	0	0	85	80	97	97	99	9	9	1	80	1	75	5	100	98	0	1	99	99	99
Aeromonas salmonicida ssp salmonicida	1	60	0	0	0	0	0	0	1	0	75	50	54	1	0	0	0	0	0	50	100	100	0	1	0	99	99
Grimontia hollisae	1	0	0	0	0	0	0	0	94	0	0	10	0	0	0	0	0	0	0	0	100	100	0	25	0	99	99
Photobacterium damselae	95	99	75	0	1	0	98	0	0	10	1	50	0	99	0	0	1	0	0	0	100	99	0	95	99	99	99
Plesiomonas shigelloides	99	99	100	100	0	0	1	0	100	0	0	99	0	99	0	0	0	99	0	0	100	99	0	95	99	94	94
Vibrio alginolyticus	0	1	98	75	60	0	1	0	100	10	75	99	100	0	1	0	100	1	10	1	100	47	0	100	96	94	94
Vibrio cholerae	98	1	94	97	75	0	0	0	99	58	92	98	98	0	1	0	94	0	5	0	100	96	0	100	96	99	99
Vibrio fluvialis	95	99	0	0	1	0	0	0	80	1	75	75	80	1	0	0	75	0	36	75	100	100	0	100	99	99	99
Vibrio mimicus	99	1	99	99	1	0	0	0	99	1	75	99	99	1	0	0	0	1	0	0	100	95	0	100	95	99	99
Vibrio parahaemolyticus	0	1	100	99	50	0	1	0	100	1	75	100	99	9	0	0	1	0	12	50	100	63	0	100	98	99	99
Vibrio vulnificus	99	0	91	0	25	0	0	0	99	1	99	99	75	97	0	1	1	1	90	0	99	54	0	100	99	99	99
Pasteurella aerogenes	99	0	0	80	0	0	99	0	99	0	0	99	1	0	99	1	99	99	0	75	75	100	0	0	100	100	100
Pasteurella multocida 1	4	0	0	0	0	0	0	0	99	0	0	29	1	0	1	0	75	1	0	0	99	90	0	0	2	23	23
Pasteurella multocida 2	7	0	0	45	0	0	0	0	99	0	0	44	99	0	99	1	0	0	0	0	89	90	0	0	2	23	23
Pasteurella pneumotropica / Mannheimia haemolytica	60	0	1	10	0	0	25	0	15	7	3	35	12	12	1	1	35	1	2	1	80	99	0	0	9	33	33
Acinetobacter baumannii/calcoaceticus	0	0	0	0	51	0	1	0	0	5	5	99	0	0	0	0	0	0	0	0	0	3	0	0	90	96	0
Bordetella/Alcaligenes/Moraxella spp *	0	0	0	0	52	0	14	0	0	0	0	0	0	0	0	0	0	0	7	20	95	62	1	75	88	97	0
Burkholderia cepacia	50	0	25	0	75	0	1	0	0	25	43	60	99	99	0	0	13	0	0	20	90	40	0	99	75	90	0
Chromobacterium violaceum	0	99	0	0	12	0	0	0	14	0	99	99	0	0	0	0	10	0	0	0	99	75	0	99	57	90	10
Chryseobacterium indologenes	5	0	0	0	12	0	90	0	75	0	80	80	0	0	0	0	0	0	0	0	99	20	0	0	48	93	6
Chryseobacterium meningosepticum	77	0	0	0	20	0	1	0	85	0	90	90	0	0	0	0	0	0	0	0	99	6	0	0	1	49	10
Eikenella corrodens	0	0	75	99	0	0	0	0	0	0	0	0	0	0	0	0	0	0	0	0	100	95	0	0	49	49	49
Myroides /Chryseobacterium indologenes	15	0	0	0	50	0	75	0	1	1	75	1	0	0	0	0	1	1	10	10	99	42	0	99	84	2	2
Ochrobactrum anthropi	0	89	0	30	92	0	25	0	0	15	0	1	1	0	0	0	0	0	25	10	90	12	60	97	99	47	0
Pseudomonas aeruginosa	0	75	0	0	92	0	25	0	0	0	75	50	0	0	0	0	0	10	1	25	97	26	56	97	99	98	0
Pseudomonas fluorescens/putida	0	75	0	0	75	0	0	0	0	10	27	25	0	0	0	0	1	0	1	1	99	30	0	100	96	93	0
Pseudomonas luteola	86	0	0	0	94	0	0	0	0	25	13	84	0	0	0	0	10	15	1	45	0	26	0	100	91	94	0
Pseudomonas oryzihabitans	0	0	0	0	89	0	0	0	0	25	0	10	0	0	0	1	0	10	1	45	0	7	0	100	99	99	0
Non-fermenter spp	1	1	0	0	37	0	1	0	15	15	9	9	0	0	0	1	1	1	1	1	93	48	35	99	85	49	49
Shewanella putrefaciens group	0	0	80	80	75	75	1	0	0	0	75	1	0	0	0	0	0	0	0	2	99	96	1	99	96	9	2
Stenotrophomonas maltophilia	70	0	75	1	75	1	0	0	0	0	90	1	0	0	0	0	0	0	0	0	91	26	1	100	91	49	0

* Brucella spp possible

Courtesy of bioMérieux, Inc.

Chart IV   Characterization of Enterobacteriaceae—The Enterotube II System

Group	Genus / Organism	GLUCOSE	GAS PRODUCTION	LYSINE	ORNITHINE	H₂S	INDOLE	ADONITOL	LACTOSE	ARABINOSE	SORBITOL	VOGES-PROSKAUER	DULCITOL	PHENYLALANINE DEAMINASE	UREA	CITRATE
ESCHERICHIEAE	Escherichia	+ 100.0	+J 92.0	d 80.6	d 57.8	−K 4.0	+ 96.3	− 5.2	+J 91.6	+ 91.3	± 80.3	− 0.0	d 49.3	− 0.1	− 0.1	− 0.2
ESCHERICHIEAE	Shigella	+ 100.0	−A 2.1	− 0.0	∓B 20.0	− 0.0	∓ 37.8	− 0.0	−B 0.3	± 67.8	∓ 29.1	− 0.0	d 5.4	− 0.0	− 0.0	− 0.0
EDWARDSIELLEAE	Edwardsiella	+ 100.0	+ 99.4	+ 100.0	+ 99.0	+ 99.6	+ 99.0	− 0.0	− 0.0	∓ 10.7	− 0.2	− 0.0	− 0.0	− 0.0	− 0.0	− 0.0
SALMONELLEAE	Salmonella	+ 100.0	+C 91.9	+H 94.6	+I 92.7	+E 91.6	− 1.1	− 0.0	− 0.8	± 89.2	± 94.1	− 0.0	dD 86.5	− 0.0	− 0.0	dF 80.1
SALMONELLEAE	Arizona	+ 100.0	+ 99.7	+ 99.4	+ 100.0	+ 98.7	− 2.0	− 0.0	d 69.8	+ 99.1	+ 97.1	− 0.0	− 0.0	− 0.0	− 0.0	+ 96.8
SALMONELLEAE	CITROBACTER — freundii	+ 100.0	+ 91.4	− 0.0	d 17.2	± 81.6	− 6.7	− 0.0	d 39.3	+ 100.0	+ 98.2	− 0.0	d 59.8	− 0.0	dw 89.4	+ 90.4
SALMONELLEAE	CITROBACTER — amalonaticus	+ 100.0	+ 97.0	− 0.0	+ 97.0	− 0.0	+ 99.0	− 0.0	± 70.0	+ 99.0	+ 97.0	− 0.0	∓ 11.0	− 0.0	± 81.0	+ 94.0
SALMONELLEAE	CITROBACTER — diversus	+ 100.0	+ 97.3	− 0.0	+ 99.8	− 0.0	+ 100.0	+ 100.0	d 40.3	+ 98.0	+ 98.2	− 0.0	± 52.2	− 0.0	dw 85.8	+ 99.7
PROTEEAE	PROTEUS — vulgaris	+ 100.0	±G 86.0	− 0.0	− 0.0	+ 95.0	+ 91.4	− 0.0	− 0.0	− 0.0	− 0.0	− 0.0	− 0.0	+ 100.0	+ 95.0	d 10.5
PROTEEAE	PROTEUS — mirabilis	+ 100.0	+G 96.0	− 0.0	+ 99.0	+ 94.5	− 3.2	− 0.0	− 2.0	− 0.0	− 0.0	∓ 16.0	− 0.0	+ 99.6	+ 89.3	± 58.7
PROTEEAE	MORGANELLA — morganii	+ 100.0	±G 86.0	− 0.0	+ 97.0	− 0.0	+ 99.5	− 0.0	− 0.0	− 0.0	− 0.0	− 0.0	− 0.0	+ 95.0	+ 97.1	−L 0.0
PROTEEAE	PROVIDENCIA — alcalifaciens	+ 100.0	dG 85.2	− 0.0	− 1.2	− 0.0	+ 99.4	+ 94.3	− 0.3	− 0.7	− 0.6	− 0.0	− 0.0	+ 97.4	− 0.0	+ 97.9
PROTEEAE	PROVIDENCIA — stuartii	+ 100.0	− 0.0	− 0.0	− 0.0	− 0.0	+ 98.6	∓ 12.4	− 3.6	− 4.0	− 3.4	− 0.0	− 0.0	+ 94.5	∓ 20.0	+ 93.7
PROTEEAE	PROVIDENCIA — rettgeri	+ 100.0	∓G 12.2	− 0.0	− 0.0	− 0.0	+ 95.9	+ 99.0	− 10.0	− 0.0	d 1.0	− 0.0	− 0.0	+ 98.0	+ 100.0	+ 96.0
KLEBSIELLEAE	ENTEROBACTER — cloacae	+ 100.0	+ 99.3	− 0.0	+ 93.7	− 0.0	− 0.0	∓ 28.0	± 94.0	+ 99.4	+ 100.0	+ 100.0	d 15.2	− 0.0	± 74.6	+ 98.9
KLEBSIELLEAE	ENTEROBACTER — sakazakii	+ 100.0	+ 97.0	− 0.0	+ 97.0	− 0.0	∓ 16.0	− 0.0	+ 100.0	+ 100.0	− 0.0	+ 97.0	− 6.0	− 0.0	− 0.0	+ 94.0
KLEBSIELLEAE	ENTEROBACTER — gergoviae	+ 100.0	+ 93.0	± 64.0	+ 100.0	− 0.0	− 0.0	− 0.0	∓ 42.0	+ 100.0	− 0.0	+ 100.0	− 0.0	− 0.0	+ 100.0	+ 96.0
KLEBSIELLEAE	ENTEROBACTER — aerogenes	+ 100.0	+ 95.9	+ 97.5	+ 95.9	− 0.0	− 0.8	+ 97.5	+ 92.5	+ 100.0	+ 98.3	+ 100.0	− 4.1	− 0.0	− 0.0	+ 92.6
KLEBSIELLEAE	ENTEROBACTER — agglomerans	+ 100.0	∓ 24.1	− 0.0	− 0.0	− 0.0	∓ 19.7	− 7.5	d 52.9	+ 97.5	d 26.3	± 64.8	d 12.9	∓ 27.6	d 34.1	d 84.2
KLEBSIELLEAE	HAFNIA — alvei	+ 100.0	+ 98.9	+ 99.6	+ 98.6	− 0.0	− 0.0	− 0.0	d 2.8	+ 99.3	− 0.0	± 65.0	− 2.4	− 0.0	− 3.0	d 5.6
KLEBSIELLEAE	SERRATIA — marcescens	+ 100.0	±G 52.6	+ 99.6	+ 99.6	− 0.0	−w 0.1	∓ 56.0	− 1.3	− 0.0	+ 99.1	+ 98.7	− 0.0	− 0.0	dw 39.7	+ 97.6
KLEBSIELLEAE	SERRATIA — liquefaciens	+ 100.0	d 72.5	± 64.2	+ 100.0	− 0.0	−w 1.8	d 8.3	d 15.6	+ 97.3	+ 97.3	∓ 49.5	− 0.0	− 0.9	dw 3.7	+ 93.6
KLEBSIELLEAE	SERRATIA — rubidaea	+ 100.0	dG 35.0	± 61.0	− 0.0	− 0.0	−w 2.0	± 88.0	+ 100.0	+ 100.0	− 8.0	+ 92.0	− 0.0	− 0.0	dw 4.0	± 88.0
KLEBSIELLEAE	KLEBSIELLA — pneumoniae	+ 100.0	+ 96.0	+ 97.2	− 0.0	− 0.0	− 0.0	± 89.0	+ 98.7	+ 99.9	+ 99.4	+ 93.7	∓ 33.0	− 0.0	+ 95.4	+ 96.8
KLEBSIELLEAE	KLEBSIELLA — oxytoca	+ 100.0	+ 96.0	+ 97.2	− 0.0	− 0.0	+ 100.0	± 89.0	∓ 98.7	+ 100.0	+ 98.0	+ 93.7	∓ 33.0	− 0.0	∓ 95.4	∓ 96.8
KLEBSIELLEAE	KLEBSIELLA — ozaenae	+ 100.0	d 55.0	∓ 35.8	− 1.0	− 0.0	− 0.0	+ 91.8	d 26.2	+ 100.0	± 78.0	− 0.0	− 0.0	− 0.0	d 14.8	± 28.1
KLEBSIELLEAE	KLEBSIELLA — rhinoscleromatis	+ 100.0	− 0.0	− 0.0	− 0.0	− 0.0	− 0.0	+ 98.0	d 6.0	+ 100.0	+ 98.0	− 0.0	− 0.0	− 0.0	− 0.0	− 0.0
YERSINIAE	YERSINIA — enterocolitica	+ 100.0	− 0.0	− 0.0	+ 90.7	− 0.0	∓ 26.7	− 0.0	− 0.0	+ 98.7	+ 98.7	− 0.1	− 0.0	− 0.0	+ 90.7	− 0.0
YERSINIAE	YERSINIA — pseudotuberculosis	+ 100.0	− 0.0	− 0.0	− 0.0	− 0.0	− 0.0	− 0.0	− 0.0	± 55.0	− 0.0	− 0.0	− 0.0	− 0.0	+ 100.0	− 0.0

E. *S. enteritidis* bioserotype Paratyphi A and some rare biotypes may be H₂S negative.

F. *S. typhi*, *S. enteritidis* bioserotype Paratyphi A and some rare biotypes are citrate-negative and *S. cholerae-suis* is usually delayed positive.

G. The amount of gas produced by *Serratia*, *Proteus* and *Providencia alcalifaciens* is slight; therefore, gas production may not be evident in the ENTEROTUBE II.

H. *S. enteritidis* bioserotype Paratyphi A is negative for lysine decarboxylase.

I. *S. typhi* and *S. gallinarum* are ornithine decarboxylase-negative.

J. The Alkalescens-Dispar (A-D) group is included as a biotype of *E. coli*. Members of the A-D group are generally anaerogenic, nonmotile and do not ferment lactose.

K. An occasional strain may produce hydrogen sulfide.

L. An occasional strain may appear to utilize citrate.

Chart V   **Reaction Interpretations for API Staph**

PROCEDURE

- Cocci
- Gram +
- Catalase +

Columbia blood agar

0.5 McF

API Staph Medium

API Staph

0 ⟶ URE

ADH
URE

Resistance to lysostaphin

18:00–24:00   36°C ± 2°C

API Staph

VP : **VP 1 + VP 2**
NIT : **NIT 1 + NIT 2**
PAL : **ZYM A + ZYM B**

+ - + - + -

Courtesy of bioMérieux, Inc.

Chart VI   Biochemistry of API Staph Tests

READING TABLE

TESTS	ACTIVE INGREDIENTS	QTY (mg/cup.)	REACTIONS / ENZYMES	RESULT	
				NEGATIVE	POSITIVE
0	No substrate		Negative control	red	—
GLU	D-glucose	1.56	(Positive control) (D-GLUcose)		
FRU	D-fructose	1.4	acidification (D-FRUctose)		
MNE	D-mannose	1.4	acidification (D-ManNosE)		
MAL	D-maltose	1.4	acidification (MALtose)		
LAC	D-lactose (bovine origin)	1.4	acidification (LACtose)	red *	yellow
TRE	D-trehalose	1.32	acidification (D-TREhalose)		
MAN	D-mannitol	1.36	acidification (D-MANnitol)		
XLT	xylitol	1.4	acidification (XyLiTol)		
MEL	D-melibiose	1.32	acidification (D-MELibiose)		
NIT	potassium nitrate	0.08	Reduction of NITrates to nitrites	NIT 1 + NIT 2 / 10 min	
				colorless-light pink	red
PAL	ß-naphthyl phosphate	0.0244	ALkaline Phosphatase	ZYM A + ZYM B / 10 min	
				yellow	violet
VP	sodium pyruvate	1.904	Acetyl-methyl-carbinol production (Voges-Proskauer)	VP 1 + VP 2 / 10 min	
				colorless-light pink	violet-pink
RAF	D-raffinose	1.56	acidification (RAFfinose)		
XYL	D-xylose	1.4	acidification (XYLose)		
SAC	D-saccharose (sucrose)	1.32	acidification (SACcharose)	red	yellow
MDG	methyl-αD-glucopyranoside	1.28	acidification (Methyl-αD-Glucopyranoside)		
NAG	N-acetyl-glucosamine	1.28	acidification (N-Acetyl-Glucosamine)		
ADH	L-arginine	1.904	Arginine DiHydrolase	yellow	orange-red
URE	urea	0.76	UREase	yellow	red-violet

The acidification tests should be compared to the negative (0) and positive (GLU) controls.

* When MNE and XLT are preceded or followed by positive tests, then an orange test should be considered negative.

• The quantities indicated may be adjusted depending on the titer of the raw materials used.

• Certain cupules contain products of animal origin, notably peptones.

Lysostaphin resistance test

Determine resistance to lysostaphin on P agar, as indicated in the following procedure or according to the manufacturer's recommendations.

To perform the test, inoculate the surface of a P agar plate, by flooding it with a bacterial suspension (approximately 10^7 organisms/ml).

Leave to dry for 10–20 minutes at 36°C ± 2°C.

Place a drop of lysostaphin solution (200 µg/ml) on the surface of the agar.

Incubate for 18–24 hrs. at 35–37°C.

Total or partial lysis of the bacterial culture indicates susceptibility to the enzyme.

This test constitutes the 21st test of the strip. It is considered positive if resistance to lysostaphin is determined.

Chart VII   **API Staph Profile Register**

IDENTIFICATION TABLE
% of reactions positive after 18–24 hrs. at 36°C ± 2°C

API STAPH V4.0	0	GLU	FRU	MNE	MAL	LAC	TRE	MAN	XLT	MEL	NIT	PAL	VP	RAF	XYL	SAC	MDG	NAG	ADH	URE	LSTR
Staphylococcus aureus	0	100	100	95	96	88	91	80	0	0	83	97	78	1	0	97	2	90	80	80	0
Staphylococcus auricularis	0	100	99	36	72	10	90	9	0	0	81	0	1	0	0	40	0	15	90	1	0
Staphylococcus capitis	0	100	99	80	43	22	2	36	0	0	86	23	90	0	0	50	0	1	85	35	0
Staphylococcus caprae	0	100	99	70	10	75	74	10	0	0	99	95	99	0	0	0	0	1	99	60	0
Staphylococcus carnosus	0	100	100	99	0	99	99	99	0	0	99	83	83	0	0	0	0	100	100	0	0
Staphylococcus chromogenes	0	100	100	99	79	100	100	13	0	0	96	96	1	0	1	100	0	31	89	95	0
Staphylococcus cohnii ssp cohnii	0	100	99	66	99	2	97	88	33	0	21	66	94	0	0	2	0	9	2	1	0
Staph.cohnii ssp urealyticum	0	100	100	99	98	98	100	94	64	0	1	94	87	0	0	0	0	98	0	99	0
Staphylococcus epidermidis	0	100	99	70	99	81	2	0	0	1	80	84	68	1	0	97	4	18	73	88	0
Staphylococcus haemolyticus	0	99	75	5	99	80	91	60	0	1	78	3	57	0	0	98	13	83	85	1	0
Staphylococcus hominis	0	98	94	41	97	50	86	28	0	1	82	27	70	1	0	97	4	50	43	84	0
Staphylococcus hyicus	0	100	99	99	0	87	99	0	0	0	90	90	15	0	0	99	2	93	100	68	0
Staphylococcus lentus	0	100	100	100	100	100	100	100	7	99	92	21	57	100	100	100	28	100	0	1	0
Staphylococcus lugdunensis	0	100	89	88	99	66	99	0	0	0	99	16	99	0	0	100	0	90	1	50	0
Staphylococcus saprophyticus	0	100	99	2	97	90	99	88	22	0	35	14	79	1	0	96	1	70	30	65	0
Staphylococcus schleiferi	0	100	80	100	0	1	71	0	0	0	99	97	99	0	0	0	0	94	99	0	0
Staphylococcus sciuri	0	99	99	99	99	70	93	98	0	0	83	67	30	0	16	95	7	68	0	0	0
Staphylococcus simulans	0	100	100	57	11	95	92	73	4	0	83	27	38	0	4	97	2	90	97	84	0
Staphylococcus warneri	0	99	99	50	98	19	96	70	0	0	23	16	90	0	0	99	0	6	77	97	0
Staphylococcus xylosus	0	100	100	92	81	85	95	90	30	9	82	75	67	11	82	87	10	80	5	90	0
Kocuria kristinae	0	99	96	99	90	9	84	3	0	0	6	3	93	0	0	90	12	0	0	0	97
Kocuria varians/rosea	0	91	92	8	1	1	8	1	0	0	75	4	8	4	8	4	0	1	1	29	95
Micrococcus spp	0	2	4	0	1	0	1	0	0	0	8	15	1	0	0	1	0	1	11	11	91

Courtesy of bioMérieux, Inc.

Reading References

History

Brock, Thomas D. *Robert Koch: A Life in Medicine and Bacteriology*. Herndon, VA: ASM Press, 1998.

Dubos, Rene. *Pasteur and Modern Science*. Herndon, VA: ASM Press, 1998.

General Information

Calandar, Richard L. *The Bacteriophages*. 2nd ed. New York: Oxford University Press, 2005.

Moat, Albert G., Foster, John W., and Spector, Michael P. *Microbial Physiology*. 4th ed. Hoboken, NJ: Wiley-Liss, 2002.

Nester, Eugene W., Anderson, Denise G., Roberts, C. Evans, Jr., and Nester, Martha T. *Microbiology: A Human Perspective*. 7th ed. New York: McGraw-Hill, 2012.

Rosenberg, Eugene, DeLong, Edward F., Lory, Stephen, Stackebrandt, Erko, and Thompson, Fabiano (eds.). *The Prokaryotes: Prokaryotic Physiology and Biochemistry*. 4th ed. New York: Springer, 2013.

Varnam, Alan H. *Environmental Microbiology*. Herndon, VA: ASM Press, 2000.

White, David, Drummond, James, and Fuqua, Clay. *The Physiology and Biochemistry of Prokaryotes*. 4th ed. New York: Oxford University Press, 2011.

Identification of Microorganisms

Garrity, G., senior editor. *Bergey's Manual of Systematic Bacteriology*. Vol. 2, *The Proteobacteria*. 2nd ed. New York: Springer, 2005.

Holt, John G., Krieg, Noel R., Sneath, Peter H. A., Staley, James T., and Williams, Stanley T. *Bergey's Manual of Determinative Bacteriology*. 9th ed. Philadelphia, PA: Lippincott Williams and Wilkins, 1993.

Laboratory Methods

Atlas, R. M. *Handbook of Microbiological Media*. 4th ed. Herndon, VA: ASM Press, 2010.

Reddy, C. A., Beveridge, T. J., Breznak, J. A., Marzluf, G., Schmidt, T. M., and Snyder, L. R. *Methods for General and Molecular Microbiology*. 3rd ed. Herndon, VA: ASM Press, 2007.

Versalovic, James (ed.). *Manual of Clinical Microbiology*. 10th ed. Herndon, VA: ASM Press, 2011.

Zimbro, Mary Jo, Power, David A., Miller, Sharon M., Willson, George E., and Johnson, Julie A. Difco and BBL Manual, *Manual of Microbiological Culture Media*. 2nd ed. Sparks, MD: Becton, Dickinson and Co., 2009.

Applied Microbiology

Downes, Frances P., and Ito, Keith. *Compendium of Methods for the Microbiological Examination of Foods*. 4th ed. Washington D.C.: American Public Health Association, 2001.

Montville, Thomas J., Mathews, Karl R., Kalmia, Nad, and Kniel, E. *Food Microbiology: An Introduction*. 3rd ed. Herndon, VA: ASM Press, 2012.

Rice, E. W., Baird, R. B., Eaton, A. D., and Clesceri, L. S. (eds.). *Standard Methods for the Examination of Water and Wastewater*. 22nd ed. Washington D.C.: American Public Health Association, 2012.

Wehr, Michael, and Frank, Joseph (eds.). *Standard Methods for the Examination of Dairy Products*. 17th ed. Washington D.C.: American Public Health Association, 2004.

Index